ALGEBRAIC STRUCTURE AND MATRICES

ALGEBRAIC STRUCTURE AND MATRICES

BY

E. A. MAXWELL, Ph.D.

Fellow of Queens' College Cambridge

BEING PART II OF
ADVANCED ALGEBRA

CAMBRIDGE
AT THE UNIVERSITY PRESS
1969

CAMBRIDGE UNIVERSITY PRESS
Cambridge, New York, Melbourne, Madrid, Cape Town, Singapore, São Paulo, Delhi

Cambridge University Press
The Edinburgh Building, Cambridge CB2 8RU, UK

Published in the United States of America by Cambridge University Press, New York

www.cambridge.org
Information on this title: www.cambridge.org/9780521109055

First published 1965
Reprinted 1969
This digitally printed version 2009

A catalogue record for this publication is available from the British Library

ISBN 978-0-521-05695-3 hardback
ISBN 978-0-521-10905-5 paperback

CONTENTS

SECTION II: MATRICES

PREFACE

Besides the general acknowledgement mentioned in the Introduction, I would also express my more specific thanks to Mr A. P. Rollett and Dr G. Matthews who gave me very valuable help and criticism at the manuscript stage. Mr W. G. Kellaway, of the Department of Education in the University of Cambridge, read the proofs and made a number of important suggestions.

This manuscript set the printers a difficult task, and I am most grateful to them for the skill and cheerfulness with which they have overcome it.

E. A. M.

March, 1964

It is probably true that this would have been a different book if written in 1969—I have learned much as the years pass, even though they have been comparatively few. But I doubt whether the basic plan would have changed greatly. This still seems to me to be a reasonable way to present the subject to the still immature, but developing, young mathematician.

I am grateful for many comments and much advice. To adopt all would have gone beyond second printing to second edition, but I have incorporated improvements where readily possible. There were also more mis-prints (or, to give them their proper names, mistakes) than I like in the first printing and I acknowledge help from a number of friends, especially the late Mr. C. V. Durell and, in kindly but persistent detail, Mr. F. Gerrish.

E. A. M.

February, 1969

INTRODUCTION

Since the peaceful days of 1958–9 when I wrote the first volume of this work, an explosion of considerable violence has struck the world of teachers of mathematics. 'In 1959, the Organisation for European Economic Cooperation convoked at Royaumont, near Paris, France, a two-week seminar on "New Thinking in School Mathematics".' This quotation, from the later O.E.E.C. book, *Synopses for Modern Secondary School Mathematics*, describes the force that gathered in preparation for the explosion. Much preliminary work had been done before 1959, but it is fair to ascribe to that gathering and to its chairman, Professor Marshall Stone of the University of Chicago, high credit for the more directed activity that has since taken place.

For my own part, I had the privilege of being present for some of the time, and of speaking and discussing the problems informally. It became clear to me that something vital was involved, though I did not feel able to go as far as many of the protagonists would wish. Whether or not I have judged correctly here, the reader must decide.

It is not possible, I think, to write at this stage a fully satisfactory book on 'modern algebra' for class-room use; there just is not sufficient teaching experience of the problems involved. I have tried to look with the eye of imagination at a beginner approaching the subject, and to judge how he can best be served as he tries, on the one hand, to see what it is all about and, on the other (should he be proceeding to further study) to prepare himself for a firm grasp of the details that will come later.

I have had in mind a pupil in the top forms at school or in the first year of a university course. In writing, I have done all I can to make myself clear, but I have not attempted to pretend that the subject is easy. It is exciting and illuminating, and very rewarding, but it is not easy, and the pupil who wishes to master it will have to make the effort.

The choice of subject-matter has been extremely difficult; many alternatives are possible both in content and in presentation. Every reader familiar with the subject as a whole will regret some omission or other. A conspicuous omission, for example, is any reference to mathematical logic. Notation, too, has proved troublesome, for many alternatives are in existence. I can only hope that the particular choices made will not prove unduly troublesome in later reading; my consolation is that other variants would probably have proved no less unpopular.

It would be hard to express my indebtedness for the material used. I have given a fairly detailed bibliography and take this opportunity of recording how much I owe to all these authors. The list is far from exhaustive, but it does indicate possible next steps for reading in a subject which I hope this book may do something to encourage.

BIBLIOGRAPHY

The following list of books for further, or alternative, reading includes many which I have consulted and to which I would pay tribute. It also contains references to exploratory work both before and after the conference at Royaumont.

1. The introduction of 'modern mathematics' into the curriculum

Beberman, M., *An emerging programme of secondary school mathematics.* Oxford, 1958.

An account of the Illinois experiment dating from 1952, one of the first serious attempts to tackle the problem.

Program for College preparatory mathematics (with appendices). 1958.

Report of a Commission on Mathematics made to the College Entrance Examination Board of America. Set up in 1955, it gives a complete 'modernization' programme for all school levels, with class-room details.

New thinking in school mathematics. 1961.

The report of the O.E.E.C. Conference at Royaumont.

School mathematics in O.E.E.C. countries.

A summary of the position in schools and training colleges for all O.E.E.C. countries.

Mathematics for physicists and engineers.

Lectures and discussions at an O.E.E.C. seminar.

Synopses for modern secondary school mathematics.

A very detailed series of suggestions for a radical re-thinking of secondary school mathematics, with considerable elaboration of syllabuses.

Report of the Gulbenkian Foundation on mathematics at A-level. 1960.

The first detailed study of the problem in the United Kingdom to suggest practical steps for class-room and examination.

On teaching mathematics. Pergamon, 1961.

A symposium held at Southampton under Professor Thwaites, with, again, detailed suggestions for syllabuses.

2. Textbooks at a fairly elementary level

Adler, I., *The new mathematics.* 1958. A 'Mentor' book; New American Library of World Literature, Inc.

Covers a wide range vividly and simply. The guiding theme is the gradual development and abstraction of the number system.

Alexandroff, P. S., *Introduction to the theory of Groups.* Blackie, 1959.

Translation of a fascinating account by a Russian master of the subject.

Coxeter, H. S. M., *Introduction to geometry.* Wiley, 1961.

Not strictly relevant, but an original and thrilling account of similar possibilities in geometry.

Davenport, H., *The higher arithmetic*. Hutchinson, 1952.
An admirable survey, indispensable for the mathematician.

Goodstein, R. L., *Fundamental concepts of mathematics*. Pergamon, 1962.
A simple but scholarly review for 'the cultivated amateur'. Excellent chapters on 'Classes and truth functions' and 'Networks and maps'.

Ledermann, W., *Introduction to the theory of finite groups*. Oliver and Boyd, 1949.
A very thorough elementary account.

Mansfield, D. E. and Thompson, D., *Mathematics, a new approach*. Chatto and Windus, 1962.
An important, and highly successful, attempt to produce a textbook at a really elementary level. One of the most original and lucid books yet produced to deal with the problem. It is based on actual class-room experience dating from the 'pre-Royaumont' era, and deals with the 'O' level of the General Certificate of Education.

Sawyer, W. W., *A concrete approach to abstract algebra*. Freeman (San Francisco), 1959.
Covers a very wide range in a thoroughly readable manner.

3. More advanced texts

Aitken, A. C., *Determinants and matrices*. Oliver and Boyd, 1939.
An extraordinarily able piece of compression, and an excellent account of the essentials. The present author is indebted to his treatment of linear systems of equations.

Albert, A. A., *Fundamental concepts of higher algebra*. Chicago, 1956.
A very thorough review, especially aimed towards the theory of finite fields.

Birkhoff, G. and MacLane, S., *A survey of modern algebra*. Macmillan, 1941.
A very widely ranging survey of the subject as a whole.

Ferrar, W. L., *Algebra*. Oxford, 1941.
A well-established standard textbook which covers the basic groundwork very readably.

Hadley, G., *Linear algebra*. Addison-Wesley, 1961.
A thorough introduction at a comparatively elementary level. It 'leads logically to its author's *Linear Programming* text'.

Halmos, P. R., *Finite-dimensional vector spaces*. Princeton, 1948.
An abstract, but clear and lucid, account.

Hohn, F. E., *Applied Boolean algebra, an elementary introduction*. Macmillan, New York, 1960.
Deals with Boolean algebra, relay circuits, and propositional logic.

Hodge, W. V. D. and Pedoe, D., *Methods of algebraic geometry*. Book I. Cambridg , 1947.
Although concerned with geometry, Book I contains a very detailed introduction to the algebraic essentials.

Mirsky, L., *An introduction to linear algebra.* Oxford, 1955.

Covers those parts of the subject which must form part of any standard course.

Shilov, G. E., *Theory of linear spaces.* Prentice-Hall, 1961.

An excellent translation of a first-rate account of the subject which will be enjoyed by all mathematicians.

Thrall, R. M. and Tornheim, L., *Vector spaces and matrices.* Chapman and Hall, 1957.

Most of the standard theory is covered, with a wealth of detail. An excellent text for the advanced student.

Turnbull, H. W. and Aitken, A. C., *Theory of Cannonical matrices.* Blackie, 1932.

A very elegant account of the basic principles.

Van der Waerden, B. L., *Modern algebra.* Ungar, 1949.

A two-volume work which has already established itself as a classic. Very readable, but, of necessity, not easy.

4. Foreign Language texts

Structures algébriques et structures topologiques. By various authors, being No. 7 of Monographies de l'Enseignement Mathématique. Paris, 1958.

A brilliant symposium, with all the lucidity that would be expected.

Elements de Mathématique. Bourbaki, 1954 etc. I. *Théorie des ensembles,* II *Algèbre.*

Perhaps the most famous work of present-day mathematics by a group of authors believed to be commutative.

Mathématiques du 20ᵉ siècle. Brussels, 1960.

Various authors, following a seminar. Gives a good insight into some modern trends.

Godement, R., *Cours d'algèbre.* Hermann, 1963.

A masterly exposition, essential for the advanced student. At 664 pages for 60 francs it is expensive but extremely good value.

5. Supplementary list

(*Texts issued while this book was in press.*)

G. Matthews, *Matrices I*; C. A. R. Bailey, *Sets and logic, I*; F. B. Lovis, *Computers I*; J. A. C. Reynolds, *Shape, size and place.* These are St. Dunston's College Booklets, published by Edward Arnold.

A. J. Moakes, *Numerical Mathematics* and *The core of Mathematics*; W. Chellingsworth, *Mathematics for circuits.* These are in the series, Introductory Monographs in Mathematics, published by Macmillan.

Georges Papy, *Groups.* Macmillan, 1964.

T. J. Fletcher, Editor, *Some lessons in mathematics.* Cambridge University Press. 1964.

The School Mathematics Project, '*Book T*'. Cambridge University Press, 1964.

SECTION I

ALGEBRAIC STRUCTURE

1

ONE-PROCESS ALGEBRA

The familiar processes of algebra, such as addition or multiplication, are usually expressed by means of a symbolism which it is our purpose to extend so as to cover many topics which seem, at first sight, to be quite unrelated.

But first a number of notations are introduced. Experience indicates that the young and developing mathematician, to whom this book is addressed, likes to have a good supply of mathematical shorthand at his disposal, and it is hoped that the new notations will soon become familiar through use.

1. Notation and definitions. The word *set* is used to denote any collection of objects subject to the sole requirement of a law to decide definitely whether any given *element* does or does not belong to it. For example, the set of two-digit powers of 2 consists precisely of the numbers 16, 32, 64 and no others.

The individual members of a set are denoted by letters such as a, b, $c, \ldots, x, \ldots$ and the set itself, regarded as a single entity, by a letter such as S or E. A set is also named by enumerating its elements within brackets; for example,

$$\{16, 32, 64\}.$$

The *symbol of inclusion* $\in$ denotes *membership of a set*; thus the statement

$$x \in S$$

reads, *x is a member of S*. For example,

$$32 \in \{16, 32, 64\}.$$

The phrase

$$x \notin S$$

reads, *x is not a member of S*. For example,

$$8 \notin \{16, 32, 64\}.$$

Example

1. State, for example in the form

$$2 \in P,$$

which of the following elements are members of the sets enumerated:

Elements: 1, 2, 3, 4, 5, 6, 7,
triangle, square, rhombus, circle,
cube, tetrahedron, sphere, cylinder,
angle, length, volume, area,
$\sqrt{2}$, $\sqrt{3}$, $\sqrt{5}$, $\sqrt{7}$, $\sqrt{9}$, $\sqrt{11}$, $\sqrt{16}$;

Sets: N, positive integers,
Q, rational numbers,
F, geometrical figures,
S, polygons,
T, measures,
U, rectilinear geometrical figures,
V, even integers,
W, irrational numbers,
X, four-sided figures.

The *symbol of consequence* $\Rightarrow$ denotes that *a logical deduction can be drawn in the 'sense' indicated by the arrow.* Thus the symbolism

$$p \Rightarrow q$$

means that a statement, indicated for brevity by the symbol p, leads inevitably to the statement indicated by q.

It does not follow that q leads to p.

For example,

$$\begin{aligned} &a \text{ and } b \text{ are positive even integers} \\ \Rightarrow\ &a+b \text{ is a positive even integer.} \end{aligned}$$

On the other hand,

$$\begin{aligned} &a+b \text{ is a positive even integer} \\ \not\Rightarrow\ &a \text{ and } b \text{ are positive even integers.} \end{aligned}$$

The *symbol of necessary and sufficient consequence* $\Leftrightarrow$ is used when *the logical deduction follows each of the two senses indicated by the arrows.*

For example, for positive integers,

$$\begin{aligned} &\text{one of } a, b \text{ is odd and the other is even} \\ \Leftrightarrow\ & a+b \text{ is odd.} \end{aligned}$$

Again, in a triangle ABC,

$$\begin{aligned} & AB = AC \\ \Leftrightarrow\ & \angle ACB = \angle ABC. \end{aligned}$$

The *existence symbol* $\exists$ is used for the words *there exist(s)* or their equivalent.

For example, if N is the set† of positive integers

$$N \equiv \{1, 2, 3, \ldots\},$$

then $\exists\ x \in N$ such that $x+3 = 7$.

The *symbol of universal inclusion* $\forall$ is used in the sense that '$\forall\ x$' means *for all (relevant) values of* x.

For example, if N is the set of positive integers,

$$\forall\ x, \quad x \in N \Rightarrow x \leqslant x^2.$$

It may be remarked, finally, that most of the sets to be considered are *discrete*, the elements being capable of having, as it were, a separate label attached to each. This is, however, by no means necessary: a set might, for example, consist of all values of x in the interval $0 < x < 1$, or of all functions bounded for values of x in the interval $a \leqslant x \leqslant b$. Sets of this type are called *continuous*.

Example

1. Criticize the following arguments:

(*a*)

$$\begin{aligned} & x = 2 \\ \Rightarrow\ & x^2 = 4 \\ \Rightarrow\ & x = +2 \quad \text{or} \quad x = -2. \end{aligned}$$

† By convention, a symbol such as

$$\{1, 2, 3, \ldots, 9\}$$

denotes the set consisting of the first 9 positive integers, whereas the symbol

$$\{1, 2, 3, \ldots\}$$

without 'terminus' denotes the 'infinite set of all positive integers'.

(*b*)
$$5(4x+1) = 4(5x+1)$$
$$\Rightarrow \quad 20x+5 = 20x+4$$
$$\Rightarrow \quad 5 = 4.$$

(*c*) In $\triangle ABC$,
$$b = 1, \quad c = \sqrt{3}, \quad B = \tfrac{1}{6}\pi$$
$$\Rightarrow A = \tfrac{1}{2}\pi.$$

2. The structure of elementary addition. The positive integers forming the set
$$N \equiv \{1, 2, 3, \ldots\}$$
are subject to laws of addition which, for present purposes, have three main features, called the *laws of structure*:

(i) The *inclusive law* or the law of *closure*, that addition is everywhere defined; so that, if a and b are any two positive integers whatever, then their sum $a+b$ is defined uniquely and is included within the set N. That is,
$$\forall\, a, b, \quad \text{where} \quad a \in N,\ b \in N,$$
$$\exists \text{ unique } a+b \text{ and also } a+b \in N.$$

(ii) The *associative law* governing the equality of two different associations when three numbers are summed; thus
$$\forall\, a, b, c \in N,$$
$$(a+b)+c = a+(b+c).$$
The three numbers can be added *either* by grouping the first two and then adding the third *or* by taking the first and then adding the sum of the second and third. The total is written simply
$$a+b+c.$$

(iii) The *commutative law*, that the order in which the summed elements are written is irrelevant. Thus
$$\forall\, a,b \in N,$$
$$a+b = b+a.$$

There are generalized operations, to be met later, for which the commutative law does not hold; for example, the operation of subtraction, where $a-b \neq b-a$. In such cases it may *not* be true that, for example, $a+(b+c) = (a+c)+b$; but, of course, the meanings of the symbols will have to be extended for this alternative to be possible.

3. Generalization of the idea of addition. The laws of structure enumerated in §2 can be extended to many diverse sets of objects, provided that a suitable *rule of combination* can be prescribed. The symbol $*$ will be used to denote such a rule, in a way exactly analogous to that in which the symbol $+$ is used for elementary addition.

The use of the symbol may be first illustrated by means of an example in which everything except the notation is familiar. As before, let

$$N \equiv \{1, 2, 3, \ldots\}$$

be the set of positive integers, and let the symbol $*$ denote the operation of elementary multiplication; thus

$$3 * 5 = 15, \quad 8 * 9 = 72.$$

[Read '$3 * 5$' as '3 star 5'.]

Then (i) *the inclusive law is obeyed*, since

$$\forall\ a, b, \quad \text{where} \quad a \in N, b \in N,$$

$$\exists \text{ unique } a * b \text{ so that } a * b \in N;$$

(ii) *the associative law is obeyed*, since

$$(a * b) * c = a * (b * c),$$

this being the usual associative law $(ab)c = a(bc)$ of multiplication;

(iii) *the commutative law is obeyed*, since

$$a * b = b * a,$$

this being the law $ab = ba$.

The Illustrations which follow are intended to give increasing familiarity with these ideas by exhibiting them in a variety of contexts.

***Illustration* 1.** *Simple examples where the laws of structure are not obeyed.* These examples illustrate very elementary rules of combination and are designed to emphasize that the laws of structure are by no means automatic. For all cases in this Illustration the set is that of positive integers,

$$N \equiv \{1, 2, 3, \ldots\}.$$

(i) *Failure of the inclusive law.* Take for combination the rule of elementary division, so that

$$a * b \text{ means } a \div b.$$

Then the quotient $a * b$ is not normally a positive integer; for example,

$$3 * 7 \equiv \tfrac{3}{7} \notin N.$$

(ii) *Failure of the associative law.* Take for combination the rule of elementary subtraction, so that

$$a * b \quad \text{means} \quad a - b.$$

Then $$(a * b) * c \equiv (a-b)-c = a-b-c,$$
whereas $$a * (b * c) \equiv a-(b-c) = a-b+c,$$
so that $$(a * b) * c \neq a * (b * c).$$

(iii) *Failure of the commutative law.* Take for combination the rule of 'exponentiation', so that
$$a * b \quad \text{means} \quad a^b.$$
Then $$a * b = a^b,$$
whereas $$b * a = b^a,$$
so that $$a * b \neq b * a.$$

These Illustrations, involving very elementary operations of common occurrence, are inserted early in the exposition to warn the reader not to be led astray by notation but to test all rules of combination carefully before assuming that they necessarily obey the laws of structure.

The remaining Illustrations of this section are concerned with cases, familiar in this type of work, where the laws are effectively obeyed.

Illustration 2. *Digital arithmetic.* This Illustration demonstrates the effects of a change in the rules of combination. The ground that it covers is, in fact, somewhat ahead of the present stage of development, but it is included here in the hope of catching the reader's attention and of widening his experience.

Consider the set of 10 elements
$$M \equiv \{0, 1, 2, 3, 4, 5, 6, 7, 8, 9\}$$
and subject them in turn to the following rules:

(i) a rule of *digital addition*, whereby the sum, denoted by $a+b$, of two numbers a, b is the *units digit* of the natural arithmetical sum; for example,
$$2+7 = 9, \quad 5+8 = 3, \quad 6+4 = 0;$$

(ii) a rule of *digital multiplication*, whereby the product, denoted by $a \times b$, of two numbers a, b is the *units digit* of the natural arithmetical product; for example,
$$2\times 3 = 6, \quad 4\times 7 = 8, \quad 5\times 6 = 0.$$

Text† Examples

1. Show that the rule of digital addition satisfies the three laws of structure.

2. Show that the rule of digital multiplication satisfies the three laws of structure.

DEVELOPMENTS. The rules of digital arithmetic lead to results quite different from some long-accepted features of elementary arithmetic. For example,

† The Text Examples are usually essential parts of the text, strengthening the standard phrase, 'It may easily be shown...'.

(i) since $5\times 4 = 0$, it is possible for the product of two numbers to be 'zero' although neither number is itself zero;

(ii) since $2\times 3 = 6$ and $2\times 8 = 6$, it is possible for a number to have two distinct 'halves';

(iii) since the relation

$$x^2+5x+4 = 0$$

is satisfied by $x = 1,4,6,9$, it is possible for a quadratic equation to have four roots.

Text Examples

1. Identify a number $g \in M$ to which you would be prepared to attach the symbol '-1'.

2. Identify the two numbers $j, k \in M$ to which you would be prepared to attach the symbol '$\sqrt{(-1)}$', and verify that $j+k = 0$.

Verify also that the equation

$$x^2+4x+5 = 0$$

has precisely the two roots

$$2g+j, \quad 2g+k$$

(corresponding to '-2'$\pm$'$\sqrt{(-1)}$' of the more familiar arithmetic).

3. Prove that the elements 2, 3, 7, 8 have no square roots, but that each element has precisely one cube root.

Illustration 3. *Polynomial expressions.* Let P denote the set whose elements are quadratic polynomials (in x) with positive integral coefficients. For convenience of notation, call the polynomials $a, b, c, \ldots$, where, for example,

$$a \equiv a_1x^2+a_2x+a_3,$$
$$b \equiv b_1x^2+b_2x+b_3.$$

As rule of combination $*$, take the rule of direct addition, so that

$$a * b \equiv (a_1x^2+a_2x+a_3)+(b_1x^2+b_2x+b_3).$$

Then (i) *the inclusive law is obeyed,* since

$$a * b \equiv (a_1+b_1)x^2+(a_2+b_2)x+(a_3+b_3),$$

a quadratic polynomial with integral coefficients;

(ii) *the associative law is obeyed,* since

$$(a * b) * c = a * (b * c),$$

each being the quadratic polynomial

$$(a_1+b_1+c_1)x^2+(a_2+b_2+c_2)x+(a_3+b_3+c_3);$$

(iii) *the commutative law is obeyed,* since

$$\begin{aligned} a * b &\equiv (a_1+b_1)x^2+(a_2+b_2)x+(a_3+b_3) \\ &\equiv (b_1+a_1)x^2+(b_2+a_2)x+(b_3+a_3) \\ &\equiv b * a. \end{aligned}$$

***Illustration* 4.** *Rotations.* Let $ABCD$ be a given square free to rotate in its own plane in the counterclockwise sense about its centre O. Starting from any given position, denote by the symbol n (where $n = 1, 2, 3, \ldots$) the rotation of the square through n right angles, and let R denote the set of rotations

$$R \equiv \{1, 2, 3, \ldots\}.$$

Note, incidentally, that two rotations whose defining symbols differ by a multiple of 4 define the same position of the square.

For rule of combination, let

$$a * b \qquad (a \in R, \quad b \in R)$$

denote the rotation b followed by the rotation a.

Text Examples

1. Prove that this rule of combination satisfies the three laws of structure.

2. Suggest a modification in the above definition of rotation to allow n to take any of the values

$$\ldots, -3, -2, -1, 0, 1, 2, 3, \ldots$$

and verify that, under the corresponding rule of combination, the three laws of structure are still obeyed.

***Illustration* 5.** *Vectors.* The idea of a vector is assumed to be familiar from mechanics, though no previous knowledge is, in fact, essential. For present purposes an alternative, but equivalent, definition is more convenient.

A line issuing from a fixed point O, having given magnitude and direction, is exactly determined when its end-point P is known. Such a line represents a *vector* $\overrightarrow{OP}$. [The name *vector* or *carrier*, indeed, suggests the carrying of a point from O to P.]

Referred to a given set of rectangular coordinate axes, with origin O, the vector $\overrightarrow{OP}$ is equally well determined by the coordinates (p_1, p_2) of P. The totality of all vectors such as $\overrightarrow{OP}$ is essentially the same as the totality of all *ordered pairs*, as we may call them, such as (p_1, p_2), and a pair like (p_1, p_2) is conveniently denoted by the single symbol $\mathbf{p}$. The elements p_1, p_2 are normally regarded, in this context, as belonging to the set of real numbers.

In the usual treatment of vectors in a mechanics course, the *vector sum* of two vectors $\overrightarrow{OP}, \overrightarrow{OQ}$ is, by the parallelogram law, the vector $\overrightarrow{OR}$, where R is the fourth vertex of the parallelogram of which OP, OQ are adjacent sides. By elementary coordinate geometry, this is exactly equivalent to the definition:

The vector sum of the two vectors

$$\mathbf{p}(p_1, p_2) \quad \text{and} \quad \mathbf{q}(q_1, q_2)$$

is the vector $\mathbf{r}(r_1, r_2)$ such that

$$r_1 = p_1 + q_1, \quad r_2 = p_2 + q_2.$$

Led by this definition, we propose the following *rule of combination:*

$$\mathbf{p} * \mathbf{q} \text{ is to be the vector } (p_1+q_1, p_2+q_2),$$

and we denote by V the set of all vectors of the type $\mathbf{p}(p_1,p_2)$ subject to this rule of combination.

Text Example

1. Prove that, if the 'components' such as p_1, p_2 are taken to be positive integers, then the three laws of structure are all satisfied.

4. The union of two sets. It is not our purpose to make a close examination of Set Theory, but there are some features which are both valuable in themselves and also interesting as further examples of algebraic structure. These are the themes of the next few paragraphs.

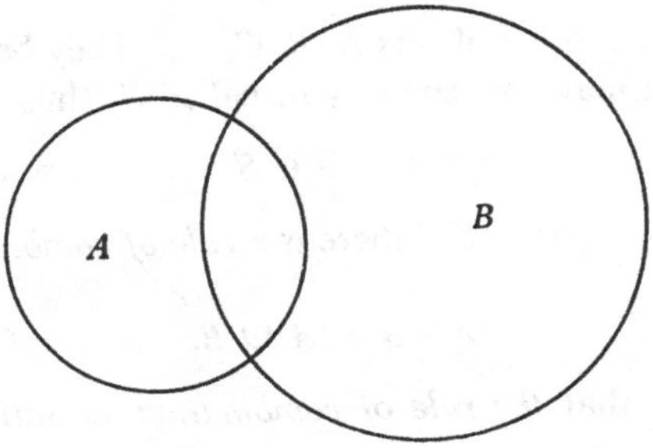

Fig. 1

Let A and B be two given sets, possibly having elements in common. They can be represented, entirely diagrammatically, by a couple of areas whose common part represents the elements common to the sets. Such a diagram is known as a *Venn Diagram*† representing the sets.

In particular, the *union* of the two given sets is defined to be *the set consisting of all those elements which belong* **either** *to A* **or** *to B* **or** *to both.*

The union of A and B is denoted by the notation

$$A \cup B,$$

read 'A cup B', so that, in precise language,

$$\begin{cases} x \in A \Rightarrow x \in A \cup B, \\ y \in B \Rightarrow y \in A \cup B. \end{cases}$$

† Named after J. Venn (1834-1923).

The set $A \cup B$ corresponding to the individual sets A and B indicated in Fig. 1 is exhibited in Fig. 2.

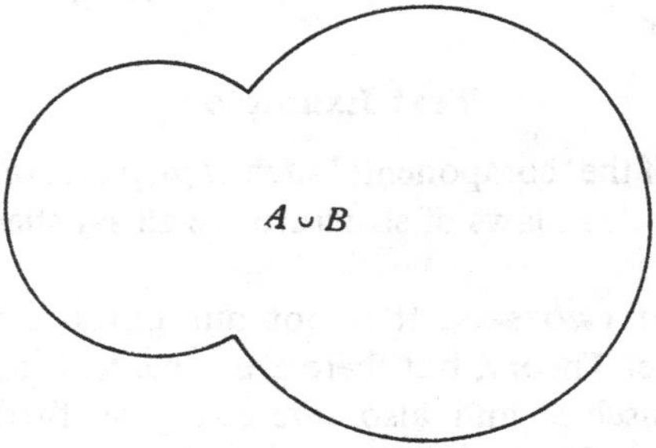

Fig. 2

Consider now a number of sets $A, B, C, \ldots$. They belong† to the 'set of all sets', which may conveniently be called S; thus

$$A \in S, \quad B \in S, \ldots.$$

Within the elements (sets) of S there is a *rule of combination* defined by the relation

$$A * B \equiv A \cup B.$$

We now prove that *the rule of combination* $\cup$ *satisfies the laws of structure:*

(i) The *inclusive law* is satisfied, since $A \cup B$ is a uniquely defined set which, by the very fact of being a set, belongs to S.

(ii) The *associative law*

$$(A \cup B) \cup C = A \cup (B \cup C)$$

is satisfied, each side of this relation denoting precisely the set whose elements belong to one at least of A, B, C.

More formally,

$$x \in A \Rightarrow x \in (A \cup B),$$
$$x \in (A \cup B) \Rightarrow x \in (A \cup B) \cup C;$$

and

$$x \in A \Rightarrow x \in A \cup (B \cup C).$$

Thus each element of A (and, similarly of B and C) is an element of each of the sets $(A \cup B) \cup C$ and $A \cup (B \cup C)$. (See Fig. 3.)

† This is perhaps less simple than it seems, but the ordinary intuitive idea is all that is intended here.

Finally, no point not in A, B or C can belong to $(A \cup B) \cup C$ or to $A \cup (B \cup C)$, as follows intuitively or by formal argument similar to that just given.

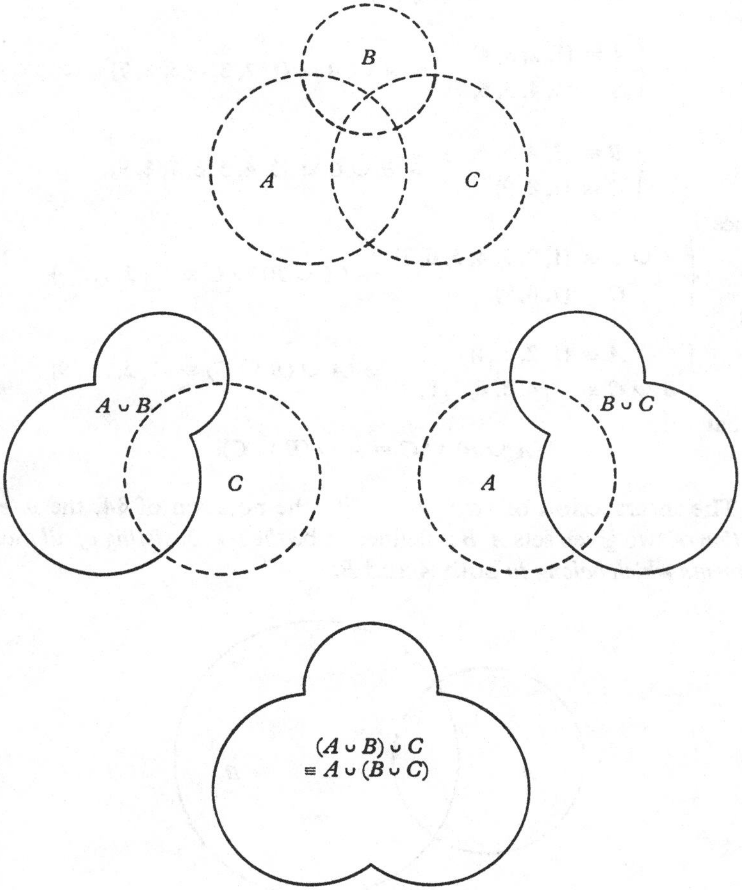

Fig. 3

(iii) The *commutative law*

$$A \cup B = B \cup A$$

is satisfied, each side, by definition, denoting the set whose elements belong either to A or to B or to both.

***Illustration* 6.** Suppose that

$$A \equiv \{1, 2, 3, 4\},$$
$$B \equiv \{3, 4, 5, 6, 7\},$$
$$C \equiv \{7, 8, 9\}.$$

Then

$$\left\{\begin{array}{l} A \equiv \{1, 2, 3, 4\} \\ B \equiv \{3, 4, 5, 6, 7\} \end{array}\right. \Rightarrow A \cup B \equiv \{1, 2, 3, 4, 5, 6, 7\}$$

and

$$\left\{\begin{array}{l} B \equiv \{3, 4, 5, 6, 7\} \\ C \equiv \{7, 8, 9\} \end{array}\right. \Rightarrow B \cup C \equiv \{3, 4, 5, 6, 7, 8, 9\}.$$

Hence

$$\left\{\begin{array}{r} A \cup B \equiv \{1, 2, 3, 4, 5, 6, 7\} \\ C \equiv \{7, 8, 9\} \end{array}\right. \Rightarrow (A \cup B) \cup C \equiv \{1, 2, \ldots, 9\}$$

and

$$\left\{\begin{array}{r} A \equiv \{1, 2, 3, 4\} \\ B \cup C \equiv \{3, 4, 5, 6, 7, 8, 9\} \end{array}\right. \Rightarrow A \cup (B \cup C) \equiv \{1, 2, \ldots, 9\}$$

so that

$$(A \cup B) \cup C \equiv A \cup (B \cup C).$$

5. The intersection of two sets. With the notation of §4, the *intersection* of two given sets A, B is defined to be *the set consisting of all those elements which belong to* **both** A **and** B.

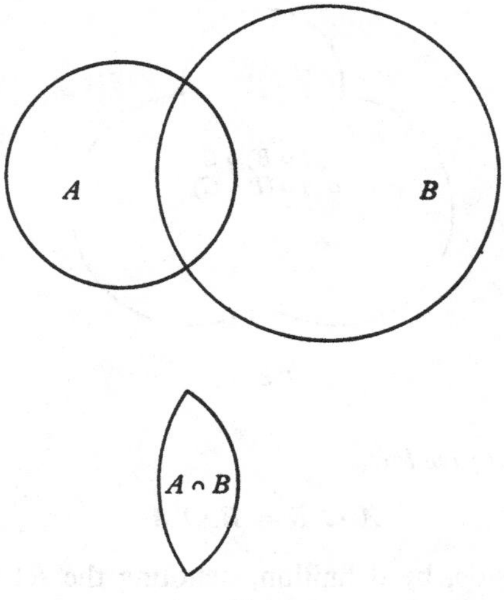

Fig. 4

The intersection of A and B is denoted by the notation

$$A \cap B,$$

read 'A cap B', so that, in precise language,

$$\left.\begin{array}{l} x \in A \\ x \in B \end{array}\right\} \Rightarrow x \in A \cap B.$$

and

See Fig. 4.

When A is the set of points of a line l and B the set of points of a line m, then $A \cap B$ is, in ordinary language, the *point of intersection* of l, m.

Text Example

1. Verify the laws of structure when the rule of combination is

$$A * B \equiv A \cap B.$$

6. The empty set. The set containing no elements is called the *empty set* and is denoted by the symbol ϕ.

There is only one empty set, distinguished from all others by the complete absence of any element—and there is no point in even attempting to decide what those elements are that it does not contain. Hence

$$\nexists\, x, \quad x \in \phi.$$

If A is any non-empty set, then

$$A \cup \phi \equiv A,$$
$$A \cap \phi \equiv \phi.$$

Text Example

1. Give a formal proof of the above two formulae.

Revision Examples

1. Determine which of the three laws of structure are obeyed by the following sets:

(i) {1, 2, 3, 4} under addition,
(ii) {2, 4, 6, 8, ...} under addition,
(iii) {1, 3, 5, 7, ...} under addition,
(iv) {1, 2, 4, 8, 16, 32, ...} under multiplication,

(v) rotations of an equilateral triangle in its plane through multiples of 120° about its centre, subject to the rule

$$\theta * \psi \equiv \text{rotation through } \psi \text{ followed by rotation through } \theta.$$

2. Prove that the set of ordered number-pairs, of which (a_1, a_2) is typical, satisfy the three laws of structure when the rule of combination is

(i) $(a_1, a_2) * (b_1, b_2) \equiv (a_1 b_2 + a_2 b_1, a_2 b_2)$,

(ii) $(a_1, a_2) * (b_1, b_2) \equiv (a_1 b_1 - a_2 b_2, a_1 b_2 + a_2 b_1)$.

3. A set consists of all positive numbers expressed in decimal form. The rule of combination is that

$$a * b \equiv \text{the integral part of the sum } a+b.$$

(For example, $2{\cdot}8 * 3{\cdot}1 = 5$.) Prove that $(a * b) * c$ is not necessarily equal to $a * (b * c)$.

4. The set

$$N \equiv \{1, 2, 3, \ldots\}$$

is subject to the rule of combination

(i) $a * b \equiv a - b$, (ii) $a * b \equiv a \div b$.

Prove that none of the laws of structure is necessarily satisfied.

5. If

$$M \equiv \{1, 2, 3, \ldots, 20\}, \quad N \equiv \{1, 3, 5, \ldots, 19\},$$

enumerate $M \cup N$ and $M \cap N$.

6. If

$$M \equiv \{1, 2, 4, 8, 16\}, \quad N \equiv \{2, 4, 6, 8, 10, 12, 14, 16\},$$
$$P \equiv \{1, 2, 3, \ldots, 16\},$$

identify

$$M \cup (N \cup P), \quad (M \cup N) \cup P, \quad M \cap (N \cap P), \quad (M \cap N) \cap P,$$
$$M \cup (N \cap P), \quad (M \cup N) \cap P, \quad M \cap (N \cup P), \quad (M \cap N) \cup P.$$

7. The set M consists of all rhombi in a plane and the set N consists of all rectangles in the same plane. Identify the set $M \cap N$.

8. A 'parent' set consists of n elements. From these elements, two sets A, B are taken (possibly having elements in common). The elements of the parent set other than those of A form a set $\overline{A}$ called the *complement* of A, and similar notation is used for the complements of other sets. Prove that

$$\overline{A \cup B} = \overline{A} \cap \overline{B}, \quad \overline{A \cap B} = \overline{A} \cup \overline{B}.$$

9. The set N consists of all positive integers; the set R consists of all real numbers; the set C consists of all complex numbers. Determine the sets to which x belongs if

(i) $x^2-3x+2=0$, (ii) $2x^2+3x-4=0$,
(iii) $x^2+2x+5=0$.

10. The set R consists of all real positive numbers (greater than zero). The law of combination is

$$a*b=\frac{ab-1}{a+b}.$$

Prove that $$(a*b)*c=a*(b*c).$$

11. In a company of 40 women, 20 wear scarves, 16 wear hats; of these 6 wear both. The rest wear neither. Illustrate this in a Venn diagram, giving the number who wear neither.

12. All the 35 boys in a certain class take one or more of the subjects E, F, G (English, French, German). The total numbers of boys taking E, F, G are respectively 15, 18, 23. Of these, 6 take at least E, F; 8 take at least E, G; 9 take at least F, G. Draw a Venn diagram representing 'the sets E, F, G' and their intersections $E\cap F$, $E\cap G$, $F\cap G$. Deduce that $E\cap F\cap G=2$, so that the number of boys taking all three subjects is 2.

13. A group of 28 men gave evidence that they had observed one or more of three events A, B, C. In all, 11 had seen A, 13 had seen B, and

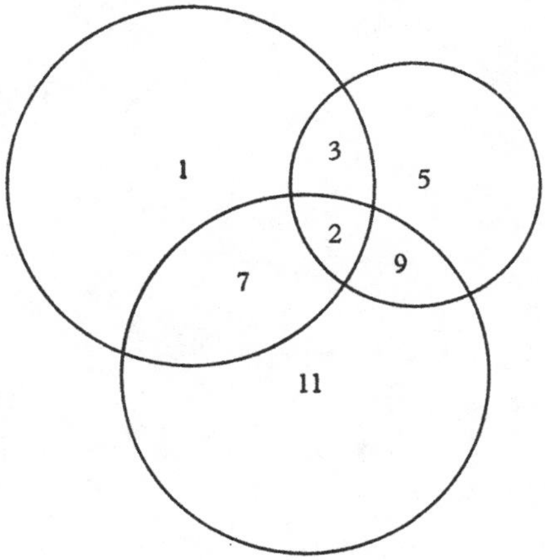

Fig. 5

15 had seen C. Of these, 4 had seen B and C, but not A, 3 had seen C and A but not B, while 1 had seen all three. Illustrate this in a Venn diagram and confirm that 2 had seen A and B but not C.

14. Invent one or more problems of the type of Questions 11, 12, 13 based on the Venn diagram indicated in the accompanying Fig. 5, where the three given sets have $1+3+2+7$ elements, $5+9+2+3$ elements, $11+7+2+9$ elements.

2

NEUTRAL ELEMENTS AND INVERSE OPERATIONS

1. Neutral elements. Let A be a given set, of which a is a typical element, subject to a given rule of combination. There may, or may not, exist a *neutral element*, temporarily called e, which has no effect in the sense that

$$a * e = e * a = a$$

for all elements a of the set.

Illustration 1. (i) The set of positive integers

$$N \equiv \{1, 2, 3, \ldots\},$$

subject to the normal laws of addition, has no neutral element. Every member is changed by any addition.

(ii) The same set N, subject to the normal laws of multiplication, has neutral element 1, since

$$x \in N$$

$$\Rightarrow x \times 1 = 1 \times x = x.$$

(iii) The set of positive integers with zero

$$M \equiv \{0, 1, 2, 3, \ldots\},$$

subject to the normal laws of addition, has neutral element 0, since

$$x \in M$$

$$\Rightarrow x + 0 = 0 + x = x.$$

2. Uniqueness of the neutral element. It is important to observe that *a set A, subject to a given rule of combination, cannot have more than one neutral element:*

Suppose that, on the contrary, e and f are both neutral elements. Then, $\forall\ x \in A$,

$$x * e = e * x = x,$$

$$x * f = f * x = x.$$

In particular, take the special cases when x in the first line is f, so that

$$e * f = f,$$

and when x in the second line is e, so that

$$e * f = e.$$

Then
$$f = e,$$

so that the two units are identical.

3. Subtraction. Let the set of positive integers, with zero, be

$$M \equiv \{0, 1, 2, 3, \ldots\},$$

subject to ordinary addition as the rule of combination. Then

$$x \in M, \quad y \in M$$
$$\Rightarrow \exists\, z \in M \quad \text{such that} \quad x+y = z.$$

When x and y are given, the 'sum' z can always be found. On the other hand, it may or may not be possible to find, say, y when x and z are given; for example,

$$\nexists\, y \in M \quad \text{such that} \quad 8+y = 2.$$

To obtain *a set in which the relation*

$$x+y = z$$

is satisfied when any two of x, y, z are given, the set M must be augmented by the negative integers in the form

$$Z \equiv \{\ldots, -3, -2, -1, 0, 1, 2, 3, \ldots\}.$$

The relation

$$8+y = 2$$

can then be satisfied for $y \in Z$ by selecting the value $y = -6$.

4. Division. Let the set of positive integers, without zero, be

$$N\{1, 2, 3, \ldots\},$$

subject to ordinary multiplication as the rule of combination. Then

$$x \in N, \quad y \in N$$
$$\Rightarrow \exists\, z \in N \quad \text{such that} \quad x \times y = z.$$

Here, again, it may not be possible to find y when x and z are given; for example,

$$\nexists\, y \in N \quad \text{such that} \quad 5 \times y = 8.$$

To obtain *a set in which the relation*

$$x \times y = z$$

is satisfied when any two of x, y, z are given, the set N must be augmented to the set Q of rational numbers whose elements are of the form a/b, where a, b are positive integers ($b \neq 0$). The relation

$$5 \times y = 8$$

can then be satisfied for $y \in Q$ by selecting the value $y = \frac{8}{5}$.

NOTE: It is important to observe that the solubility of the relations

$$8 + y = 2,$$
$$5 \times y = 8$$

depends not so much on the relations themselves as on the set of values from which y can be drawn. In the notation of the preceding paragraphs, the first relation has no solution in M and the second relation has no solution in Z.

5. Inverse operations. Subtraction and division are familiar examples of *inverse processes* for addition and multiplication respectively. More generally, let

$$S \equiv \{a, b, c, \ldots\}$$

be any set whose elements satisfy the three laws of structure for the relevant rule of combination. Then

$$x \in S, \quad y \in S$$
$$\Rightarrow \exists\, z \in S \quad \text{such that} \quad x * y = z.$$

It does not, however, follow by any means that this relation can be satisfied within S when, say, x and z are given. The examples of the preceding paragraphs demonstrate this clearly. On the other hand, such sets certainly do exist, and their properties must now be investigated.

Suppose, then, that S is a typical set with the property that, for a given rule of combination, the relation

$$x * y = z$$

can be solved within it when *any two* of x, y, z are given.

Certain important results follow (associativity being assumed):

(i) *The set S contains a neutral element.* By hypothesis, members e, f of S can be found such that, for *given* $a \in S$,

$$a * e = a,$$

$$f * a = a.$$

Suppose now that x is an *arbitrary* element of S. Then

$$\exists\, u, v \in S$$

such that

$$a * u = x, \quad v * a = x.$$

But, since the associative law is valid for the operation $*$, we have

$$x * e = (v * a) * e = v * (a * e) = v * a = x$$

and $$f * x = f * (a * u) = (f * a) * u = a * u = x,$$

so that, although e, f were defined in the first instance with reference to the particular element a, *the corresponding relations*

$$x * e = x, \quad f * x = x$$

are true for all $x \in S$.

In the relation $x * e = x$, put $x = f$; in the relation $f * x = x$, put $x = e$. Then

$$f * e = f, \quad f * e = e,$$

so that $$f = e.$$

Hence *there exists an element* $e \in S$ *with the property that, for arbitrary* x,

$$x * e = e * x = x,$$

so that, by definition, e is a neutral element, often called a *unit*.

It has already been proved (§2) that *the neutral element is unique*.

(ii) *Every element* $x \in S$ *has an inverse* $y \in S$ *defined by the relations*

$$x * y = e, \quad y * x = e.$$

The existence of y for these two relations severally is an automatic consequence of the hypothesis that each relation can be solved when two of its elements are known. The fresh fact is that y is the same in each case.

Suppose that y, z are defined by the relations

$$x * y = e, \quad z * x = e.$$

Then

$$\begin{aligned} y &= e * y \\ &= (z * x) * y \\ &= z * (x * y) \\ &= z * e \\ &= z. \end{aligned}$$

(iii) *The inverse of a given element* a *is unique.* If not, suppose that a has two distinct inverses u and v, so that

$$a * u = u * a = e,$$
$$a * v = v * a = e.$$

In particular,

$$a * u = e, \quad v * a = e.$$

Hence, by (ii), with a mere change of notation,

$$u = v.$$

NOTATION. The inverse of an element a may be denoted by the symbol a'. Thus

$$a * a' = a' * a = e.$$

(iv) *The relation*

$$x * a = b$$

defines x *uniquely when* a, b *are given.*

The element a has the unique inverse a'. Now

$$\begin{aligned} & x * a = b \\ \Rightarrow\ & (x * a) * a' = b * a' \\ \Rightarrow\ & x * (a * a') = b * a' \\ \Rightarrow\ & x * e = b * a' \\ \Rightarrow\ & x = b * a'. \end{aligned}$$

The only possible value for x is thus $b * a'$, and substitution in the given relation $x * a = b$ shows that it is indeed a solution.

In the same way,

$$\begin{aligned} & a * x = b \\ \Rightarrow\ & x = a' * b, \end{aligned}$$

which, again, is actually a solution.

IMPORTANT NOTE. When the operation of combination is not commutative, the two solutions

$$x = b * a', \quad x = a' * b$$

are usually distinct. We have not yet given much attention to non-commutative operations, but the fact should be registered at once.

(v) *The inverse of a product: The inverse of the element* $a * b$ *is* $b' * a'$ (*in that order*), so that

$$(a * b)' = b' * a'.$$

[Here, again, the importance of the order in which the elements are taken refers to the non-commutative case.]

By the associative law, followed by the definitions of inverse and neutral elements,

$$\begin{aligned}(b' * a') * a &= b' * (a' * a)\\ &= b' * e\\ &= b',\end{aligned}$$

and so

$$\begin{aligned}(b' * a') * (a * b) &= b' * b\\ &= e,\end{aligned}$$

showing that $b' * a'$ is the inverse of $a * b$.

Text Examples

1. Prove that the inverse of a' is a itself.

2. Prove that the inverse of $a * b * c$ is $c' * b' * a'$.

3. A set S is subject to a rule of combination that satisfies the laws of structure. It has a neutral element e, such that

$$\forall\, a \in S, \quad a * e = e * a = a.$$

Each element $a \in S$ has an inverse $a' \in S$ such that

$$a * a' = a' * a = e.$$

Prove that the equation

$$x * y = z$$

can be satisfied within S when any two of x, y, z are given. Is the commutative law necessary to this problem?

4. Prove that the neutral element (when it exists) is its own inverse.

Give an example of a set with a law of structure under which an element other than the neutral element is its own inverse.

5. Prove that the set of positive even integers, under multiplication, has no neutral element; and that in the set of positive odd integers, under multiplication, no element other than the neutral element itself has an inverse.

6. Determine, when they exist, the neutral elements and the inverses of arbitrary elements for each of the following sets, defined in Chapter 1, extending the set in the obvious way when that is necessary:

(i) Digital arithmetic on the set

$$\{0, 1, 2, 3, 4, 5, 6, 7, 8, 9\}$$

for (*a*) digital addition, (*b*) digital multiplication;

(ii) Polynomial expressions

$$a_1 x^2 + a_2 x + a_3$$

for addition;

(iii) Rotations, where

$$a * b$$

means rotation b followed by rotation a.

(iv) Vectors, where

$$\mathbf{p} * \mathbf{q}$$

is the vector $(p_1 + q_1, p_2 + q_2)$.

3

THE GROUP STRUCTURE

1. First ideas. A group is, speaking informally, a set S of elements subject to a rule of combination displaying properties discussed in the first two chapters:

(i) obedience to the laws of structure (but the commutative law not being essential),

(ii) solubility of the relation

$$x * y = z$$

when any two of these elements are given.

Illustration 1. *Addition modulo* 4. Let

$$S \equiv \{0, 1, 2, 3\}$$

be a set of four given numbers subject to the rule of combination that

$$a * b$$

means the remainder after dividing the ordinary arithmetical sum $a+b$ by 4. There is thus a *set* and a *rule*. In detail, the 'sums' are

$$\begin{array}{llll} 0*0=0, & 0*1=1, & 0*2=2, & 0*3=3, \\ 1*0=1, & 1*1=2, & 1*2=3, & 1*3=0, \\ 2*0=2, & 2*1=3, & 2*2=0, & 2*3=1, \\ 3*0=3, & 3*1=0, & 3*2=1, & 3*3=2. \end{array}$$

The three laws of structure (including, in this case, the commutative law) are all satisfied; in particular,

$$(a * b) * c = a * (b * c),$$

since each expression is the remainder on dividing the arithmetical sum $a+b+c$ by 4.

Further, the relation

$$x * y = z$$

is satisfied within S when any two of x, y, z are given. This can be verified directly from the table; alternatively, the set has a neutral element 0 and the elements

$$0, 1, 2, 3$$

have inverses

$$0, 3, 2, 1$$

respectively. The relations (p. 23)

$$x = z * y', \quad y = x' * z$$

then give x, y in turn.

These four elements subject to this rule therefore form a *group*.

Illustration 2. *The fourth roots of unity.* Let

$$T \equiv \{1, i, -1, -i\},$$

where i is a complex root of -1, be a set of four given numbers subject to the rule of combination that

$$a * b$$

means the ordinary product of a by b. Here, again, there is a *set* and a *rule*. In detail, the 'products' are

$$\begin{array}{llll} 1 * 1 = 1, & 1 * i = i, & 1 * -1 = -1, & 1 * -i = -i, \\ i * 1 = i, & i * i = -1, & i * -1 = -i, & i * -i = 1, \\ -1 * 1 = -1, & -1 * i = -i, & -1 * -1 = 1, & -1 * -i = i, \\ -i * 1 = -i, & -i * i = 1, & -i * -1 = i, & -i * -i = -1. \end{array}$$

Text Example

1. Verify that the set, subject to this rule of combination, forms a group. Identify the neutral element and the inverse of each of the four given elements.

2. Group tables and isomorphism. The detailed 'sums' and 'products' given in the preceding Illustrations are displayed clumsily; their appeal can be greatly enhanced by the use of tabular form. For this, the 'sums' of the first Illustration are exhibited:

	0	1	2	3
0	0	1	2	3
1	1	2	3	0
2	2	3	0	1
3	3	0	1	2

and the products of the second Illustration are exhibited:

	1	i	-1	$-i$
1	1	i	-1	$-i$
i	i	-1	$-i$	1
-1	-1	$-i$	1	i
$-i$	$-i$	1	i	-1

In each case, *an element $a * b$ is located by finding the intersection of the row containing the element a with the column containing the element b.*

From these tables a startling fact emerges. Consider a third table which, with developments immediately in view, we write in the form

	e	a	b	c
e	e	a	b	c
a	a	b	c	e
b	b	c	e	a
c	c	e	a	b

The first table is, very obviously, a particular case of it, obtained by writing

$$e = 0, \quad a = 1, \quad b = 2, \quad c = 3.$$

But *the second table is also a particular case*, obtained by writing

$$e = 1, \quad a = i, \quad b = -1, \quad c = -i$$

In other words, *the tables corresponding to the groups given in Illustrations* 1 and 2 *are identical except for notation.* It is this possibility that gives to groups their position of fundamental importance in mathematics. They pick out, as it were, the essentially identical structures of sets and rules of combination which seem, at first sight, to have little in common. On this basis, generalization can proceed in many unexpected directions.

Without wishing to emphasize either the word or the idea unduly at present, we add that groups which can be exhibited by means of identical tables (subject only to notation) are said to be *isomorphic*—that is, having the same form. (See later, p. 285.)

3. More formal statement of group structure. Let there be given a set S whose elements satisfy the *first two* laws of structure defined on p. 6. (The third law is held in abeyance for the present.) Thus

$$a \in S, \quad b \in S$$

$\Rightarrow \exists$ unique $c \in S$ such that

$$a * b = c,$$

and, further,

$$a \in S, \quad b \in S, \quad c \in S$$

$$\Rightarrow \quad (a * b) * c = a * (b * c).$$

Finally, let the set be subject to the further condition that the relation

$$a * b = c$$

can be satisfied *uniquely* within S when any two of a, b, c are given.

DEFINITION. The set S, subject to a rule of combination satisfying these conditions, is said to form a *group*.

When the commutative law is also satisfied, so that

$$a \in S, \quad b \in S \Rightarrow a * b = b * a,$$

the group is said to be *commutative* or *Abelian*.

[The Norwegian mathematician Niels Henrik Abel was born at Findoë in 1802 and died at Arendal in 1829.]

The fact that the relation

$$a * b = c$$

can be satisfied within S when any two of a, b, c are given leads at once (p. 22) to the following properties of groups:

(i) there is a neutral element e,

(ii) every element $a \in S$ has a unique inverse a' such that

$$a * a' = a' * a = e.$$

These two properties are often used instead of the $a * b = c$ property to define a group. They form, in any event, a convenient test—without unit and inverses there cannot be a group.

Attention should also be drawn to the unique solution of the relations

$$x * a = b, \quad a * x = b$$

in the form

$$x = b * a', \quad x = a' * b$$

respectively.

THEOREM. *To prove that the inverse* $(a * b)'$ *of an element* $a * b$ *is the element* $b' * a'$. *(Compare also p. 24.)*

By the fundamental rules,

$$(a * b) * (b' * a') = a * (b * b') * a' = a * a' = e,$$
$$(b' * a') * (a * b) = b' * (a' * a) * b = b' * b = e,$$

as required.

A group that contains precisely n elements is said to be a *finite* group of *order n*.

A group may contain an 'infinite' number of elements: for example, the integers, positive, negative or zero under addition. It may also be *discrete* or *continuous* according as its elements form a discrete set (for example, the integers) or a continuous set (for example, the values of x in the interval $0 \leqslant x \leqslant 1$).

Most of the work in this book is concerned with sets and groups whose elements are discrete and, usually, finite in number.

NOTATION. From now on, the *group inverse* of an element a will be denoted by the notation a^{-1}.

4. The structure table of a group. We have already given particular examples of structure tables. Suppose, more generally, that a group S has a finite number of discrete elements $a, b, c, \dots$. The products $a * a$, $a * b, \dots$ may be exhibited in a table in the form

	a	b	c	d	...
a	$a*a$	$a*b$	$a*c$	$a*d$	...
b	$b*a$	$b*b$	$b*c$	$b*d$	...
c	$c*a$	$c*b$	$c*c$	$c*d$	...
...	...	...	...	...	...

where the product $p * q$ is found at the intersection of the row through p with the column through q.

The following properties, though simple, are very important:

(i) *The 'product' elements in any row* (*column*) *are all distinct.*

If, for example,

$$a * c = k$$

and

$$a * g = k,$$

then the equation

$$a * x = k$$

has two distinct solutions, which is impossible since the relation must be uniquely soluble.

(ii) *Each row* (*column*) *contains all the elements of S.*

If, for example, the element k is missing from the row defined by a, then the equation

$$a * x = k$$

has no solution, which is impossible since the relation must be soluble.

It is easy to verify that these two properties are satisfied by the tables given in §2 (p. 27).

5. Recognition of a group. Given a set of elements S and a rule of combination, it is not always easy to decide whether they form a group.

A possible first step is to form a structure table. If it can be filled in at all, that will at least ensure that

$$x \in S, \quad y \in S$$
$$\Rightarrow \exists\, x * y \in S.$$

In addition, there must be a neutral element and also each row (column) must contain each element exactly once; alternatively, each element must have a unique inverse.

The existence of a proper table establishes all the properties required for a group, with the exception of the associative law

$$(x * y) * z = x * (y * z).$$

This may often be proved directly from the very nature of the rule of combination. In stubborn cases the products may have to be evaluated individually; here the structure table is, of course, very helpful.

Illustration 3. *The structure table for 'multiplication modulo* 4'; *disproof of group*. Let

$$S \equiv \{0, 1, 2, 3\}$$

be a set of four given numbers subject to the rule of combination that

$$a * b$$

means the remainder after dividing the ordinary arithmetical product by 4. The structure table is

	0	1	2	3
0	0	0	0	0
1	0	1	2	3
2	0	2	0	2
3	0	3	2	1

It follows at once from this table that *the operation does not define a group*. In fact, rows (columns) headed by 0 or 2 do not contain all four elements, which would be essential for a group.

The group definitions break down, since the equations

$$0 * x = k \qquad (k = 1, 2, 3),$$
$$2 * x = k \qquad (k = 1, 3)$$

cannot be satisfied.

Illustration 4. *Reflexions*.

For convenience, this work is expressed in terms of rectangular Cartesian coordinates.

Given an arbitrary point (x,y) in a plane, its reflexions in the origin, in the y-axis and in the x-axis are respectively the points $(-x,-y)$, $(-x,y)$, $(x,-y)$.

Denote by u, v, w the *operations* corresponding to these reflexions in turn, and by e the 'identity operation' which leaves (x,y) unaltered.

Write

$$u(x,y), \quad v(x,y), \quad w(x,y)$$

to denote the results of operating on (x,y) by u, v, w, so that

$$u(x,y) = (-x,-y),$$
$$v(x,y) = (-x,y),$$
$$w(x,y) = (x,-y).$$

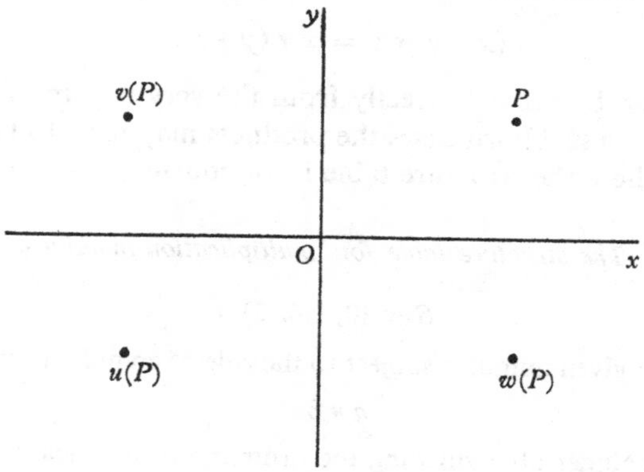

Fig. 6

Further, let notation such as

$$u * w$$

be used to denote the operation whereby w acts on (x,y) and then u acts on the result. Thus, in detail,

$$\begin{aligned} w(x,y) &= (x,-y), \\ u * w(x,y) &= u(x,-y) \\ &= \{-x,-(-y)\} \\ &= (-x,y). \end{aligned}$$

The point of this is that we recognize $(-x,y)$ as the reflexion $v(x,y)$, so that

$$u * w(x,y) = v(x,y).$$

It then becomes natural to say that *the operation $u * w$ is identical in effect with the operation v*, so that, as an 'algebra of operators',

$$u * w = v.$$

The reader is now in a position to verify the following examples.

Text Examples

1. Prove that the structure table for products analogous to those just defined is the following:

	e	u	v	w
e	e	u	v	w
u	u	e	w	v
v	v	w	e	u
w	w	v	u	e

2. Verify, with the help of the above table or otherwise, that the products of operations obey the associative law.

[For example,

$$(u * v) * w = w * w = e,$$

$$u * (v * w) = u * u = e,$$

so that $$(u * v) * w = u * (v * w).]$$

3. Deduce that the four operations, subject to the above rule of combination, form a group.

REMARKS. (i) The group with this structure is a very famous one, known as *Klein's four-group*.

(ii) The top left-hand corner of the table is

	e	u
e	e	u
u	u	e

which is itself a group, a *subgroup* of the larger group. It has e as unit and the other element u obeys the law

$$u^2 = e \text{ (where } u^2 \equiv u * u).$$

If we care to interpret e as 'even' and u as 'odd', the table may, should we wish, be read:

even + even = even; even + odd = odd,
odd + even = odd; odd + odd = even.

6. Permutations. The three numbers 1, 2, 3 can be written down in six distinct orders,

123, 231, 312; 132, 213, 321,

corresponding to the six *permutations* of the three numbers. Each of these permutations may be regarded as an *operation* on the triplet to bring it to an assigned order. The six corresponding *operators* may be called e, a, b, u, v, w, where

$$e(123) = 123, \quad a(123) = 231, \quad b(123) = 312,$$
$$u(123) = 132, \quad v(123) = 213, \quad w(123) = 321.$$

The *product* of two of these operations is defined as that permutation which is obtained when the numbers 1, 2, 3 are subjected to them in succession.

For example (omitting the asterisks for convenience),

$$au(123) = a\{u(123)\} = a(132),$$

and the effect of the operation by a is to change 1, 2, 3 into 2, 3, 1 respectively, so that, finally,

$$au(123) = (213)$$
$$= v(123).$$

Hence

$$au = v.$$

In the same way,

$$w^2(123) = w\{w(123)\} = w(321),$$

where, now, w changes 1 to 3, 2 to 2, 3 to 1, so that

$$w^2(123) = (123)$$
$$= e(123).$$

Hence

$$w^2 = e.$$

Text Examples

1. Verify the following structure table:

	e	a	b	u	v	w
e	e	a	b	u	v	w
a	a	b	e	v	w	u
b	b	e	a	w	u	v
u	u	w	v	e	b	a
v	v	u	w	a	e	b
w	w	v	u	b	a	e

2. Verify that the products obey the associative law:
(i) by appeal to the law of combination,

(ii) by argument such as

$$(a * u) * w = v * w = b,$$

$$a * (u * w) = a * a = b,$$

so that $$(a * u) * w = a * (u * w).$$

3. Prove that the six permutations form a group.

4. Prove that the three permutations e, a, b form a subgroup, but that u, v, w do not.

5. Prove that the group of six permutations is not Abelian.
(For example, $v * w = b$, $w * v = a$. Find other examples.)

6. Verify that the inverses of e, a, b, u, v, w are respectively e, b, a, u, v, w.

7. Establish the relations $a^3 = e$, $b^3 = e$.

8. Verify directly the formulae for inverses:

$$(uvw)^{-1} = w^{-1}v^{-1}u^{-1},$$

$$(ubc)^{-1} = c^{-1}b^{-1}u^{-1},$$

$$(auv)^{-1} = v^{-1}u^{-1}a^{-1}$$

***Illustration* 5.** *The Klein four-group again.* Consider the four permutations on four numbers 1, 2, 3, 4:

$$e(1234) = 1234,$$

$$u(1234) = 4321,$$

$$v(1234) = 3412,$$

$$w(1234) = 2143,$$

where e is the identity and u, v, w are found by interchanging 14, 23; 24, 31; 34, 12 respectively.

By direct calculation,

$$u^2(1234) = u\{u(1234)\} = u(4321)$$
$$= 1234,$$

on interchanging 1, 4 and 2, 3 for the last step.

Hence $$u^2 = e.$$

Similarly $$v^2 = e, \quad w^2 = e.$$

Also $$vw(1234) = v\{w(1234)\} = v(2143)$$
$$= 4321,$$

on interchanging 2, 4 and 3, 1 for the last step. Hence

$$vw = u.$$

Similarly $$wu = v, \quad uv = w.$$

The structure table is

	e	u	v	w
e	e	u	v	w
u	u	e	w	v
v	v	w	e	u
w	w	v	u	e

which we recognize as that of Klein's four-group (p. 33).

***Illustration* 6.** *The cross-ratio group.* The familiar set of six cross-ratios of four parameters can, surprisingly, be made the basis of a group whose structure table is identical with that of the six permutations on three numbers (p. 33).

Let x be any given number and denote by e, a, b, u, v, w the six operations which, acting on x, replace it in turn by the values x, $1/(1-x)$, $(x-1)/x$, $1-x$, $1/x$, $x/(x-1)$; thus

$$ex = x, \qquad ax = \frac{1}{1-x}, \qquad bx = \frac{x-1}{x},$$

$$ux = 1-x, \quad vx = \frac{1}{x}, \qquad wx = \frac{x}{x-1}.$$

The products are to be defined by successive operations. For example,

$$aux = a\{ux\} = a(1-x).$$

Now the process a acting on any number ξ changes it to $1/(1-\xi)$, so that

$$a(1-x) = 1/\{1-(1-x)\} = 1/x.$$

Hence

$$aux = vx,$$

so that

$$au = v.$$

Similarly

$$w^2x = w\{w(x)\} = w\left\{\frac{x}{x-1}\right\}$$

$$= \frac{\left(\frac{x}{x-1}\right)}{\left(\frac{x}{x-1}\right)-1} = \frac{x}{x-(x-1)} = x$$

$$= ex,$$

so that

$$w^2 = e.$$

Text Example

1. Prove that the structure table for this group is the same as that given on p. 34.

7. Cyclic groups. Let S be any group, with neutral element e. If a is any other element, then, by the group property, S contains each of the 'powers'

$$a * a, (a * a) * a, \ldots,$$

denoted by

$$a^2, a^3, \ldots$$

respectively, where, inductively,

$$a^{n+1} = a^n * a;$$

thus, if k is any positive integer,

$$a^k \in S.$$

Further the element a^k has an inverse for each k, and it is natural to denote that inverse by the notation a^{-k}, so that

$$a^k * a^{-k} = a^{-k} * a^k = e.$$

It follows that, *if p, q are integers (positive, negative or zero), then*

$$a^p * a^q = a^{p+q},$$

with the interpretation

$$a^0 = e.$$

Text Example

1. Prove the statement just given in italics when
(i) $p > 0, q > 0$; (ii) $p > 0, q < 0$;
(iii) $p < 0, q > 0$; (iv) $p < 0, q < 0$.

Suppose now that S is a group of finite order n, that is, having n elements. The sequence of elements

$$e, a, a^2, a^3, \ldots, a^k, \ldots$$

all belong to S, so that not more than n of them can be distinct. Suppose that t is the first positive integer for which a^t is equal to some earlier member of the sequence; say

$$a^t = a^u \qquad (t > u \geqslant 0).$$

This relation ensures that, a^{-u} being the element inverse to a^u,

$$a^t * a^{-u} = a^u * a^{-u} = e,$$

so that, as proved in the preceding example,

$$a^{t-u} = e.$$

Hence *there is a positive integer s with the property that*

$$a^s = e.$$

When referring to this integer hereafter we shall assume that s is the *least* positive integer with this property.

Then, incidentally,

$$a^{p+qs} = a^p$$

for all integral q.

Hence *associated with each element a* ($\neq e$) *of a group there is an integer s (dependent on the particular element) such that*

$$a, a^2, a^3, \ldots, a^{s-1}$$

are all distinct members of S.

Definition. The number s is called the *order* of the element a in the group S.

Definition. A group whose elements consist entirely of the set

$$e, a, a^2, \ldots, a^{s-1} \qquad (a^s = e)$$

is called *a cyclic group of order s generated by a.*

Text Examples

1. Show that no two of the elements

$$e, a, a^2, \ldots, a^{s-1}$$

just defined can be identical.

2. Prove that, in the case when s is a prime number, if $\alpha(\neq e)$ is any one of these elements, then the elements can equally be expressed in the form

$$\alpha, \alpha^2, \alpha^3, \ldots, \alpha^{s-1},$$

where $\alpha^s = e$.

3. Prove that a cyclic group is necessarily Abelian (commutative).

4. Prove that the set whose elements are the n roots of the equation $x^n = 1$, subject to multiplication as the law of combination, is a cyclic group of order n.

Is the same true for the equation $x^n = -1$?

5. Prove that the group of four elements 0, 1, 2, 3, subject to addition modulo 4 (see p. 26), is a cyclic group.

***Illustration* 7.** (Demonstrating some abstract methods.)

(i) To prove that, *if each element x of a group satisfies the relation* $x^2 = e$, *where e is the unit, then the group is necessarily commutative.*

Let a, b be any two elements, and write $ab = c$. Then

$$ab = c$$
$$\Rightarrow a(ab) = ac \Rightarrow b = ac \qquad (a^2 = e)$$
$$\Rightarrow bc = (ac)c \Rightarrow bc = a \qquad (c^2 = e)$$
$$\Rightarrow b(bc) = ba \Rightarrow c = ba \qquad (b^2 = e).$$

Hence
$$ab = c = ba.$$

(ii) To prove that *a group of four distinct elements is necessarily commutative.* Let e, a, b be three distinct elements, so chosen that e is the unit but b is not the inverse of a. Then

$$ab \neq e \qquad (a, b \text{ not inverse}),$$
$$ab \neq a \qquad (b \neq e),$$
$$ab \neq b \qquad (a \neq e).$$

Hence the four elements are

$$e, a, b, ab.$$

But, once again,

$$ba \neq e \qquad (a, b \text{ not inverse}),$$
$$ba \neq a \qquad (b \neq e),$$
$$ba \neq b \qquad (a \neq e).$$

Hence the fourth element is also ba, so that

$$ab = ba.$$

The rest follows easily; for example,

$$a(ab) = a(ba) = (ab)a.$$

Revision Examples

1. Prove that the following Illustrations in Chapter 1 define groups:

(i) Illustration 2 (p. 8), digital arithmetic under addition but not under multiplication,

(ii) Illustration 3 (p. 9), polynomial expressions,

(iii) Illustration 4 (p. 10), rotations,

(iv) Illustration 5 (p. 10), vectors.

2. Prove that the following sets, with corresponding rules of combination, do not define groups:

(i) the odd integers subject to

(*a*) ordinary addition; (*b*) ordinary multiplication;

(ii) the set $\{0, 1, 2, 3, 4, 5, 6, 7, 8, 9, 10\}$ subject to

(*a*) ordinary addition, (*b*) ordinary multiplication;

(iii) the quadratic polynomials $\{a \equiv a_1x^2 + a_2x + a_3\}$ subject to ordinary multiplication.

3. Determine which of the following sets, with corresponding rules of combination, form groups:

(i) the complex numbers subject to
(*a*) addition, (*b*) multiplication;

(ii) the positive rational numbers, exclusive of zero, subject to
(*a*) addition, (*b*) multiplication;

(iii) the rational numbers, positive, negative or zero, subject to
(*a*) addition, (*b*) multiplication;

(iv) the positive integral powers of 3

$$\{1, 3, 3^2, 3^3, \ldots\}$$

subject to
(*a*) addition, (*b*) multiplication.

4. Construct the structure table for a group of three elements e, a, b, showing that, basically, there is only one possibility. Prove that the group is commutative.

5. Four operations e, a, b, c act on the 'ordered quadruplet' $(1,2,3,4)$; the operation e leaves the quadruplet unaltered; a changes 1,2,3,4 to 2,3,4,1; b changes 1,2,3,4 to 3,4,1,2; c changes 1,2,3,4 to 4,1,2,3. (For example, $ab(1,2,3,4) = a(3,4,1,2) = (4,1,2,3) = c(1,2,3,4)$, so that the operation ab is the same as c.) Prove that the operations e, a, b, c form a group. Construct the structure table and show that it is essentially the same as that given on page 28.

6. Four numbers $e \equiv 1$, $a = \omega$, $b \equiv \omega^2$, $c = \omega^3$ are given, where $\omega^4 = 1$, and they are subjected to the operation of normal multiplication (for complex numbers). Prove that they form a group and construct the structure table.

Repeat the analogous work for the group of order 5 when $\omega^5 = 1$ $(\omega \neq 1)$.

7. Prove that in a group of even order there must be at least one element x such that $x^2 = e$, $x \neq e$.

8. In a given group, there are a number of elements x all of which commute with a given element a, so that, for each such x,

$$ax = xa.$$

Prove that the elements x form a group.

9. The eight elements

$$I, (A), (B), (C), (BC), (CA), (AB), (ABC)$$

combine in pairs according to the rule that *the combination of any pair is that one of the eight elements containing those letters A, B, C which occur once, and only once, in the pair.* Thus $(AB)(BC) = (AC)$. If there are no such letters, the combination is I. The order of the letters in the elements is irrelevant, and I is an identity element. Show that they form a commutative group and write out the group table. List the subgroups.

Show that the group is isomorphic (p. 28) to a subgroup of the permutation group on six symbols.

10. The operation R_n takes the point (x,y) to the position (x',y') given by the relations

$$x' = x\cos\beta + y\sin\beta, \quad y' = -x\sin\beta + y\cos\beta,$$

where $\beta = 2\pi n/N$. Show, by geometrical considerations, or otherwise, that these operations, with $n = 0, 1, 2, \ldots, N-1$, form a group.

11. Show that the symmetry operations of a rhombus with unequal diagonals form a group, and obtain the group table expressing the relations between the operations.

12. Prove that, if A, X, Y belong to a group and if either $AX = AY$ or $XA = YA$, then $X = Y$. Deduce that no member of the group can appear more than once in the same row (or column) of the group multiplication table.

A non-cyclic group has exactly 4 members. Show that it must consist of I, P, Q, PQ, where $P^2 = Q^2 = I$. Give the group multiplication table and state whether the group is commutative.

Show that such a group is generated by operations P and Q on a complex number z, where P and Q are defined by the relations

$$Pz = z^{-1}, \quad Qz = \bar{z}.$$

13. The complex number z represents an arbitrary point in the Argand diagram. The transformations S and P take the point z to the points $z' = ze^{\frac{2}{3}\pi i}$ and $z' = \bar{z}$ respectively.

(i) If the transformations are applied in any order (for example, $SSPS$) how many different results are possible? Show that the resulting transformations form a non-commutative group G.

(ii) Show that this group is isomorphic with the group of permutations of 3 things.

(iii) Find the subgroups of G.

14. A card in the form of an equilateral triangle lies on a horizontal table. The various operations of rotation that leave its appearance unchanged include two rotations, R and R^2, about a vertical axis and three rotations, A, B, C, that turn it upside down. Construct the multiplication table of the group that they define and give examples of subgroups.

15. The identical operation and seven other operations A, B, C, A^{-1}, B^{-1}, C^{-1}, S constitute a group which has the property that S is the square of every element except itself and I. Prove that the square of S can only be I.

For any element X, other than I and S, show that the product XS is the inverse of X and is equal to SX.

If $AB = C$, show that $BC = A$ and construct a multiplication table for the group.

16. Show that S, the operation of subtracting a number x from unity, and D, the operation of dividing unity by x, generate a group of 6 operators. Set up the multiplication table, and show that it is like that for the 6 permutations of 3 objects.

Find 3 numbers x_1, x_2, x_3 which the group generated from S and D actually permutes among themselves.

17. A transformation of the type

$$x' = \frac{ax-b}{cx-d} \qquad (ad-bc \neq 0)$$

is written $x' = Tx$. Show that the inverse transformation, written $x = T^{-1}x'$, is of the same type.

Show also that, if T_1x and T_2x are two such transformations, then $T_1(T_2x)$ is also of the same type.

T^2x and T^3x are written for $T(Tx)$ and $T(T^2x)$. Find the conditions satisfied by a, b, c, d if, identically in x,

$$\text{(i) } T^2x = x, \qquad \text{(ii) } T^3x = x.$$

Show that the transformations of this type with $(a,b,c,d) = (1,0,0,-1)$, $(5,7,3,4)$, $(4,7,3,5)$, $(2,-1,3,2)$, $(11,19,6,11)$, $(13,18,9,13)$ form a group, and determine the subgroups.

18. Show that in any group with an even number of elements, at least one element other than the identity is its own inverse.

Show that any group of 4 elements in which only one element other than the identity is its own inverse is cyclic, and that in any non-cyclic group of 4 elements each element is its own inverse.

Show that the group of symmetry operations of a rhombus is non-cyclic.

19. The elements of a set S consist of the number 1 together with all positive integers that are expressible as products of distinct primes. A law of combination is defined for elements $a, b \in S$ by putting

$$a * b = ab/d_{ab}^2,$$

where d_{ab} is the highest common factor of a, b. Prove that, with this law of combination, S forms a group.

20. Show that a group G, in which all the elements, except the unit element, have order 2 (p. 37), is Abelian.

If all the elements of a group G, except the unit element, have order 3, show that, for each given a in G,

$$(g^{-1}ag)a = a(g^{-1}ag)$$

for all $g \in G$.

21. Prove that the set $\{2, 4, 6, 8\}$ subject to the rule of digital multiplication (p. 8) is a group isomorphic with the group $\{1, i, -1, -i\}$ subject to ordinary multiplication.

4

EUCLIDEAN SPACES

The work of this chapter is essentially a re-statement in vector language of ideas familiar in Cartesian coordinate geometry, together with an attempt to generalize them in a reasonably simple and natural way.

We indicated in Chapter 1 (p. 10) reasons for regarding vectors as *number-pairs* manipulated according to rules designed to agree with the vectors of applied mathematics. This is now our starting point.

It is assumed throughout this chapter that all arithmetic and algebraic symbols denote real numbers manipulated according to the normal rules of elementary mathematics.

1. Vectors. In a given rectangular coordinate system let a point P have coordinates (p_1,p_2). These coordinates form an *ordered number-pair* conveniently designated as a single unit by the symbol **p** in **bold** type†, so that

$$\mathbf{p} \equiv (p_1,p_2).$$

This confronts us at once with two sets of symbols: those denoting the 'numbers' of elementary arithmetic and algebra, and those denoting the ordered number-pairs. The first are combined among themselves by the usual *two* rules of addition and multiplication. The second are now defined to obey the *single* rule of combination

$$\mathbf{p}+\mathbf{q} \equiv (p_1+q_1, p_2+q_2),$$

and hence to obey the laws of structure studied in Chapter 1; in particular,

$$(\mathbf{p}+\mathbf{q})+\mathbf{r} = \mathbf{p}+(\mathbf{q}+\mathbf{r}).$$

DEFINITIONS. In this context, the ordinary algebraic elements, usually denoted by symbols like $a, b, c, \ldots$, or by arithmetical numbers, are called *scalars*. The ordered number-pairs, denoted by symbols like $\mathbf{p}, \mathbf{q}, \mathbf{r}, \ldots$, are called *vectors*; the individual elements p_1, p_2 of a vector $\mathbf{p}$ are called

† In writing, bold type may be indicated by a 'squiggly' line under the letter: $\underset{\sim}{p}$.

its *components* and are usually, when regarded as individuals, ordinary arithmetical and algebraic numbers also. The totality of all vectors $\mathbf{p} \equiv (p_1, p_2)$, subject to the law

$$\mathbf{p}+\mathbf{q} \equiv (p_1+q_1, p_2+q_2)$$

and to rules of combination with scalars to be given shortly, is called a *vector space*, often denoted by V.

2. Properties of vectors. One further rule is required, showing how scalars and vectors may be combined; a number of properties may then be deduced. The status of these properties should be noted carefully: they can be *proved* for the vectors considered here and in § 1, but they will reappear later (Chapter 5) as *axioms* once the idea of a vector has been generalized. In other words, these properties will serve to guide us in extending the conceptions to more general constructs.

1. The *rule of combination:* The *product of a vector* $\mathbf{p}$ *by a scalar* a is defined by the relation

$$a\mathbf{p} \equiv \mathbf{p}a \equiv (ap_1, ap_2).$$

The following properties may now be deduced:

2. The *three distributive laws,* which should be proved formally for vectors of type $\mathbf{p}(p_1, p_2)$:

(i) $$(a+b)\mathbf{p} = a\mathbf{p}+b\mathbf{p},$$

(ii) $$a(b\mathbf{p}) = (ab)\mathbf{p},$$

(iii) $$a(\mathbf{p}+\mathbf{q}) = a\mathbf{p}+a\mathbf{q}.$$

3. The *zero vector* and *the negative* $-\mathbf{p}$ of $\mathbf{p}$: The zero vector $\mathbf{0}$ is defined by the relation

$$\mathbf{0} \equiv (0, 0),$$

and the negative, $-\mathbf{p}$, of $\mathbf{p}$ by the relation

$$-\mathbf{p}+\mathbf{p} = \mathbf{0}.$$

4. The following *consequences* should be verified:

(i) $$-\mathbf{p} \equiv (-p_1, -p_2) \equiv (-1).\mathbf{p},$$

(ii) $$0.\mathbf{p} = \mathbf{0},$$

(iii) $$a.\mathbf{0} = \mathbf{0},$$

(iv) $$a(-\mathbf{p}) = -(a\mathbf{p}),$$

(v) $$1.\mathbf{x} = \mathbf{x}.$$

Text Examples

1. Prove that the vectors of V form an Abelian group in which the neutral element is the zero vector $\mathbf{0}$ and the inverse of a vector $\mathbf{p}$ is its negative $-\mathbf{p}$.

2. Mark in a diagram the position of the vectors

$$\mathbf{a} \equiv (3, 1), \quad \mathbf{b} \equiv (1, 2),$$

$\mathbf{a}+\mathbf{b}$, $2\mathbf{a}$, $2\mathbf{b}$, $2(\mathbf{a}+\mathbf{b})$, $-\mathbf{a}$, $-\mathbf{b}$, $-(\mathbf{a}+\mathbf{b})$.

Repeat the example when

(i) $\mathbf{a} \equiv (1, 0)$, $\mathbf{b} \equiv (0, 2)$; (ii) $\mathbf{a} \equiv (-1, -1)$, $\mathbf{b} \equiv (1, 0)$.

3. The scalar product (inner product) of two vectors. Although vectors are combined by the single law, called addition, there are two important functions associated with them and having something of the nature of products. This paragraph concerns the first, called the *scalar*, or *inner*, *product.*

DEFINITION. The *scalar* (*inner*) *product* of two vectors $\mathbf{p}$, $\mathbf{q}$ is the function, denoted by $\mathbf{p}.\mathbf{q}$, given by the relation

$$\mathbf{p}.\mathbf{q} \equiv p_1q_1+p_2q_2.$$

When p, q are the same function, the product $\mathbf{p}.\mathbf{p}$ is sometimes written as $\mathbf{p}^2$, so that

$$\mathbf{p}^2 \equiv \mathbf{p}.\mathbf{p} \equiv p_1^2+p_2^2.$$

There are five basic *scalar product properties.* These are almost immediate for the particular case under discussion, but they assume vital importance in more abstract or more general cases.

(i) *The scalar product is a bilinear form:* that is, the function

$$\mathbf{p}.\mathbf{q} \equiv p_1q_1+p_2q_2$$

is linear both in the elements p_1, p_2 and also in the elements q_1, q_2.

[Other examples of bilinear forms are

$$2p_1q_2-3p_2q_1,$$
$$p_1q_1+2p_1q_2+3p_2q_1+4p_2q_2.]$$

(ii) *The scalar product is commutative:* that is,

$$\mathbf{p}.\mathbf{q} = \mathbf{q}.\mathbf{p}.$$

(iii) *The scalar product obeys the distributive law*

$$\mathbf{p}.(\mathbf{q}+\mathbf{r}) = \mathbf{p}.\mathbf{q}+\mathbf{p}.\mathbf{r}.$$

(iv) *The scalar product obeys the law*

$$(a\mathbf{p}.\mathbf{q}) = a(\mathbf{p}.\mathbf{q})$$

for any scalar a.

(v) *The scalar product* $\mathbf{p}.\mathbf{p} \equiv p_1^2+p_2^2$ *of a vector with itself is always greater than zero, except that it is zero when* $\mathbf{p}$ *is the zero vector.* Thus

$$\mathbf{p}.\mathbf{p} > 0 \Leftrightarrow \mathbf{p} \neq \mathbf{0},$$
$$\mathbf{p}.\mathbf{p} = 0 \Leftrightarrow \mathbf{p} = \mathbf{0}.$$

4. Metrical properties of 'number-pair' vectors. The work in this paragraph is little more than a re-statement of ideas familiar in elementary plane trigonometry and coordinate geometry. Its importance, once again, lies in the extensions which the use of the language foreshadows.

DEFINITIONS. (i) The magnitude

$$+\sqrt{(\mathbf{p}.\mathbf{p})} \equiv +\sqrt{(p_1^2+p_2^2)}$$

is called the *length* of the vector $\mathbf{p}$ and is denoted by the symbol

$$|\mathbf{p}|.$$

(ii) The magnitude

$$\frac{\mathbf{p}.\mathbf{q}}{|\mathbf{p}||\mathbf{q}|}$$

(which, as is proved almost immediately, lies between -1 and $+1$) is the cosine of a definite angle θ lying between 0 and π. This angle θ is called *the angle between the vectors* $\mathbf{p}$, $\mathbf{q}$.

To prove that the function lies between -1 and $+1$, note that

$$\begin{aligned}
&|\mathbf{p}|^2|\mathbf{q}|^2-(\mathbf{p}.\mathbf{q})^2\\
&\quad = (p_1^2+p_2^2)(q_1^2+q_2^2)-(p_1q_1+p_2q_2)^2\\
&\quad = p_1^2q_2^2-2p_1q_2p_2q_1+p_2^2q_1^2\\
&\quad = (p_1q_2-p_2q_1)^2\\
&\quad \geqslant 0,
\end{aligned}$$

since all components are real. Hence

$$(\mathbf{p}.\mathbf{q})^2 \leqslant |\mathbf{p}|^2|\mathbf{q}|^2,$$

so that $\mathbf{p}.\mathbf{q}$ is numerically less than $|\mathbf{p}||\mathbf{q}|$, and so

$$-1 \leqslant \frac{\mathbf{p}.\mathbf{q}}{|\mathbf{p}||\mathbf{q}|} \leqslant 1.$$

(iii) The vectors **p**, **q** are said to be *orthogonal* when

$$\mathbf{p}.\mathbf{q} \equiv p_1q_1+p_2q_2 = 0.$$

In that case, $\theta = \frac{1}{2}\pi$.

Typical orthogonal pairs are given by

$$\mathbf{p} \equiv (1, 0), \quad \mathbf{q} \equiv (0, 1)$$

or

$$\mathbf{p} \equiv (3, 4), \quad \mathbf{q} \equiv (8, -6).$$

These definitions are, of course, selected to give for vectors the formulae familiar in elementary coordinate geometry. They also lead consistently to other well-known results.

Suppose, for example, that components of vectors are interpreted as rectangular Cartesian coordinates. Mark the points **p**, **q**, **p**+**q**; then *the 'sum'* **p**+**q** *gives the fourth vertex of the parallelogram of which adjacent sides join O*, **p** and *O*, **q** *respectively.*

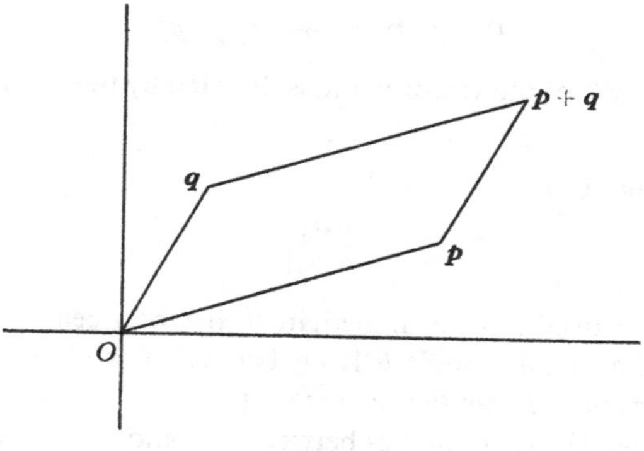

Fig. 7

Further, the square of the length of **p**+**q** is given by

$$(\mathbf{p}+\mathbf{q}).(\mathbf{p}+\mathbf{q}),$$

which, by the distributive and commutative laws, is

$$\mathbf{p}.\mathbf{p}+2\mathbf{p}.\mathbf{q}+\mathbf{q}.\mathbf{q}.$$

Thus

$$|\mathbf{p}+\mathbf{q}|^2 = |\mathbf{p}|^2+|\mathbf{q}|^2+2|\mathbf{p}||\mathbf{q}|\cos\theta,$$

and this, suitably adjusted, is the familiar *cosine formula*

$$a^2 = b^2+c^2-2bc\cos A$$

of elementary trigonometry. When **p**, **q** are orthogonal, the result is equivalent to the *theorem of Pythagoras.*

5. The elements of three-dimensional metrical geometry. It may be useful to give at this stage a brief note on how ordinary coordinate geometry in three-dimensional space is set up, a basic knowledge of the framework of 'solid geometry' being assumed. For fuller details a special textbook must be consulted; the present account does little beyond extending elementary ideas of plane coordinate geometry.

If three mutually perpendicular planes are given, the *position of a point P* is determined by its distances, with suitable sign conventions, from them. These distances, say p_1, p_2, p_3, are combined to give the *position vector* $\mathbf{p}(p_1, p_2, p_3)$.

When two points P, Q are given, the *distance between them* is calculated, by two appeals to the theorem of Pythagoras, to give

$$PQ^2 = (p_1-q_1)^2+(p_2-q_2)^2+(p_3-q_3)^2.$$

By direct extension of the work given for a plane in §4, the *scalar product* of two vectors $\mathbf{u}$, $\mathbf{v}$ is defined by the relation

$$\mathbf{u}.\mathbf{v} = u_1v_1+u_2v_2+u_3v_3,$$

so that, here,

$$PQ^2 = (\mathbf{p}-\mathbf{q}).(\mathbf{p}-\mathbf{q})$$
$$= (\mathbf{p}-\mathbf{q})^2.$$

The *length* $|\mathbf{p}|$ of a vector $\mathbf{p}$ is the positive magnitude given by the formula

$$|\mathbf{p}|^2 = \mathbf{p}.\mathbf{p} = p_1^2+p_2^2+p_3^2,$$

agreeing with the more general formula just given.

The *angle between two vectors* $\mathbf{p}$, $\mathbf{q}$ is the angle θ between 0 and π such that

$$\cos\theta = \frac{\mathbf{p}.\mathbf{q}}{|\mathbf{p}||\mathbf{q}|}.$$

When $\theta = \frac{1}{2}\pi$, so that

$$\mathbf{p}.\mathbf{q} = 0,$$

the vectors are said to be *orthogonal.*

(These results are *proved* in books on coordinate geometry as consequences, ultimately, of the axioms and theorems of Euclid. In a more abstract treatment they are regarded as *definitions*, selected to agree with the Euclidean language.)

A vector of unit length is called a *unit vector*. If $\mathbf{p}$, $\mathbf{q}$ are unit vectors, then

$$\mathbf{p}^2 = 1, \quad \mathbf{q}^2 = 1,$$
$$\cos\theta = \mathbf{p}.\mathbf{q},$$

where θ is the angle between them.

There are many occasions when it is convenient to express vectors in terms of the three *orthogonal unit coordinate vectors*

$$\mathbf{i} \equiv (1, 0, 0), \quad \mathbf{j} \equiv (0, 1, 0), \quad \mathbf{k} \equiv (0, 0, 1).$$

The name emphasizes the relations

$$\mathbf{i}^2 = \mathbf{j}^2 = \mathbf{k}^2 = 1,$$
$$\mathbf{j}.\mathbf{k} = \mathbf{k}.\mathbf{i} = \mathbf{i}.\mathbf{j} = 0.$$

Then

$$\mathbf{p} \equiv (p_1, p_2, p_3) \equiv p_1(1, 0, 0)+p_2(0, 1, 0)+p_3(0, 0, 1)$$
$$\equiv p_1\mathbf{i}+p_2\mathbf{j}+p_3\mathbf{k}.$$

Note, as a matter of consistency, that

$$\mathbf{p}^2 = p_1^2\mathbf{i}^2+p_2^2\mathbf{j}^2+p_3^2\mathbf{k}^2+2p_2p_3\mathbf{j}.\mathbf{k}+2p_3p_1\mathbf{k}.\mathbf{i}+2p_1p_2\mathbf{i}.\mathbf{j}$$
$$= p_1^2+p_2^2+p_3^2.$$

Text Example

1. Establish the relation

$$\mathbf{p}.\mathbf{q} = p_1q_1+p_2q_2+p_3q_3$$

for the vectors

$$\mathbf{p} \equiv p_1\mathbf{i}+p_2\mathbf{j}+p_3\mathbf{k}, \quad \mathbf{q} \equiv q_1\mathbf{i}+q_2\mathbf{j}+q_3\mathbf{k}.$$

Suppose now that l is a given line *passing through the origin.* The coordinates p_1, p_2, p_3 of all points P upon it are constant in ratio, so that there exists a *fixed* vector $\mathbf{l}$ (l_1, l_2, l_3) such that

$$\mathbf{p} = k\mathbf{l}$$

where (i) $\mathbf{l}$ is a vector defining the direction of the line, and (ii) k is a scalar defining the position of the point on the line.

For definiteness, $\mathbf{l}$ is often chosen to be a *unit* vector, and, in that case,

$$\mathbf{l}^2 = 1,$$

or

$$l_1^2+l_2^2+l_3^2 = 1.$$

When this is done,

$$\mathbf{p}^2 = (k\mathbf{l})^2 = k^2\mathbf{l}^2 = k^2,$$

so that k is the actual distance of $P(p_1, p_2, p_3)$ from the origin $O(0,0,0)$. The 'normalized' vector $\mathbf{l}$, with $\mathbf{l}^2 = 1$, is called the *direction vector of the line.*

More generally, if *any given line l* is taken, not necessarily passing through the origin, its direction is the same as that of the parallel line through the origin and so that direction may be described by means of a unit vector $\mathbf{l}$. If P, Q are two points on the line l, a simple calculation by 'change of origin', exactly analogous to that used in plane geometry, shows that the vector $\mathbf{p}-\mathbf{q}$ is proportional to $\mathbf{l}$; hence there is a scalar k such that

$$\mathbf{p}-\mathbf{q} = k\mathbf{l}.$$

For unit vector $\mathbf{l}$, the scalar k is the distance between P and Q.

To complete this brief résumé, suppose that π is a *given plane*. Draw from the origin the line OA perpendicular to π, meeting it at A, and let $\mathbf{l}$ be the unit direction vector of OA, so that, if d (positive or negative) is the actual length of OA, then

$$\mathbf{a} = d\mathbf{l}.$$

If X is any point in the plane π, the vector AX is $\mathbf{x}-\mathbf{a}$, and this is perpendicular to OA, so that

$$(\mathbf{x}-\mathbf{a}).\mathbf{l} = 0,$$

or

$$\mathbf{x}.\mathbf{l} = \mathbf{a}.\mathbf{l} = d\mathbf{l}.\mathbf{l}$$
$$= d \qquad (\mathbf{l}.\mathbf{l} = 1).$$

Hence *the coordinates of all points in π satisfy the relation*

$$\mathbf{x}.\mathbf{l} = d,$$

or

$$l_1 x_1 + l_2 x_2 + l_3 x_3 = d.$$

Revision Examples

1. Find the lengths of the sides of the triangle ABC with vertices

$$\mathbf{a} \equiv (1, 1, 1), \quad \mathbf{b} \equiv (4, 5, 13), \quad \mathbf{c} \equiv (13, 5, 4).$$

2. Prove that the middle point of AB has position vector $\frac{1}{2}(\mathbf{a}+\mathbf{b})$.

3. A tetrahedron has vertices A, B, C, D. The middle points of BC, CA, AB, AD, BD, CD are P, Q, R, L, M, N respectively. Prove that PL, QM, RN have the same middle point U, where

$$\mathbf{u} = \tfrac{1}{4}(\mathbf{a}+\mathbf{b}+\mathbf{c}+\mathbf{d}).$$

4. Given two points A, B, identify geometrically the point with position vector $\mathbf{a}-\mathbf{b}$.

Interpret geometrically the equation

$$(\mathbf{b}-\mathbf{c})+(\mathbf{c}-\mathbf{a})+(\mathbf{a}-\mathbf{b}) = \mathbf{0}.$$

5. Find the angles of the triangle whose vertices are

$$O(0, 0, 0), \quad A(1, -1, 1), \quad B(2, 2, -2).$$

6. Prove that

$$\mathbf{a}^2+\mathbf{b}^2+\mathbf{c}^2+(\mathbf{a}+\mathbf{b}+\mathbf{c})^2 = (\mathbf{b}+\mathbf{c})^2+(\mathbf{c}+\mathbf{a})^2+(\mathbf{a}+\mathbf{b})^2$$

and interpret this result for a parallelepiped of which OA, OB, OC are adjacent sides.

7. Show that the points $\mathbf{p}$, $\mathbf{q}$, $\lambda\mathbf{p}+(1-\lambda)\mathbf{q}$ are collinear.

8. The points A, B, C, P, Q, R have coordinate vectors $\mathbf{a}, \mathbf{b}, \mathbf{c}, \mathbf{p}, \mathbf{q}, \mathbf{r}$, where

$$\mathbf{p} = \lambda\mathbf{a}, \quad \mathbf{q} = \mu\mathbf{b}, \quad \mathbf{r} = \nu\mathbf{c},$$

λ, μ, ν being real and distinct scalars. The lines BC, QR meet at L; CA, RP meet at M; AB, PQ meet at N, corresponding coordinate vectors being $\mathbf{l}$, $\mathbf{m}$, $\mathbf{n}$. Prove that

$$(\mu-\nu)\mathbf{l} = \mu(1-\nu)\mathbf{b}-\nu(1-\mu)\mathbf{c},$$
$$(1-\lambda)(\mu-\nu)\mathbf{l}+(1-\mu)(\nu-\lambda)\mathbf{m}+(1-\nu)(\lambda-\mu)\mathbf{n} = \mathbf{0}.$$

Prove also that $\mathbf{l}-\mathbf{n}$ and $\mathbf{m}-\mathbf{n}$ have the same direction and interpret this result geometrically.

9. Find a vector $\mathbf{p}$ equally inclined to the vectors

$$(1, 2, 2), \quad (2, 3, 6), \quad (0, 3, 4).$$

10. Define the orthogonal unit coordinate vectors $\mathbf{i}$, $\mathbf{j}$, $\mathbf{k}$ and find an expression for the angle between the two vectors

$$\overrightarrow{OP} = p_1\mathbf{i}+p_2\mathbf{j}+p_3\mathbf{k}, \quad \overrightarrow{OQ} = q_1\mathbf{i}+q_2\mathbf{j}+q_3\mathbf{k}.$$

A regular tetrahedron $OP_1P_2P_3$ is determined as follows:

(i) the position vector of P_1 is $\mathbf{i}+\mathbf{j}$,

(ii) the position vector of P_2 is of the form $-p\mathbf{i}+q\mathbf{j}$, where p,q are positive,

(iii) the position vector of P_3 is of the form $u\mathbf{i}+v\mathbf{j}+w\mathbf{k}$, where w is positive.

Find the position vectors of P_2 and P_3.

11. A variable $\mathbf{r}$ is given by the relation

$$\mathbf{r} = u\mathbf{i}+v\mathbf{j}+w\mathbf{k},$$

where u, v, w are given functions of a parameter t. A vector $\mathbf{s}$ is defined by

$$\mathbf{s} = u'\mathbf{i}+v'\mathbf{j}+w'\mathbf{k},$$

where u', v', w' are the differential coefficients of u, v, w. If $u = 2at$,

$v = b(t^2+1)$, $w = c(t^2-1)$, prove that there are three vectors **r** such that **r** and **s** are othogonal, provided that

$$c^2 > 2a^2+b^2.$$

12. The position vectors of P, Q, R are

$$\mathbf{p} \equiv \mathbf{i}-2\mathbf{j}+\mathbf{k}, \quad \mathbf{q} \equiv 4\mathbf{i}-4\mathbf{j}+7\mathbf{k}, \quad \mathbf{r} \equiv 6\mathbf{i}-10\mathbf{j}+10\mathbf{k}.$$

Show that the lengths of PQ, QR are equal and find the position vector of the other vertex S of the rhombus $PQRS$.

Calculate in terms of **i**, **j**, **k** a unit vector perpendicular to the plane $PQRS$.

13. Define the scalar product **a.b** of two vectors **a** and **b**. State *all* the conditions under which $\mathbf{a}.\mathbf{b} = 0$.

Let α be a scalar quantity and **a**, **b**, **c** three vectors such that **a** and **b** are orthogonal. Prove that, if

$$\alpha\mathbf{x}+(\mathbf{x}.\mathbf{b})\mathbf{a} = \mathbf{c},$$

then, if $\alpha \neq 0$,

$$\mathbf{x}.\mathbf{b} = \frac{\mathbf{c}.\mathbf{b}}{\alpha}.$$

Hence find **x** in terms of α, **a**, **b**, **c** when $\alpha \neq 0$.

If $\alpha = 0$, $\mathbf{c} = \mathbf{0}$ and **a**, **b** are non-zero vectors, what non-zero vectors **x** satisfy the equation?

14. Find a formula for a vector coplanar with **a** and **b**, and perpendicular to **a**. Evaluate this vector and the angle between **a** and **b** when these have components $(1,1,-2)$, $(1,-2,1)$. Give a vector perpendicular to both **a** and **b**.

15. If **r** denotes the vector from the origin to the point (x,y,z) and **a**, **b**, **c** are constant vectors, prove that the relation

$$\mathbf{r}.\mathbf{a} = \mathbf{b}.\mathbf{a}$$

implies that the point (x,y,z) lies in a certain plane.

Derive expressions for the magnitude of the projection of **c** on this plane and for the distance from the plane of the point whose position vector is **c**.

6. The Euclidean space R^n. The reader will have recognized what has preceded in this chapter as a re-writing of some of the elementary parts of coordinate geometry, and will know that the proofs, in a normal school course, depend ultimately on the study known as *Euclidean geometry*. The plane or space to which they are conceived as referring is called the

Euclidean plane or space. It is for this reason that the name of Euclid is also attached to the abstract extensions which follow, and the reader will observe (as he has been amply forewarned) that the whole point of the particular way in which these extensions are cast is to emphasize their origins in the familiar Euclidean framework.

Once again, *all 'scalars' and 'components of vectors' are assumed to be real.*

DEFINITIONS. The 'ordered n-tuple' of real numbers

$$(p_1, p_2, p_3, \ldots, p_n)$$

is called an *n-dimensional vector* and denoted by the single symbol $\mathbf{p}$. The words 'n-dimensional' are omitted when the implication is clear. The numbers of ordinary arithmetic and algebra are again called *scalars.* The totality of all these vectors, subject to the rules to be given, is called an *n-dimensional Euclidean space R^n.*

The vectors are combined according to the *rule of addition*

$$\mathbf{p}+\mathbf{q} = (p_1+q_1, p_2+q_2, \ldots, p_n+q_n)$$

which, in its turn, obeys the *associative law*

$$(\mathbf{p}+\mathbf{q})+\mathbf{r} = \mathbf{p}+(\mathbf{q}+\mathbf{r}).$$

The *rule of combination* for scalars and vectors is given by the formula

$$a\mathbf{p} \equiv (ap_1, ap_2, \ldots, ap_n).$$

The *zero vector* is given by the formula

$$\mathbf{0} \equiv (0, 0, \ldots, 0)$$

and the *negative* $-\mathbf{p}$ of $\mathbf{p}$ by

$$-\mathbf{p}+\mathbf{p} = \mathbf{0},$$

so that

$$-\mathbf{p} \equiv (-p_1, -p_2, \ldots, -p_n).$$

Text Example

1. Prove for n-dimensional vectors the following properties:

(i) The *distributive laws*

$$(a+b)\mathbf{p} = a\mathbf{p}+b\mathbf{p},$$
$$a(b\mathbf{p}) = (ab)\mathbf{p},$$
$$a(\mathbf{p}+\mathbf{q}) = a\mathbf{p}+a\mathbf{q};$$

(ii) the *zero rules*

$$0.\mathbf{p} = \mathbf{0},$$
$$a.\mathbf{0} = \mathbf{0}.$$

7. The scalar product (inner product) for *n*-dimensional vectors.

DEFINITION. The *scalar product* (*inner product*) of two vectors **p**, **q** is the function **p**.**q** such that

$$\mathbf{p}.\mathbf{q} = p_1q_1+p_2q_2+\ldots+p_nq_n.$$

In particular,

$$\mathbf{p}^2 = \mathbf{p}.\mathbf{p} = p_1^2+p_2^2+\ldots+p_n^2.$$

Text Example

1. Establish for *n*-dimensional scalar products the following properties:

(i) *The scalar product is a bilinear form;*

(ii) *The scalar product is commutative,*

$$\mathbf{p}.\mathbf{q} = \mathbf{q}.\mathbf{p};$$

(iii) *The scalar product obeys the distributive law*

$$\mathbf{p}.(\mathbf{q}+\mathbf{r}) = \mathbf{p}.\mathbf{q}+\mathbf{p}.\mathbf{r};$$

(iv) *The scalar product obeys the law*

$$(a\mathbf{p}.\mathbf{q}) = a(\mathbf{p}.\mathbf{q});$$

(v) *The scalar product of* **p** *with itself is always greater than or equal to zero:*

$$\mathbf{p}.\mathbf{p} > 0 \Leftrightarrow \mathbf{p} \neq \mathbf{0},$$
$$\mathbf{p}.\mathbf{p} = 0 \Leftrightarrow \mathbf{p} = \mathbf{0}.$$

8. Metrical properties for *n*-dimensional vectors.

DEFINITIONS. The magnitude

$$+\sqrt{(\mathbf{p}.\mathbf{p})} \equiv +\sqrt{(p_1^2+p_2^2+\ldots+p_n^2)}$$

is called the *length* of the vector **p** and is denoted by the symbol

$$|\mathbf{p}|.$$

The magnitude

$$\frac{\mathbf{p}.\mathbf{q}}{|\mathbf{p}||\mathbf{q}|}$$

(which, as is proved below, lies between -1 and $+1$) is the cosine of an angle θ ($0 \leqslant \theta \leqslant \pi$) called *the angle between the vectors* $\mathbf{p}$, $\mathbf{q}$.

To prove that the function lies between -1 and $+1$ we appeal to the known basic properties of scalar products. Since $\mathbf{p}$, $\mathbf{q}$ are vectors, so also is $\mathbf{p}+k\mathbf{q}$ for any scalar k, and this vector has a scalar product for which

$$(\mathbf{p}+k\mathbf{q}).(\mathbf{p}+k\mathbf{q}) \geqslant 0,$$

with equality if, and only if, $\mathbf{p}+k\mathbf{q} = 0$. By the distributive and commutative laws, we have

$$\mathbf{p}.\mathbf{p}+2k\mathbf{p}.\mathbf{q}+k^2\mathbf{q}.\mathbf{q} \geqslant 0.$$

This is a quadratic polynomial in k which is always greater than or equal to zero, whatever the value of k, and so (Part I, p. 129)

$$(\mathbf{p}.\mathbf{p})(\mathbf{q}.\mathbf{q}) \geqslant (\mathbf{p}.\mathbf{q})^2,$$

so that, numerically,

$$\mathbf{p}.\mathbf{q} \leqslant |\mathbf{p}||\mathbf{q}|.$$

Hence

$$-1 \leqslant \frac{\mathbf{p}.\mathbf{q}}{|\mathbf{p}||\mathbf{q}|} \leqslant 1.$$

DEFINITION. Two vectors $\mathbf{p}$, $\mathbf{q}$ for which

$$\mathbf{p}.\mathbf{q} = 0$$

are said to be *orthogonal.*

The n vectors

$$(1,0,0,\ldots,0),\ (0,1,0,\ldots,0),\ (0,0,1,\ldots,0),\ \ldots,\ (0,0,0,\ldots,1)$$

are so related that all pairs are orthogonal.

Two orthogonal vectors satisfy the '*Pythagoras*' *relation*

$$|\mathbf{p}+\mathbf{q}|^2 = |\mathbf{p}|^2+|\mathbf{q}|^2.$$

9. Two basic inequalities. The preceding work has, by implication, covered two well-known inequalities:

(i) THE SCHWARZ INEQUALITY. Let

$$a_1, a_2, \ldots, a_n$$

and

$$b_1, b_2, \ldots, b_n$$

be two sets of *n positive* numbers. Then *the Schwarz inequality states that*

$$a_1 b_1+a_2 b_2+\ldots+a_n b_n \leqslant \sqrt{\{(a_1^2+a_2^2+\ldots+a_n^2)(b_1^2+b_2^2+\ldots+b_n^2)\}}.$$

This is merely the cosine inequality

$$\mathbf{a}.\mathbf{b} \leqslant |\mathbf{a}||\mathbf{b}|.$$

The case of equality occurs when, and only when, $a_i = b_i$ for $i = 1, 2, \ldots, n$.

(ii) THE TRIANGLE INEQUALITY. Reference to the diagram of Fig. 7, interpreted now for n-dimensional vectors, shows the existence of triangles with sides of lengths $|\mathbf{p}|$, $|\mathbf{q}|$, $|\mathbf{p}+\mathbf{q}|$. *The triangle inequality states that the sum of any two of these magnitudes is greater than the third* (with equality only when the points are collinear):

By the 'cosine' inequality,

$$-|\mathbf{p}||\mathbf{q}| \leqslant \mathbf{p}.\mathbf{q} \leqslant |\mathbf{p}||\mathbf{q}|$$
$$\Rightarrow |\mathbf{p}|^2-2|\mathbf{p}||\mathbf{q}|+|\mathbf{q}|^2 \leqslant \mathbf{p}.\mathbf{p}+2\mathbf{p}.\mathbf{q}+\mathbf{q}.\mathbf{q} \leqslant |\mathbf{p}|^2+2|\mathbf{p}||\mathbf{q}|+|\mathbf{q}|^2$$
$$\Rightarrow \{|\mathbf{p}|-|\mathbf{q}|\}^2 \leqslant (\mathbf{p}+\mathbf{q}).(\mathbf{p}+\mathbf{q}) \leqslant \{|\mathbf{p}|+|\mathbf{q}|\}^2$$
$$\Rightarrow \{|\mathbf{p}|-|\mathbf{q}|\}^2 \leqslant |\mathbf{p}+\mathbf{q}|^2 \leqslant \{|\mathbf{p}|+|\mathbf{q}|\}^2.$$

Hence

(i) $$|\mathbf{p}+\mathbf{q}| \leqslant |\mathbf{p}|+|\mathbf{q}|$$

since each side is positive by definition;

(ii) if $|\mathbf{p}| \geqslant |\mathbf{q}|$, then, taking positive square roots,

$$|\mathbf{p}|-|\mathbf{q}| \leqslant |\mathbf{p}+\mathbf{q}|,$$

so that

$$|\mathbf{p}| \leqslant |\mathbf{q}|+|\mathbf{p}+\mathbf{q}|,$$

and also, automatically for this case,

$$|\mathbf{q}| \leqslant |\mathbf{p}|+|\mathbf{p}+\mathbf{q}|.$$

The three triangle inequalities thus hold when $|\mathbf{p}| \geqslant |\mathbf{q}|$ and so, by similar argument, when $|\mathbf{p}| \leqslant |\mathbf{q}|$. They therefore hold generally.

Examples

1. Find the lengths of the following vectors and the cosines of the angles between them in pairs:

$$(1,1,1,1), \quad (1,-1,-1,1), \quad (0,1,2,3), \quad (1,0,3,-2).$$

2. Find the values of k which make the following vectors of unit lengths:

$$(k,2k,3k), \quad (k,-k,k,-k,2k), \quad (0,0,3k,4k,12k).$$

3. Find the value of k for which the vectors

$$(1, 3, 5, 7), \quad (2, 4, 6, k)$$

are orthogonal.

4. Find p, q, r if the vector

$$(p, q, r, 1, 2)$$

is orthogonal to each of the vectors

$$(0, 1, 2, 3, 4) \quad (4, 3, 2, 1, 0), \quad (1, 1, 1, 1, 1).$$

10. The vector product in a Euclidean space of three dimensions. There is a second product, familiar in applied mathematics, and a brief account may be given here although it is not, strictly speaking, in the direct line of argument. Attention is confined to the Euclidean space R^3, where a typical vector is $\mathbf{p} \equiv (p_1, p_2, p_3)$, as in §5.

Let $\mathbf{p}, \mathbf{q}$ be two given vectors, and suppose that, if possible, *a vector* $\mathbf{x}$ *is selected orthogonal to each.* Then (p. 49)

$$\mathbf{x}.\mathbf{p} = 0, \quad \mathbf{x}.\mathbf{q} = 0,$$

so that

$$p_1 x_1 + p_2 x_2 + p_3 x_3 = 0,$$
$$q_1 x_1 + q_2 x_2 + q_3 x_3 = 0.$$

These equations can (in general) be solved in the form

$$\frac{x_1}{p_2 q_3 - p_3 q_2} = \frac{x_2}{p_3 q_1 - p_1 q_3} = \frac{x_3}{p_1 q_2 - p_2 q_1},$$

and any vector with components proportional to these denominators will be perpendicular to both $\mathbf{p}$ and $\mathbf{q}$. For simplicity, the coefficient of proportionality is taken to be 1, and the resulting vector, known as the *vector product of* $\mathbf{p}$ *by* $\mathbf{q}$, is denoted by the symbol $\mathbf{p} \wedge \mathbf{q}$ (or $\mathbf{p} \times \mathbf{q}$), so that

$$\mathbf{p} \wedge \mathbf{q} \equiv (p_2 q_3 - p_3 q_2, p_3 q_1 - p_1 q_3, p_1 q_2 - p_2 q_1).$$

COROLLARIES, to be proved as *Examples.*

(i) *The vector product is antisymmetrical (skew-symmetrical)*, so that

$$\mathbf{p} \wedge \mathbf{q} = -\mathbf{q} \wedge \mathbf{p}$$

(ii) $$\mathbf{p} \wedge \mathbf{p} = \mathbf{0},$$

(iii) $$\mathbf{p}.(\mathbf{p} \wedge \mathbf{q}) = 0, \quad \mathbf{q}.(\mathbf{p} \wedge \mathbf{q}) = 0,$$

(iv) $$(a\mathbf{p}) \wedge \mathbf{q} = a(\mathbf{p} \wedge \mathbf{q}),$$

(v) $$\mathbf{p} \wedge (\mathbf{q} + \mathbf{r}) = \mathbf{p} \wedge \mathbf{q} + \mathbf{p} \wedge \mathbf{r}.$$

Scalar multiplication and vector multiplication can be combined to give *scalar triple products* and *vector triple products.* We are not greatly concerned with them here, so we give the definitions and state the fundamental properties, leaving the proofs to be worked as examples.

DEFINITION. The *scalar triple product* of three vectors **p**, **q**, **r**, *in that order*, is the scalar product of the vector **p** with the vector product $\mathbf{q} \wedge \mathbf{r}$. It is denoted by the symbol $(\mathbf{pqr})$, so that

$$(\mathbf{pqr}) = \mathbf{p}.(\mathbf{q} \wedge \mathbf{r}).$$

[Sometimes commas are inserted for clarity: $(\mathbf{p},\mathbf{q},\mathbf{r})$.]

Properties

1. The scalar triple product is not changed by cyclic interchange of its vectors:

$$(\mathbf{pqr}) = (\mathbf{qrp}) = (\mathbf{rpq})$$

2. The scalar triple product is changed in sign if the order of its vectors is reversed:

$$(\mathbf{pqr}) = -(\mathbf{rqp}).$$

3. The scalar triple product can be expressed as a determinant in the form

$$(\mathbf{pqr}) = \begin{vmatrix} p_1 & p_2 & p_3 \\ q_1 & q_2 & q_3 \\ r_1 & r_2 & r_3 \end{vmatrix}.$$

4. The scalar triple product obeys the rules:

$$(a\mathbf{p},\mathbf{q},\mathbf{r}) = a(\mathbf{pqr}),$$
$$(\mathbf{p},\mathbf{q},\mathbf{r}+\mathbf{s}) = (\mathbf{pqr})+(\mathbf{pqs}).$$

5. If the vectors **a**, **b**, **c** are coplanar [that is, linearly dependent, so that there is a relation

$$\lambda\mathbf{a}+\mu\mathbf{b}+\nu\mathbf{c} = \mathbf{0}],$$

then $(\mathbf{abc}) = 0$.

DEFINITION. The *vector triple product* of the vector **p** with the vector $\mathbf{q} \wedge \mathbf{r}$ is defined to be the vector

$$\mathbf{p} \wedge (\mathbf{q} \wedge \mathbf{r}).$$

This vector can, in fact, be expressed in terms of the two vectors **q**, **r** with suitable scalar products in the form:

$$\mathbf{p} \wedge (\mathbf{q} \wedge \mathbf{r}) = (\mathbf{p}.\mathbf{r})\mathbf{q}-(\mathbf{p}.\mathbf{q})\mathbf{r}.$$

To prove this by direct computation, note that

$$\mathbf{q} \wedge \mathbf{r} \equiv (q_2 r_3 - q_3 r_2, q_3 r_1 - q_1 r_3, q_1 r_2 - q_2 r_1),$$

so that the *first component* of $\mathbf{p} \wedge (\mathbf{q} \wedge \mathbf{r})$ is

$$\begin{aligned} & p_2(q_1 r_2 - q_2 r_1) - p_3(q_3 r_1 - q_1 r_3) \\ &= (p_2 r_2 + p_3 r_3) q_1 - (p_2 q_2 + p_3 q_3) r_1 \\ &= (p_1 r_1 + p_2 r_2 + p_3 r_3) q_1 - (p_1 q_1 + p_2 q_2 + p_3 q_3) r_1 \\ &= (\mathbf{p}.\mathbf{r}) q_1 - (\mathbf{p}.\mathbf{q}) r_1. \end{aligned}$$

Similarly the second and third components are

$$(\mathbf{p}.\mathbf{r}) q_2 - (\mathbf{p}.\mathbf{q}) r_2, \quad (\mathbf{p}.\mathbf{r}) q_3 - (\mathbf{p}.\mathbf{q}) r_3$$

respectively, and the result is established.

Properties

1. The vector triple product obeys the rules:

$$\mathbf{p} \wedge (\mathbf{q} \wedge \mathbf{r}) = -(\mathbf{q} \wedge \mathbf{r}) \wedge \mathbf{p},$$
$$(\mathbf{p}+\mathbf{s}) \wedge (\mathbf{q} \wedge \mathbf{r}) = \mathbf{p} \wedge (\mathbf{q} \wedge \mathbf{r}) + \mathbf{s} \wedge (\mathbf{q} \wedge \mathbf{r}),$$
$$\{\mathbf{p} \wedge (\mathbf{q} \wedge \mathbf{r})\} + \{\mathbf{q} \wedge (\mathbf{r} \wedge \mathbf{p})\} + \{\mathbf{r} \wedge (\mathbf{p} \wedge \mathbf{q})\} = \mathbf{0}.$$

Examples

1. Prove that

$$(\mathbf{a} \wedge \mathbf{b}).(\mathbf{c} \wedge \mathbf{d}) + (\mathbf{b} \wedge \mathbf{c}).(\mathbf{a} \wedge \mathbf{d}) + (\mathbf{c} \wedge \mathbf{a}).(\mathbf{b} \wedge \mathbf{d}) = 0,$$
$$(\mathbf{a} \wedge \mathbf{b}) \wedge (\mathbf{b} \wedge \mathbf{c}) = (\mathbf{abc})\mathbf{b}.$$

2. Given that

$$\mathbf{a} \equiv (1,2,3), \quad \mathbf{b} \equiv (2,3,1), \quad \mathbf{c} \equiv (3,1,2),$$

find $\mathbf{b}.\mathbf{c}$, $\mathbf{c}.\mathbf{a}$, $\mathbf{a}.\mathbf{b}$; $\mathbf{b} \wedge \mathbf{c}$, $\mathbf{c} \wedge \mathbf{a}$, $\mathbf{a} \wedge \mathbf{b}$, $(\mathbf{abc})$.

3. Given that $\mathbf{a} \equiv (1,2,3)$ and that $\mathbf{x}$ satisfies the relation

$$\mathbf{x} = 3\mathbf{a} + \mathbf{x} \wedge \mathbf{a},$$

find $\mathbf{x}$.

4. Prove that, if $\mathbf{a}, \mathbf{b}, \mathbf{c}$ are non-coplanar, then

$$\mathbf{b} \wedge \mathbf{c}, \quad \mathbf{c} \wedge \mathbf{a}, \quad \mathbf{a} \wedge \mathbf{b}$$

are also non-coplanar.

Prove that an arbitrary vector **x** can be expressed in the form

$$\mathbf{x} = \lambda(\mathbf{b} \wedge \mathbf{c}) + \mu(\mathbf{c} \wedge \mathbf{a}) + \nu(\mathbf{a} \wedge \mathbf{b}),$$

where, for example,

$$\lambda = \mathbf{a}.\mathbf{x}/(\mathbf{abc}).$$

5. Prove that the vectors

$$\mathbf{a} \wedge \mathbf{p}, \quad \mathbf{a} \wedge \mathbf{q}, \quad \mathbf{a} \wedge \mathbf{r}$$

are coplanar.

6. Solve for **x** the equation

$$\mathbf{x} \wedge \mathbf{a} = \mathbf{b} - \mathbf{x}.$$

7. The unit vectors **a**, **b** are perpendicular and the unit vector **c** is inclined at an angle θ to each of **a** and **b**. Prove that

$$\mathbf{c} = (\mathbf{a}+\mathbf{b}) \cos\theta + \mathbf{a} \wedge \mathbf{b}\sqrt{(-\cos 2\theta)}.$$

Comment on the case $\theta = \frac{1}{4}\pi$.

8. The coordinate vectors of three points A, B, C are **a**, **b**, **c**. Find the condition for the point D with position vector $\mathbf{d} \equiv \lambda\mathbf{a} + \mu\mathbf{b} + \nu\mathbf{c}$ to lie in the plane ABC.

If AB, CD meet in E, and AC, BD meet in F, find the coordinate vectors of E, F and show that the point of intersection of AD, EF has coordinate vector $\dfrac{1}{1+\lambda}(\lambda\mathbf{a}+\mathbf{d})$.

9. The points P, Q have coordinate vectors **p**, **q**; a plane α passes through the origin, the perpendiculars to α being in the direction **n**. Find the position vectors **u**, **v** of the feet of the perpendiculars from P, Q to α. Express the angle subtended at the origin by the feet, giving your answer in terms of **p**, **q**, **n**.

Prove that, when

$$\mathbf{p} \equiv (1, 2, -2), \quad \mathbf{q} \equiv (5, 2, 2), \quad \mathbf{n} \equiv (2, 1, -1),$$

then the cosine of the angle is $-5/7$.

10. Show that the most general solution of the equation

$$\mathbf{x} \wedge \mathbf{a} = \mathbf{b}$$

is

$$\mathbf{x} = \lambda\mathbf{a} + (\mathbf{a} \wedge \mathbf{b})/a^2,$$

where λ is an arbitrary scalar.

11. Prove that, if **u** is a unit vector, so that $\mathbf{u}^2 = 1$, then

$$\mathbf{u} \wedge (\mathbf{r} \wedge \mathbf{u}) = \mathbf{r} - (\mathbf{u}.\mathbf{r})\mathbf{u}.$$

12. Prove that, if

$$\mathbf{p} = \frac{\mathbf{b} \wedge \mathbf{c}}{(\mathbf{abc})}, \quad \mathbf{q} = \frac{\mathbf{c} \wedge \mathbf{a}}{(\mathbf{abc})}, \quad \mathbf{r} = \frac{\mathbf{a} \wedge \mathbf{b}}{(\mathbf{abc})},$$

then

$$(\mathbf{pqr})(\mathbf{abc}) = 1.$$

13. Prove that

$$(\mathbf{p} \wedge \mathbf{q}).(\mathbf{p} \wedge \mathbf{r}) = \det\begin{pmatrix} \mathbf{p}^2 & \mathbf{p}.\mathbf{r} \\ \mathbf{q}.\mathbf{p} & \mathbf{q}.\mathbf{r} \end{pmatrix},$$

the determinant whose elements are $\mathbf{p}^2$, $\mathbf{p}.\mathbf{r}$, $\mathbf{q}.\mathbf{p}$, $\mathbf{q}.\mathbf{r}$.

14. Prove that

$$(\mathbf{p} \wedge \mathbf{q}) \wedge (\mathbf{p} \wedge \mathbf{r}) = (\mathbf{pqr})\mathbf{p}.$$

15. Vectors **a**, **b**, **c** and a non-zero scalar k are given. It is required to find a vector

$$\mathbf{x} \equiv \lambda(\mathbf{b} \wedge \mathbf{c}) + \mu(\mathbf{c} \wedge \mathbf{a}) + \nu(\mathbf{a} \wedge \mathbf{b})$$

such that

$$\mathbf{x}.\mathbf{a} = k, \quad \mathbf{x} \wedge \mathbf{b} = \mathbf{c}.$$

Determine any conditions between **a**, **b**, **c**, k for these relations to be consistent, and find the general solution.

16. Solve for **x** the equation

$$k\mathbf{x} + (\mathbf{b}.\mathbf{x})\mathbf{a} = \mathbf{c}$$

when (i) $k + \mathbf{a}.\mathbf{b} \neq 0$, (ii) $k + \mathbf{a}.\mathbf{b} = 0$.

17. Given three non-coplanar vectors **a**, **b**, **c**, find three vectors **u**, **v**, **w** such that, for any vector **x**,

$$\mathbf{x} = (\mathbf{x}.\mathbf{u})\mathbf{a} + (\mathbf{x}.\mathbf{v})\mathbf{b} + (\mathbf{x}.\mathbf{w})\mathbf{c}.$$

18. Given that **i**, **j**, **k** are mutually orthogonal unit vectors and that

$$\mathbf{a} = 2\mathbf{i} + 3\mathbf{j} - 5\mathbf{k}, \quad \mathbf{b} = 4\mathbf{i} - 7\mathbf{j} + 2\mathbf{k},$$

find the vector sum $\mathbf{a} + \mathbf{b}$, the scalar product $\mathbf{a}.\mathbf{b}$ and the vector product $\mathbf{a} \wedge \mathbf{b}$.

Find all the vectors **c** such that

$$\mathbf{a} \wedge \mathbf{c} = \mathbf{a} \wedge \mathbf{b}$$

and solve for **x** the vector equation

$$\mathbf{x} = 2\mathbf{a} + (\mathbf{x} \wedge \mathbf{a}).$$

19. Three concurrent lines OA, OB, OC are such that no two coincide or are perpendicular. The plane α passes through OA and is orthogonal to the plane OBC; β, γ are defined similarly. Prove that the planes α, β, γ have a line in common.

5

TWO-PROCESS ALGEBRA: VECTOR SPACES

The set of positive integers

$$N \equiv \{1, 2, 3, \ldots\}$$

is, in normal practice, subjected not to one law of combination, but to two. We have, indeed, used these two laws when operating with the scalars of the previous chapter and have also used the two laws for scalars in combination with the single law (addition) for vectors. The problem now is to give increasing abstraction to the idea of vectors, leading to conceptions of considerable generality.

We begin by recalling the formal laws under which, usually unnoticed, elementary arithmetic and algebra are constrained to operate.

1. The two laws for elementary arithmetic. The positive integers belonging to the set

$$N \equiv \{1, 2, 3, \ldots\}$$

are subject to two rules of combination:

(i) *addition*, denoted by $+$, whereby

$$x \in N, \quad y \in N, \quad z \in N$$

$$\Rightarrow \begin{cases} \exists \text{ unique } x+y \in N, \\ (x+y)+z = x+(y+z), \\ \qquad x+y = y+x; \end{cases}$$

(ii) *multiplication*, denoted by $\times$, whereby

$$x \in N, \quad y \in N, \quad z \in N$$

$$\Rightarrow \begin{cases} \exists \text{ unique } x \times y \in N, \\ (x \times y) \times z = x \times (y \times z), \\ \qquad x \times y = y \times x. \end{cases}$$

For multiplication, the product is often written $x . y$ or even xy.

As we have seen earlier, the commutative law does not always hold for generalizations. It will, however, be assumed throughout this book that *operations denoted by the symbol + are to be commutative.*

The new feature induced by the existence of two rules of combination is the need to provide for their simultaneous use. This is effected by the *distributive law*:

$$x \in N, \quad y \in N, \quad z \in N$$
$$\Rightarrow \begin{cases} x \times (y+z) = x \times y + x \times z, \\ (y+z) \times x = y \times x + z \times x, \end{cases}$$

the two bracketed rules being the same when the multiplication is commutative.

In simpler notation, the distributive law is

$$\begin{cases} x(y+z) = xy+xz, \\ (y+z)x = yx+zx. \end{cases}$$

This law is the justification for the normal manipulation of brackets in elementary algebra. It can be extended in obvious ways to formulae like

$$(x+y)(x+y) = x^2+xy+yx+y^2.$$

Such modifications will usually be incorporated without comment.

The following treatment of vector spaces forms a link between one-process algebra of the first three chapters and fully two-process algebra to be considered later.

2. Generalization of the idea of a vector space. The work now to be considered involves two sets: a set $S \equiv \{a, b, c, \ldots\}$ of *scalars* and a set $V \equiv \{\mathbf{x}, \mathbf{y}, \mathbf{z}, \ldots\}$ of *vectors*.

The *scalars*, when expressed algebraically, are named in *italic* type and consist of the ordinary numbers of arithmetic subject to the two rules of addition and multiplication set out in § 1. It will normally be necessary to give an explicit statement of the set from which these elements are taken—positive integers, rational numbers, complex numbers, etc. They will be assumed to include a zero 0 and a unit 1.

The *vectors* are named in **bold** type. Their properties are stated almost immediately, followed by the rules combining them with the scalars.

There is a single rule of combination for the vectors of V, denoted by the symbol +. The effect of the laws now to be stated is that V forms a *commutative group*:

$$\mathbf{x} \in V, \quad \mathbf{y} \in V, \quad \mathbf{z} \in V$$
$$\Rightarrow \quad \exists \text{ unique } \mathbf{x}+\mathbf{y} \in V,$$
$$(\mathbf{x}+\mathbf{y})+\mathbf{z} = \mathbf{x}+(\mathbf{y}+\mathbf{z}),$$
$$\mathbf{x}+\mathbf{y} = \mathbf{y}+\mathbf{x},$$

and, finally, given a relation

$$\mathbf{x}+\mathbf{y} = \mathbf{z},$$

it can be satisfied within V when *any two* of $\mathbf{x}, \mathbf{y}, \mathbf{z}$ are given.

As demonstrated in Chapter 3, the last rule is equivalent to the existence of a zero $\mathbf{0}$ such that

$$\mathbf{0}+\mathbf{x} = \mathbf{x}+\mathbf{0} = \mathbf{x}$$

and an inverse $-\mathbf{x}$ for every $\mathbf{x}$ such that

$$(-\mathbf{x})+\mathbf{x} = \mathbf{x}+(-\mathbf{x}) = \mathbf{0}.$$

The rules combining S and V are, in effect, designed to ensure that, to the naked eye, the manipulations familiar in elementary algebra may continue to apply. They are:

$$\begin{aligned} & a, b \in S, \quad \mathbf{x}, \mathbf{y} \in V \\ \Rightarrow \quad & \exists\, a\mathbf{x} \in V, \\ & a(\mathbf{x}+\mathbf{y}) = a\mathbf{x}+a\mathbf{y}, \\ & (a+b)\mathbf{x} = a\mathbf{x}+b\mathbf{x}, \\ & a(b\mathbf{x}) = (ab)\mathbf{x}, \\ & 1.\mathbf{x} = \mathbf{x}. \end{aligned}$$

DEFINITION. When all the above conditions are satisfied, V is said to form a *vector space* (*linear space*) with scalars in S (or over S).

***Illustration* 1.** *Some basic identities.* These properties are important in themselves and also demonstrate some manipulations of the definitions and rules. *The reader should provide the appropriate reference for each step.*

(i) *To prove that*

$$0.\mathbf{x} = \mathbf{0}.$$

We have

$$\begin{aligned} 0.\mathbf{x} &= (a-a)\mathbf{x} \\ &= a\mathbf{x}+(-a)\mathbf{x} \\ &= a\mathbf{x}+(-1)(a)\mathbf{x} \\ &= a\mathbf{x}+(-1)(a\mathbf{x}) \\ &= \mathbf{0}. \end{aligned}$$

(ii) *To prove that*

$$a.\mathbf{0} = \mathbf{0}.$$

We have

$$\begin{aligned} a.\mathbf{x}+a.\mathbf{0} &= a(\mathbf{x}+\mathbf{0}) \\ &= a\mathbf{x} \\ \Rightarrow a.\mathbf{0} &= \mathbf{0}. \end{aligned}$$

(iii) *To prove that*

$$(-a)\mathbf{x} = a(-\mathbf{x}) = -(a\mathbf{x}).$$

We have $(-a)\mathbf{x} = (-1)(a)\mathbf{x} = -1.(a\mathbf{x}) = -(a\mathbf{x})$.

Also
$$a\{(-\mathbf{x})+\mathbf{x}\} = a.\mathbf{0} = \mathbf{0}$$
$$\Rightarrow a(-\mathbf{x})+a\mathbf{x} = \mathbf{0}$$
$$\Rightarrow a(-\mathbf{x}) = -(a\mathbf{x}).$$

Text Examples

1. Prove that the zero vector $\mathbf{0}$ is *unique*.
2. Prove that every vector $\mathbf{x}$ has a *unique* inverse.

***Illustration* 2.** *The solutions of homogeneous linear equations.* Consider the two equations
$$x+y+z+t = 0,$$
$$x+2y+3z+4t = 0.$$
There are only two equations, but four variables (three ratios), and so an infinity of solutions may be expected. Suppose, for example, that z, t are given arbitrary values λ, μ respectively; then
$$x+y = -\lambda-\mu,$$
$$x+2y = -3\lambda-4\mu,$$
so that
$$x = \lambda+2\mu, \quad y = -2\lambda-3\mu.$$
Thus the general solution of the two equations can be expressed in the form
$$x = \lambda+2\mu, \quad y = -2\lambda-3\mu, \quad z = \lambda, \quad t = \mu$$
for arbitrary values of λ, μ. For example, $\lambda = 5$, $\mu = 1$ gives the solution
$$x = 7, \quad y = -13, \quad z = 5, \quad t = 1.$$

Now exhibit these solutions as *ordered quadruplets* in the form
$$(\lambda+2\mu, -2\lambda-3\mu, \lambda, \mu).$$
Symbolically this may be written
$$\lambda(1, -2, 1, 0)+\mu(2, -3, 0, 1)$$
and our purpose now is to probe this symbolism deeper.

Denote by V the set of quadruplets typified by
$$\mathbf{p} \equiv (x, y, z, t)$$
subject to the condition that the components x, y, z, t satisfy the two given equations. It is to be proved that *the elements of V can belong to a vector space where the scalars are the numbers of ordinary arithmetic.*

If V is to be a vector space, a rule of combination must be prescribed for the vectors. A natural rule is that, if
$$\mathbf{p} \equiv (x_1, y_1, z_1, t_1), \quad \mathbf{q} \equiv (x_2, y_2, z_2, t_2)$$
are two quadruplets within V, then the sum $\mathbf{p}+\mathbf{q}$ is to be given by the relation
$$\mathbf{p}+\mathbf{q} \equiv (x_1+x_2, y_1+y_2, z_1+z_2, t_1+t_2).$$

The important point is that, with this rule of combination, the quadruplet $\mathbf{p}+\mathbf{q}$ belongs to V:

For

$$\begin{aligned}(x_1+x_2)+(y_1+y_2)+(z_1+z_2)+(t_1+t_2)\\ &= (x_1+y_1+z_1+t_1)+(x_2+y_2+z_2+t_2) = 0+0\\ &= 0,\end{aligned}$$

and

$$\begin{aligned}(x_1+x_2)+2(y_1+y_2)+3(z_1+z_2)+4(t_1+t_2)\\ &= (x_1+2y_1+3z_1+4t_1)+(x_2+2y_2+3z_2+4t_2) = 0+0\\ &= 0,\end{aligned}$$

so that the components of $\mathbf{p}+\mathbf{q}$, as defined, satisfy the two given equations. Hence

$$\mathbf{p}+\mathbf{q} \in V.$$

We can now check that the vectors of V form a commutative group under the given rule of combination; for the requirements

$$(\mathbf{p}+\mathbf{q})+\mathbf{r} = \mathbf{p}+(\mathbf{q}+\mathbf{r})$$

and

$$\mathbf{p}+\mathbf{q} = \mathbf{q}+\mathbf{p}$$

are readily verified, and it is equally easy to see that the equation

$$\mathbf{p}+\mathbf{q} = \mathbf{r}$$

can be satisfied within V when any two of $\mathbf{p}$, $\mathbf{q}$, $\mathbf{r}$ are given. Incidentally, the zero is the quadruplet

$$\mathbf{0} \equiv (0, 0, 0, 0)$$

and the inverse of $\mathbf{p} \equiv (x, y, z, t)$ is the quadruplet

$$-\mathbf{p} \equiv (-x, -y, -z, -t).$$

The rule required for combining the elements of S and V is easily proposed:

$$a\mathbf{p} \equiv (ax, ay, az, at).$$

Then it is immediate that

$$a\mathbf{p} \in V.$$

With this definition of the product, the further requirements

$$\begin{aligned} a(\mathbf{p}+\mathbf{q}) &= a\mathbf{p}+a\mathbf{q},\\ (a+b)\mathbf{p} &= a\mathbf{p}+b\mathbf{p},\\ a(b\mathbf{p}) &= (ab)\mathbf{p},\\ 1.\mathbf{p} &= \mathbf{p}\end{aligned}$$

can be established by direct computation, and so it is proved that V, subject to the stated law of combination, is a vector space with scalars in S.

***Illustration* 3.** *Vector space or not?* Consider the simultaneous equations

$$\begin{aligned} xyz &= 1,\\ xy^2z^3 &= 1.\end{aligned}$$

These are *two* equations for *three* unknowns, so, once again, an infinite number of solutions may be expected. Suppose, for example, that z is given an arbitrary value λ; then λ cannot be zero since that value would make the left-hand sides of the equations zero also. Then $x = \lambda$, $y = \lambda^{-2}$, so that the general solution of the equations is

$$x = \lambda, \quad y = \lambda^{-2}, \quad z = \lambda,$$

or, in triplet form,

$$(\lambda, \lambda^{-2}, \lambda)$$

for arbitrary λ.

Suppose that W (not named V, in order to avoid prejudice) is the set of ordered triplets of which a typical member is

$$\mathbf{p} \equiv (p, p^{-2}, p) \qquad (p \neq 0)$$

for an arbitrary value of p. If the rule of combination for the triplets

$$\mathbf{p} \equiv (p, p^{-2}, p), \quad \mathbf{q} \equiv (q, q^{-2}, q)$$

were proposed to be

$$\mathbf{p}+\mathbf{q} \equiv (p+q, p^{-2}+q^{-2}, p+q),$$

the 'sum' $\mathbf{p}+\mathbf{q}$ would not belong to W, since $p^{-2}+q^{-2}$ is not the square of $1/(p+q)$. Hence *the elements of W, subject to the law proposed, cannot form a vector space.*

Consider, however, an alternative rule of combination, which had better be denoted by an alternative symbol, say $*$. Let the 'sum' $\mathbf{p} * \mathbf{q}$ be defined by the relation

$$\mathbf{p} * \mathbf{q} \equiv (pq, p^{-2}q^{-2}, pq).$$

Then

$$\mathbf{p} * \mathbf{q} \in W,$$

since $p^{-2}q^{-2}$ is the square of $1/(pq)$. Hence *the inclusive law*

$$\mathbf{p} * \mathbf{q} \in W$$

is satisfied.

Moreover,

$$\begin{aligned}(\mathbf{p} * \mathbf{q}) * \mathbf{r} &\equiv (pq, p^{-2}q^{-2}, pq) * (r, r^{-2}, r) \\ &\equiv (pqr, p^{-2}q^{-2}r^{-2}, pqr) \\ &\equiv \mathbf{p} * (\mathbf{q} * \mathbf{r}),\end{aligned}$$

so that *the associative law is satisfied.* Also, by similar argument,

$$\begin{aligned}\mathbf{p} * \mathbf{q} &\equiv (pq, p^{-2}q^{-2}, pq) \equiv (qp, q^{-2}p^{-2}, qp) \\ &\equiv \mathbf{q} * \mathbf{p},\end{aligned}$$

so that *the commutative law is satisfied.*

The *zero triplet* $\mathbf{0}$ is defined by the relation

$$\mathbf{0} \equiv (1, 1, 1),$$

since then

$$\begin{aligned}\mathbf{p} * \mathbf{0} &\equiv (p.1, p^{-2}.1^{-2}, p.1) \equiv (p, p^{-2}, p) \\ &\equiv \mathbf{p},\end{aligned}$$

and, similarly,

$$\mathbf{0} * \mathbf{p} \equiv \mathbf{p}.$$

The *inverse* $-\mathbf{p}$ of $\mathbf{p}$ is defined by the relation

$$-\mathbf{p} \equiv (p^{-1}, p^2, p^{-1}),$$

since then

$$-\mathbf{p} * \mathbf{p} \equiv (p^{-1}p, p^2p^{-2}, p^{-1}p) \equiv (1, 1, 1) \equiv \mathbf{0},$$

and, similarly,

$$\mathbf{p}*(-\mathbf{p}) \equiv \mathbf{0}.$$

It follows that *the triplets*

$$\mathbf{p} \equiv (p, p^{-2}, p) \qquad (p \neq 0)$$

subject to the rule of combination

$$\mathbf{p} * \mathbf{q} \equiv (pq, p^{-2}q^{-2}, pq)$$

form a commutative group. The equation

$$\mathbf{p} * \mathbf{q} \equiv \mathbf{r}$$

is satisfied within the set of triplets whenever any two of $\mathbf{p}$, $\mathbf{q}$, $\mathbf{r}$ are given.

It is therefore possible, so far, for W to form a vector space with this rule of combination. To complete the work, it is necessary to define a rule for scalar multiplication to give triplets, denoted by $a\mathbf{p}$, with a within a given set S, such that

$$\begin{aligned} a(\mathbf{p} * \mathbf{q}) &= a\mathbf{p} * a\mathbf{q}, \\ (a+b)\mathbf{p} &= a\mathbf{p} * b\mathbf{p}, \\ a(b\mathbf{p}) &= (ab)\mathbf{p}, \\ 1.\mathbf{p} &= \mathbf{p}. \end{aligned}$$

Text Examples

1. Prove that the definition

$$a\mathbf{p} \equiv (ap, ap^{-2}, ap)$$

does *not* satisfy the conditions.

In order to satisfy these conditions, the *law of multiplication* of a vector $\mathbf{p}$ by a scalar a is taken to be

$$a\mathbf{p} \equiv (p^a, p^{-2a}, p^a).$$

Then

(i)
$$\begin{aligned} a(\mathbf{p} * \mathbf{q}) &\equiv a(pq, p^{-2}q^{-2}, pq) \\ &\equiv (p^a q^a, p^{-2a}q^{-2a}, p^a q^a) \\ &\equiv (p^a p^{-2a}, p^a) * (q^a, q^{-2a}, q^a) \\ &\equiv a\mathbf{p} * a\mathbf{q}; \end{aligned}$$

(ii)
$$\begin{aligned} (a+b)\mathbf{p} &\equiv (p^{a+b}, p^{-2a-2b}, p^{a+b}) \\ &\equiv (p^a p^b, p^{-2a}p^{-2b}, p^a p^b) \\ &\equiv (p^a, p^{-2a}, p^a) * (p^b, p^{-2b}, p^b) \\ &\equiv a\mathbf{p} * b\mathbf{p}; \end{aligned}$$

(iii)
$$\begin{aligned} a(b\mathbf{p}) &\equiv a(p^b, p^{-2b}, p^b) \\ &\equiv \{(p^b)^a, (p^{-2b})^a, (p^b)^a\} \\ &\equiv (p^{ab}, p^{-2ab}, p^{ab}) \\ &\equiv ab\mathbf{p}; \\ 1.\mathbf{p} &\equiv 1(p, p^{-2}, p) \equiv (p^1, p^{-2\times 1}, p^1) \\ &\equiv (p, p^{-2}, p) \\ &= \mathbf{p}. \end{aligned}$$

Hence W, *subject to the rules defined, does form a vector space.*

COMMENT. The importance of this example is in its emphasis that a set of elements, of itself, *neither* does *nor* does not form a vector space. It is essential that rules be clearly enunciated and that those rules be carefully verified as satisfying all the necessary conditions.

The other point to be emphasized is the very wide scope of the elements and rules which generate vector spaces.

Illustration* 4. *The solutions of a linear differential equation.

Consider the differential equation

$$\frac{d^2y}{dx^2} - 5\frac{dy}{dx} + 6y = 0.$$

[The general solution is

$$y = Ae^{2x} + Be^{3x},$$

but that fact is not actually required.]

Let u,v (functions of x) be any two solutions, so that, dashes denoting differentiations with respect to x,

$$u'' - 5u' + 6u = 0,$$
$$v'' - 5v' + 6v = 0.$$

Write
$$w = u + v.$$

Then
$$\begin{aligned} &w'' - 5w' + 6w \\ &= (u'' - 5u' + 6u) + (v'' - 6v' + 6v) \\ &= 0, \end{aligned}$$

so that w is also a solution.

The solutions are to be exhibited as the vectors of a vector space. For that reason, they will now be named in **bold** type **u**, **v**, **w**, ... Then, since the equation is satisfied by $y = 0$, there is a zero function **0** and, since the equation if satisfied by $y = u$ is also satisfied by $y = -u$, there is an inverse $-\mathbf{u}$ of every solution **u**.

Examples

1. Prove that the set of elements **u**, **v**, ..., as defined above, form a vector space under rules to be clearly enunciated and with scalars clearly defined.

2. Obtain rules of combination under which the following sets (with suitable scalars) can be the vectors of a vector space, and verify in each case that the conditions are satisfied:

(i) the quadratic polynomials

$$a_1x^2+a_2x+a_3;$$

(ii) the circles as defined by equations of the form

$$a_1(x^2+y^2)+a_2x+a_3y+a_4 = 0;$$

(iii) the set of functions that are continuous in the interval (0,1);

(iv) the ordered triplets

$$(x_1, x_2, x_3)$$

whose elements are subject to the restriction

$$x_1+x_2+x_3 = 0;$$

(v) the circles touching the x-axis at the origin;

(vi) the circles through the points (1,0), (−1,0).

3. Suggest laws of vector addition and multiplication by scalars under which the set of cubic polynomials

$$a_1x^3+a_2x^2+a_3x+a_4$$

(x real; a_1, a_2, a_3, a_4 complex) form a vector space.

Do the laws that you have suggested give a vector space when a_1, a_2, a_3, a_4 are restricted to be powers of 2?

4. Exhibit the solutions of the equations

(i)
$$2x+3y+4z = 0,$$
$$x+2y+5z = 0$$

and (ii)
$$x+y+2z+t = 0,$$
$$x+y+z+3t = 0$$

as vector spaces after the manner of Illustration 2.

5. Prove that the system of lines, in a plane, passing through the point (1, −2) belong to a vector space.

6. Prove that, under a law

$$\mathbf{p}+\mathbf{q} \equiv (p_1+q_1, p_2+q_2),$$

the points $(at^2, 2at)$ do not belong to a vector space.

7. Exhibit the solutions of the differential equation

$$\frac{d^2y}{dx^2}-3\frac{dy}{dx}+2y = 0$$

as a vector space.

8. Exhibit the set of rectangular hyperbolas through the vertices of a triangle and its orthocentre as a vector space.

9. The triplets $\mathbf{x}(x_1, x_2, x_3)$ are proposed for a vector space under the rules

$$\mathbf{x}+\mathbf{y} \equiv (x_1+y_1, x_2+y_2, x_3+y_3),$$
$$a\mathbf{x} \equiv (ax_1, ax_2, ax_3)$$

where the elements x_i and the scalars a are to be selected from some given set S. Determine whether such a vector space is possible when the set S consists of

(i) all integers (positive, negative, zero),
(ii) all powers of 2,
(iii) all complex numbers,
(iv) all rational numbers,
(v) all rational numbers in the interval $-1 \leqslant k \leqslant 1$,
(vi) the numbers 0, 1, ..., 9 in 'digital arithmetic' (p. 8).

6

TWO-PROCESS ALGEBRA: SINGLE SET, RINGS AND FIELDS

The two processes of addition and multiplication, whose rules have already been recalled, can be extended to cover many sets other than the familiar numbers of arithmetic. The earlier work has separated the processes to some extent; they must now be brought into closer contact.

1. Rings. Let

$$A \equiv \{a, b, c, \ldots\}.$$

be a set whose elements $a, b, c, \ldots$ are subjected to *two rules of combination*:

(i) a rule called *addition* and denoted by the symbol $+$;

(ii) a rule called *multiplication* and denoted by the symbol $\times$ (which may be omitted, as in elementary algebra, for convenience of notation).

The individual properties governing these rules are:

(i) *the set A forms a commutative group with respect to addition*, so that

$$\begin{aligned} & a \in A, \quad b \in A, \quad c \in A \\ \Rightarrow\ & \exists \text{ unique } a+b \in A, \\ & (a+b)+c = a+(b+c), \\ & a+b = b+a, \end{aligned}$$

and the relation

$$a+b = c$$

can be satisfied within A when any two of a, b, c are given;

(ii) *the set A obeys the inclusive and the associative laws with respect to multiplication*, so that

$$\begin{aligned} & a \in A, \quad b \in A, \quad c \in A \\ \Rightarrow\ & \exists \text{ unique } a \times b \in A, \\ & (a \times b) \times c = a \times (b \times c). \end{aligned}$$

REMARKS. The fact that A is a commutative group for addition implies

$$\exists\ 0 \text{ such that } \quad a+0 = 0+a = a,$$

and $$\exists\ -a \quad \text{such that} \quad -a+a = a+(-a) = 0.$$

It is not assumed that multiplication necessarily obeys the commutative law; nor need it define a group, commutative or otherwise.

Rules are now required to govern the simultaneous operation of addition and multiplication. These are afforded by the *distributive law*

$$a(b+c) = ab+ac,$$
$$(b+c)a = ba+ca,$$

the two relations being the same when multiplication is commutative.

DEFINITION. A set of elements subject to all these rules is called a *ring*. When the multiplication is commutative, the ring is called *commutative*.

2. Some properties of rings. Rings have a clear affinity with the integers of ordinary arithmetic, but some differences are possible:

(i) *A ring may or may not have a unity* 1 (*or* e) *with the property*

$$1.a = a.1 = a.$$

Consider, for example, the set of positive and negative *even* integers, with zero,

$$A \equiv \{\ldots, -4, -2, 0, 2, 4, \ldots\},$$

subject to the usual rules of arithmetic.

Text Example

1. Prove that A is a ring.

There is, however, no unity element. It is not possible to find $e \in A$ such that

$$ae = ea = a.$$

DEFINITION. A ring which does contain a unity element is called a *ring with unity.*

(ii) *The relation*

$$a.0 = 0.a = 0$$

is true for all $a \in A$:

By definition,

$$\begin{aligned} b+0 &= b \\ \Rightarrow \quad a(b+0) &= ab \\ \Rightarrow \quad ab+a.0 &= ab \\ \Rightarrow \quad a.0 &= 0 \end{aligned}$$

since the zero is unique. (See Chapter 2, §2.) Similarly

$$0.a = 0.$$

(iii) *If $a = 0$ or if $b = 0$, then (as in (ii))*

$$ab = 0;$$

but there are rings in which the converse is not true:

$$ab = 0$$

$$\not\Rightarrow \textit{ either } a = 0 \quad \text{or} \quad b = 0.$$

Consider, for example, the set

$$A \equiv \{0, 1, 2, 3, 4, 5\},$$

where the laws of addition and multiplication are those of ordinary arithmetic, save that all answers are reduced by multiples of 6 to bring them within the set A.

Text Examples

1. Prove that A is a ring.
2. Verify that $a = 0 \Rightarrow ab = 0$.

But *there exist pairs of non-zero numbers whose product is zero*; for

$$2 \times 3 = 0.$$

DEFINITION. When

$$ab = 0, \quad a \neq 0, \quad b \neq 0,$$

then a, b are called *divisors of zero*; a is a left-hand divisor of zero, and b a right-hand.

Examples

1. Prove that, in this ring, the equation

$$x^2 + 3x + 2 = 0$$

has four solutions and the equation

$$x^2 + 3x + 3 = 0$$

none.

2. Prove that the set $\{0, 1, 2, 3, 4\}$, subject to ordinary arithmetic with answers reduced by multiples of 5, is a ring without divisors of zero.

3. Does the set $\{1, 3, 5\}$, subject to ordinary arithmetic with answers reduced by multiples of 7, form a ring?

4. Does the set of odd numbers $\{\ldots, -3, -1, 1, 3, \ldots\}$, subject to ordinary arithmetic, form a ring?

5. Prove that the rational numbers (positive, negative or zero) form a ring when subject to ordinary arithmetic.

***Illustration* 1.** *The ring of polynomials.* (To be omitted at first if found hard.)

Let A be a given ring whose elements it is convenient to name in the form

$$\{a_0, a_1, a_2, \ldots, \quad b_0, b_1, b_2, \ldots, \quad c_0, c_1, c_2, \ldots\}.$$

Construct polynomials typified by

$$f(x) \equiv a_0 + a_1 x + a_2 x^2 + a_3 x^3 + \ldots$$

where $a_i \in A$, $x \notin A$. Each polynomial consists of a *finite* number of terms, varying from polynomial to polynomial. The element x is called an *indeterminate* and is a mere symbol, whose duty is to record the position of the relevant *coefficient* corresponding to its power. The indeterminate x is assumed to *commute* with all elements of A, so that

$$k \in A \Rightarrow kx = xk.$$

The polynomials form a set P, and they are to be combined according to the following rules:

(i) $$f(x) = g(x)$$

$\Leftrightarrow$ all corresponding coefficients are equal; in particular, the polynomials are of the same order.

[In this context, the symbol of equality = replaces the symbol of identity ≡ used in Part 1.]

(ii) If
$$f(x) \equiv a_0 + a_1 x + a_2 x^2 + \ldots,$$
$$g(x) \equiv b_0 + b_1 x + b_2 x^2 + \ldots,$$

their *sum*
$$f(x) + g(x)$$
is defined to be
$$(a_0 + b_0) + (a_1 + b_1)x + (a_2 + b_2)x^2 + \ldots.$$

(iii) The *product*
$$f(x)\,g(x)$$
is defined (following the 'schoolboy' rule for the multiplication of polynomials) to be the polynomial
$$h(x) \equiv c_0 + c_1 x + c_2 x^2 + \ldots,$$
where
$$\begin{aligned} c_0 &= a_0 b_0, \\ c_1 &= a_0 b_1 + a_1 b_0, \\ c_2 &= a_0 b_2 + a_1 b_1 + a_2 b_0, \\ c_3 &= a_0 b_3 + a_1 b_2 + a_2 b_1 + a_3 b_0, \\ &\ldots\ldots\ldots\ldots\ldots\ldots\ldots\ldots \end{aligned}$$

with zeros in the a's and b's where necessary. It will be recalled that $a_i b_j$ is not necessarily equal to $a_j b_i$.

As an illustration of a product, the rule gives (for coefficients in the set of integers subject to 'ordinary' arithmetic),

$$\begin{aligned}(2+3x)(4+5x+6x^2) &= 2.4\\ &\quad+(2.5+3.4)\,x\\ &\quad+(2.6+3.5)\,x^2\\ &\quad+\qquad 3.6\;x^3\\ &= 8+22x+27x^2+18x^3.\end{aligned}$$

It is to be proved that *the polynomials, when combined according to these rules, form a ring.*

To prove this, the definitive properties must be established in turn. This will be done here for the harder cases, leaving the reader to verify as examples those which are more obvious. For ease of comparison with the properties as stated in the definition of a ring, the polynomials will be named according to the notation:

$$a \equiv a_0+a_1x+a_2x^2+\ldots,$$
$$b \equiv b_0+b_1x+b_2x^2+\ldots.$$

Text Examples

Prove the following properties;

1. $$a \in P, \quad b \in P, \quad c \in P$$ $$\Rightarrow \exists \text{ unique } a+b \in P.$$

2. $$(a+b)+c = a+(b+c).$$

3. $$a+b = b+a.$$

4. The relation $a+b = c$ is satisfied within P when any two of a, b, c are given.

5. $\exists$ zero 0 and inverse $-a$ of a, both to be identified.

The *associative law for multiplication*

$$(ab)\,c = a(bc)$$

is harder, and is now proved. Write

$$p_k \equiv a_0b_k+a_1b_{k-1}+a_2b_{k-2}+\ldots+a_kb_0.$$

Then, by definition,

$$ab = p_0+p_1x+p_2x^2+\ldots.$$

Now write

$$q_m = p_0c_m+p_1c_{m-1}+p_2c_{m-2}+\ldots+p_mc_0.$$

Then

$$(ab)\,c = q_0+q_1x+q_2x^2+\ldots,$$

where the coefficient of x^m is

$$\begin{aligned} p_0 c_m + p_1 c_{m-1} + p_2 c_{m-2} + \ldots + p_m c_0 \\ = (a_0 b_0) c_m \\ + (a_0 b_1 + a_1 b_0) c_{m-1} \\ + (a_0 b_2 + a_1 b_1 + a_2 b_0) c_{m-2} \\ + (a_0 b_3 + a_1 b_2 + a_2 b_1 + a_3 b_0) c_{m-3} \\ + \ldots\ldots\ldots\ldots\ldots\ldots \end{aligned}$$

Arrange this summation in order of ascending suffixes in a (by 'going down' the columns, as printed, in turn):

$$\begin{aligned} q_m = a_0(b_0 c_m + b_1 c_{m-1} + b_2 c_{m-2} + b_3 c_{m-3} + \ldots), \\ + a_1(b_0 c_{m-1} + b_1 c_{m-2} + b_2 c_{m-3} + \ldots) \\ + a_2(b_0 c_{m-2} + b_1 c_{m-3} + \ldots) \\ + \ldots\ldots\ldots\ldots\ldots\ldots \end{aligned}$$

Text Examples

1. Complete the argument to prove that

$$(ab)c = a(bc).$$

2. Prove similarly the distributive laws

$$a(b+c) = ab+ac,$$
$$(b+c)a = ba+ca.$$

DEFINITION. The ring P is called the *polynomial ring associated with the given ring.*

Revision Examples

1. Give a meaning to the sign $-$ for use in the theory of a ring $A\{a, b, c, \ldots\}$, and prove that

$$a(b-c) = ab-ac, \quad (b-c)a = ba-ca, \quad (-a)(-b) = ab.$$

2. Prove that, if the ring A has a unity element (p. 74), so also has the polynomial ring P associated with it.

3. Prove that a ring cannot have two distinct unities.

4. Prove that the set of integers

$$\{\ldots, -2, -1, 0, 1, 2, \ldots\},$$

under the normal rules of addition and multiplication, forms a ring with unity.

5. Definition. The *inverse* a^{-1} of an element a in a ring with a unity element e is defined by the relation

$$aa^{-1} = a^{-1}a = e.$$

Prove that no element, other than e itself, can have two distinct inverses.

6. Definition. An element a of a ring with unity is said to be *regular* when it has an inverse a^{-1}; a regular element is also called a *unit* of the ring. Is it possible for every non-zero element in a ring to be a unit?

7. Prove that, in the ring

$$\{\ldots, -2, -1, 0, 1, 2, \ldots\}$$

under ordinary addition and multiplication, the only regular elements (units) are $+1$ and -1.

8. Prove that, if a is a unit of a ring, so also is a^{-1}. Prove that the zero element cannot be a unit.

9. A set consists of ordered pairs of real numbers

$$a = (a_1, a_2), \quad b = (b_1, b_2), \ldots$$

subject to the two rules of combination

$$a+b = (a_1+b_1, a_2+b_2), \quad ab = (a_1 b_1, a_2 b_2).$$

Prove that the elements of the set form a ring.

Verify that, for all real k, the elements $(k,0)$, $(0,k)$ are divisors of zero: thus

$$\exists\ (p,q) \quad \text{so that} \quad (p,q)\times(k,0) = (0,0).$$

10. Two elements a, b in a finite ring R are such that the equation $ax = b$ has no solution in R. Prove that there is an element c of R such that the equation $ax = c$ has more than one solution in R.

11. The elements $a_1, a_2, \ldots, a_n$ of a ring are all different from the zero element 0 of the ring but are such that

$$a_1 a_2 \ldots a_n = 0.$$

Prove that there are indices r, s, where

$$1 \leqslant r < s \leqslant n,$$

such that a_r is a left divisor of zero and a_s is a right divisor of zero. (An element a is, for example, a left divisor if $ax = 0$ for at least one x, where $x \neq 0$.)

12. Prove that the functions that are continuous in the interval $(-1, 1)$ form a ring under the obvious rules.

The function $f(x)$ is defined to be the greater of $0, x$ and the function $g(x)$ is defined to be the greater of $0, -x$. Prove that

$$f(x)g(x) = 0$$

although neither of $f(x)$, $g(x)$ is the zero function.

13. Prove that the even integers form a ring under the normal rule for addition but with the rule of multiplication replaced by

$$ab = \tfrac{1}{2}(\text{normal product of } a \text{ and } b).$$

3. Fields. Among the rings are some particular examples closely analogous to the rational numbers of ordinary arithmetic:

DEFINITION. A *field F* is a ring whose *non-zero* elements form a group with respect to the rule of *multiplication* governing the ring.

Thus

$$a \in F, \quad b \in F \qquad (a \neq 0, b \neq 0)$$

$\Rightarrow \exists\, x, y \in F$ so that

$$ax = b, \quad ya = b.$$

In particular, *a field necessarily has a unity*, denoted by 1 or e, *and each element a (other than zero) has an inverse* a^{-1}.

A field is called *commutative* when all pairs of elements commute for multiplication.

It is an immediate consequence of the definition that *a field has no divisors of zero*, since, if $a \neq 0$, so that a^{-1} exists,

$$\begin{aligned} ab &= 0 \\ \Rightarrow a^{-1}(ab) &= 0 \Rightarrow (a^{-1}a)b = 0 \\ \Rightarrow eb &= 0 \Rightarrow b = 0. \end{aligned}$$

Revision Examples

1. Prove that the set $\{0, 2, 4, 6, 8\}$, subject to normal addition and multiplication with answers reduced by multiples of 10 to bring them within the set, is a field in which 6 is the unity element.

2. Verify that (i) the rational numbers, (ii) the real numbers, (iii) the complex numbers, subject to the normal rules for addition and multiplication, are all fields.

3. Verify that the integers, subject to the normal rules of addition and multiplication, do *not* form a field.

4. Verify that the set $\{0, 1, 2, 3, 4, 5, 6\}$, subject to normal addition and multiplication with answers reduced by multiples of 7 to bring them within the set, is a field.

Identify the inverses of the non-zero elements, and solve the equations

$$\text{(i) } 3x = 5, \qquad \text{(ii) } 5x = 3.$$

Prove also that each of the equations

$$x^2+3x+2 = 0, \quad x^2+3x+3 = 0$$

has two solutions in this field; that the equation

$$x^2+3x+4 = 0$$

has one solution; and that the equation

$$x^2+3x+5 = 0$$

has no solutions.

5. Prove that, in a field,

$$ab = ac \qquad (a \neq 0)$$
$$\Rightarrow b = c.$$

6. Prove that the set $\{a_1+a_2\sqrt{2}\}$, subject to the normal rules of addition and multiplication, is (i) a ring if the elements such as a_1, a_2 are integers, (ii) a field if they are rational numbers.

In case (i), find *non-zero* numbers a_1, a_2 such that $a_1+a_2\sqrt{2}$ multiplied by another element of the ring is equal to 2.

[Thus the element 2 has a factor $a_1+a_2\sqrt{2}$ within the ring.]

7. Show that the set of all numbers of the form $a+b\sqrt{3}$ is a field, where a, b are rational.

8. Verify which of the rules

$$a * (b+c) = (a * b)+(a * c),$$
$$(b+c) * a = (b * a)+(c * a)$$

holds when the symbol $*$ stands for

(i) division $\div$, so that $p * q$ means $p \div q$;
(ii) subtraction $-$, so that $p * q$ means $p-q$;
(iii) addition $+$, so that $p * q$ means $p+q$;
(iv) multiply by the square of, so that $p * q$ means pq^2.

9. Give an example of a field F which has the property that

$$a = -a$$

for every element $a \in F$. Prove that, for every field of this type,

$$(a+b)^n = a^n+b^n,$$

where a, b are any elements of F and n is any power of 2.

10. Prove that there exists no field having exactly six elements, but that there are fields having exactly seven elements.

11. Show that the numbers of the form

$$a+b.2^{\frac{1}{3}},$$

where a, b are rational, *do not* form a field under ordinary addition and multiplication, but that the numbers of the form

$$a+b.2^{\frac{1}{3}}+c.2^{\frac{2}{3}},$$

where c is also rational, *do* form a field.

12. Prove that the union and the intersection of sets obey each of the *distributive laws*

$$A \cup (B \cap C) = (A \cup B) \cap (A \cup C),$$
$$A \cap (B \cup C) = (A \cap B) \cup (A \cap C).$$

[Note, in contrast, that whereas in ordinary arithmetic it is true that

$$a \times (b+c) = (a \times b)+(a \times c),$$

it is *not* true that

$$a+(b \times c) = (a+b) \times (a+c).]$$

7

SUBSETS

This chapter may be omitted or postponed if desired. Its aim is to introduce, to an elementary extent, ideas which are very important in a more detailed study of the subject.

1. The basic idea. A brief account has now been given of the structures of groups, vector spaces, rings and fields. A fundamental feature of all these sets is the possible existence of *subsets* consisting, in each case, of part only of the relevant set, but having the property of behaving under the appropriate rule of combination in the same way as the set itself. The subsets, called *subgroups, subspaces, subrings, subfields*, give point to many of the characteristics of the structures.

2. Subgroups. Let S be a given set of elements, together with a rule of combination $*$ satisfying the group conditions (p. 28):

$$a \in S, \quad b \in S, \quad c \in S$$

$\Rightarrow$ $\exists$ unique $a * b \in S$,

and
$$(a * b) * c = a * (b * c);$$

also the relation
$$a * b = c$$

can be satisfied within S when any two of a, b, c are given.

Equivalently, $\exists$ a neutral element e such that

$$a * e = e * a = a$$

and $\exists$ a unique inverse a' of a such that

$$a * a' = a' * a = e.$$

A *subgroup* T of S is a set whose elements all belong to S with the property that, under the same rule of combination $*$, each of the above conditions holds when S is replaced by T.

In particular, *T must contain the neutral element e and also the inverse a' of each element $a \in T$.*

There are two *improper subgroups* of each group: the group S itself and the group whose sole member is the neutral element e. The other subgroups are called *proper*.

Examples

1. Verify that each of the *improper subgroups* just described does, in fact, satisfy the necessary conditions.

2. Prove that the set

$$\{\ldots, -4, -2, 0, 2, 4, \ldots\}$$

of even integers, under ordinary addition as the rule of combination, is a subgroup of the group

$$\{\ldots, -2, -1, 0, 1, 2, \ldots\}$$

of integers.

Is the same true of the *positive* even integers?

3. Prove that, if T is a subgroup of a group S and if U is another subgroup of S, then the set $T \cap U$ whose elements are common to T and U is also a subgroup of S.

4. A subset T of a group S has the property that, for all $a, b \in T$, the element $a * b^{-1}$ is also in T, where b^{-1} is the inverse of b. Prove that T is a subgroup of S.

5. From a finite group S are selected the elements

$$e, a, a^2, a^3, \ldots$$

defined inductively by relations of the type

$$a^2 = a * a; \quad a^3 = a^2 * a; \quad a^4 = a^3 * a; \ldots.$$

Prove that there is an integer k such that

$$a^k = e,$$

and that, if k is the least such integer, then the set with elements

$$e, a, a^2, \ldots, a^{k-1}$$

is a subgroup of S.

6. Let S be the set

$$\{1, \omega, \omega^2, -1, -\omega, -\omega^2\} \qquad (\omega^3 = 1, \omega \neq 1)$$

subject to ordinary multiplication. Verify the table:

	1	ω	ω^2	-1	$-\omega$	$-\omega^2$
1	1	ω	ω^2	-1	$-\omega$	$-\omega^2$
ω	ω	ω^2	1	$-\omega$	$-\omega^2$	-1
ω^2	ω^2	1	ω	$-\omega^2$	-1	$-\omega$
-1	-1	$-\omega$	$-\omega^2$	1	ω	ω^2
$-\omega$	$-\omega$	$-\omega^2$	-1	ω	ω^2	1
$-\omega^2$	$-\omega^2$	-1	$-\omega$	ω^2	1	ω

Show that S is a group and pick out (i) a subgroup of two elements, (ii) a subgroup of three elements.

7. Prove that every subgroup of a cyclic group is cyclic.

8. Prove that, if H is a subgroup of G, then the set of elements $\{k\}$, such that

$$k^{-1}hk = h$$

for all $h \in H$, is a subgroup of G.

9. Prove that, if g is a fixed element of a group G, then the set of all elements in G that commute with g is a subgroup of G.

3. Lagrange's theorem on the order of a subgroup. This theorem, which is of fundamental importance, is introduced here chiefly to illustrate some points of more abstract argument in group theory.

Let S be a group of finite order n and T a given subgroup of order k. It is required to prove that *k is a factor of n*.

Take an arbitrary element $a \in S$ and denote by Ta the set of all elements ta for which $t \in T$. Such a set is called a *coset* of the subgroup T in S. Another element b gives rise similarly to a set Tb.

The two sets Ta, Tb may or may not be the same. We can, however, prove that

$$ab^{-1} \in T \Rightarrow \text{set } Ta = \text{set } Tb.$$

To prove this, consider the relation

$$ta = tab^{-1}b = (tab^{-1})b.$$

This shows that the element $ta \in Ta$ is the same as the element $(tab^{-1})b \in Tb$. Further, the relation†

$$tb = tba^{-1}a = t(ab^{-1})^{-1}a$$

† The inverse $(pq)^{-1}$ of an element pq is $q^{-1}p^{-1}$; see p. 29.

shows that the element $tb \in Tb$ is the same as the element

$$\{t(ab^{-1})^{-1}\}a \in Ta.$$

Thus every element of Ta is an element of Tb, and every element of Tb is an element of Ta. The sets are therefore the same.

The next point to be made is that, *if the two sets Ta, Tb are not the same, then they cannot have any element in common.*

For

$$t_1 a = t_2 b \qquad (t_1, t_2 \in T)$$
$$\Rightarrow ab^{-1} = t_1{}^{-1}t_2 \in T,$$

and this contradicts the hypothesis $Ta \neq Tb$ in virtue of the property just established.

Hence the sets Ta, Tb, Tc, ... derived from various elements $a, b, c, \ldots \in S$ are either, pair for pair, the same or entirely different.

Now examine a particular set Ta. In the first place, the members of that set defined as 'products' ta, where $t \in T$, are all distinct, for

$$t_1 a = t_2 a \Rightarrow t_1 = t_2.$$

Hence *the set Ta has the same number of elements as T itself.* In the second place, *every element of the whole group S is included in some set Tu for appropriate choice of u*; for T is a subgroup, and so T contains the neutral element e, so that any *given* element $p \in S$ belongs to the set Tp (since $p = ep$ where $e \in T$) and u can therefore be chosen to be the given element p.

To summarize: the n elements of S are divided into sets Ta, Tb, Tc, ... of k elements each, in such a way that each element of S lies in one, and only one, of the sets. If the number of sets is j, then

$$n = kj,$$

so that k is a factor of n.

***Illustration* 1.** The argument may become clearer by reference to the structure table, given on p. 34:

	e	*a*	*b*	*u*	*v*	*w*
e	*e*	*a*	*b*	*u*	*v*	*w*
a	*a*	*b*	*e*	*v*	*w*	*u*
b	*b*	*e*	*a*	*w*	*u*	*v*
u	*u*	*w*	*v*	*e*	*b*	*a*
v	*v*	*u*	*w*	*a*	*e*	*b*
w	*w*	*v*	*u*	*b*	*a*	*e*

Denote by T the subgroup

$$T \equiv \{eab\}.$$

Then

$$\begin{aligned} Te &= \{eab\} \\ Ta &= \{abe\} \\ Tb &= \{bea\} \\ Tu &= \{uvw\} \\ Tv &= \{vwu\} \\ Tw &= \{wuv\}. \end{aligned}$$

There are therefore *two* distinct sets, $\{eab\}$ and $\{uvw\}$, of *three* elements each.

Note that $\{eab\}$ is a subgroup, whereas $\{uvw\}$ is not.

Examples

1. Prove that a group whose order n is a prime number cannot have any proper subgroups.

2. If a is an element of a group S of order n such that (compare p. 37) $a^k = e$, prove that k is a factor of n. Prove also that

$$a \in S \Rightarrow a^n = e.$$

3. Prove that a finite group of prime order n is necessarily cyclic.

4. A group S has a subgroup T. Prove that, in the notation of §3,

$$a \in T \Rightarrow Ta = T,$$
$$a \notin T \Rightarrow Ta \text{ is not a subgroup.}$$

5. T is a subgroup of a group S. Prove that

$$t \in T, \quad a \in S \Rightarrow \{a^{-1}ta\} \text{ is a subgroup of } S.$$

4. Subspaces. Let

$$V \equiv \{\mathbf{x}, \mathbf{y}, \mathbf{z}, \ldots\}$$

be a vector space (p. 65) with scalars in

$$S \equiv \{a, b, c, \ldots\}.$$

It is understood that the scalars are 'ordinary' numbers subject to the normal rules of addition and multiplication, greater precision being specified if necessary. The rules governing V are

$$\begin{aligned} &\mathbf{x} \in V, \quad \mathbf{y} \in V, \quad \mathbf{z} \in V \\ &\Rightarrow \exists \text{ unique } \mathbf{x}+\mathbf{y} \in V, \\ &(\mathbf{x}+\mathbf{y})+\mathbf{z} = \mathbf{x}+(\mathbf{y}+\mathbf{z}), \\ &\mathbf{x}+\mathbf{y} = \mathbf{y}+\mathbf{x}, \\ &\exists\, \mathbf{0}, \quad \exists\, -\mathbf{x}, \end{aligned}$$

the equation

$$\mathbf{x}+\mathbf{y} = \mathbf{z}$$

being soluble within V when any two of $\mathbf{x}, \mathbf{y}, \mathbf{z}$ are given;

$$\exists\, a\mathbf{x} \in V,$$
$$a(\mathbf{x}+\mathbf{y}) = a\mathbf{x}+a\mathbf{y},$$
$$(a+b)\mathbf{x} = a\mathbf{x}+b\mathbf{x},$$
$$a(b\mathbf{x}) = (ab)\mathbf{x},$$
$$1.\mathbf{x} = \mathbf{x}.$$

A *subspace* U of V is a set of vectors of V with the property that each of the above conditions holds when V is replaced, where necessary, by U. The essentially new conditions are

$$\mathbf{x} \in U, \quad \mathbf{y} \in U, \quad a \in S$$
$$\Rightarrow \mathbf{x}+\mathbf{y} \in U,$$
$$a\mathbf{x} \in U.$$

For example, the ordered triplets

$$\mathbf{x} \equiv (x_1, x_2, x_3)$$

under the rules

$$\mathbf{x}+\mathbf{y} \equiv (x_1+y_1, x_2+y_2, x_3+y_3),$$
$$a\mathbf{x} \equiv (ax_1, ax_2, ax_3)$$

form a vector space. A subspace is given by the vectors

$$(x_1, x_2, 0)$$

in which the third component is always zero.

A familiar instance of subspaces is found in the 'space in which we live' of Euclidean geometry. The subspaces are the planes, the straight lines and the points.

Examples

1. Prove that, if two subspaces P, Q of V have a common part U, then U is a subspace of V.

2. Obtain rules under which the quartic polynomials

$$a_1 x^4 + a_2 x^3 + a_3 x^2 + a_4 x + a_5$$

form a vector space, and prove that the cubic polynomials form a subspace.

Prove that the quadratic polynomials also form a subspace, and that this is itself a subspace of the space formed by the cubic polynomials.

3. Obtain rules under which the circles of equation

$$a_1(x^2+y^2)+a_2x+a_3y+a_4 = 0$$

form a vector space.

Prove that the straight lines form a subspace; and also that the circles of a radical system form a subspace.

4. Obtain rules under which the solutions of the differential equation

$$\frac{d^3y}{dx^3}-6\frac{d^2y}{dx^2}+11\frac{dy}{dx}-6y = 0$$

form a vector space.

Prove that the solutions of the equation

$$\frac{d^2y}{dx^2}-3\frac{dy}{dx}+2y = 0$$

form a subspace.

5. Exhibit the solutions of the equations

$$x+y+z+t+u = 0,$$
$$x+2y+3z+4t+5u = 0$$

as a vector space, and prove that the solutions of the equations

$$y+2z+3t+4u = 0,$$
$$2x+3y+4z+5t+6u = 0,$$
$$x-y+z-t-u = 0$$

form a subspace.

6. Exhibit the straight lines in a plane as a vector space and those of them which pass through a fixed point as a subspace.

5. Subrings. Let

$$A \equiv \{a, b, c, \ldots\}$$

be a ring (p. 73). The two operations + and × are subject to the rules:

$$a \in A, \quad b \in A, \quad c \in A$$
$$\Rightarrow \exists \text{ unique } a+b \in A,$$
$$(a+b)+c = a+(b+c),$$
$$a+b = b+a,$$
$$\exists\, 0, \quad \exists\, (-a),$$

the equation

$$a+b = c$$

being satisfied within A when any two of a, b, c are given;

$$\exists \text{ unique } ab \in A,$$
$$(ab)c = a(bc);$$
$$a(b+c) = ab+ac; \quad (b+c)a = ba+ca.$$

A *subring* is a subset U of A with the property that all the above rules are obeyed within it. In particular,

$$a \in U, \quad b \in U$$
$$\Rightarrow a+b \in U, \quad ab \in U$$

and

$$0 \in U, \quad -a \in U.$$

Examples

1. Prove that

$$a-b \in U$$

for all a, b of U

$$\Rightarrow 0 \in U, \quad -a \in U.$$

Deduce that the conditions for a subring can be expressed in the form

$$a \in U, \quad b \in U$$
$$\Rightarrow a-b \in U, \quad ab \in U.$$

2. Prove that the ring of even integers is a subring of the ring of integers, under the ordinary rules for addition and multiplication.

3. Prove that, under the ordinary rules of addition and multiplication, the real numbers form a subring of the complex numbers.

6. Ideals. Among the subrings there is a particular class whose importance justifies brief mention even in an introductory account.

For any subring, by §5,

$$a \in U, \quad b \in U \Rightarrow a-b \in U.$$

Instead of the condition on a product, however, a wider alternative is now selected: take any element, a, of the subring U and adjoin to it *any element r of the whole ring A*; the product ar is to be required to lie within U.

DEFINITIONS. A *right ideal* of a ring A is a subset of A subject to the conditions

$$a \in U, \quad b \in U, \quad r \in A$$
$$\Rightarrow a-b \in U, \quad ar \in U.$$

A *left ideal* is defined similarly:

$$a \in U, \quad b \in U, \quad r \in A$$
$$\Rightarrow a-b \in U, \quad ra \in U.$$

When the ring is commutative the two ideals are the same and the corresponding subring is called simply an *ideal.*

Illustration* 2. *The ideal generated by an element $a \in A$.

Let a be any given element of a ring A. An ideal can be built from it according to the following pattern:

Form the sum $a+a$ and call it $2a$; then form the sum $2a+a$ and call it $3a$; then form the sum $3a+a$ and call it $4a$, proceeding in this way to the element na. Note that the symbol na does not mean 'n times the element a', for that phrase has of itself no meaning; na is, by definition, the result of the inductive process just described. We have, then,

$$\begin{aligned} a+a &= 2a, \\ 2a+a &= 3a, \\ &\cdots\cdots \\ (n-1)a+a &= na. \end{aligned}$$

Text Examples

1. Show how to extend these definitions to include elements of type νa, where ν is a negative integer.

2. Verify from your definition and from the preceding work that, for all integers m, n, positive, negative or zero,

$$ma+na = (m+n)a.$$

3. Prove that, if $b \in A$, then

$$b(na) = n(ba).$$

Now take a typical member $r \in A$ and consider the element

$$ra+na$$

for given $a \in A$, variable $r \in A$ and variable integer n, positive or negative. It is to be proved that *the set*

$$U \equiv \{ra+na\}$$

so described constitutes an ideal†: this ideal is called *the ideal generated by the element a.*

† Strictly, *a left* ideal.

The definition of an ideal (taking left ideal for this case) is (p. 90)

$$x \in U, \quad y \in U, \quad z \in A$$
$$\Rightarrow \qquad x-y \in U, \quad zx \in U.$$

Suppose, then, that (x,y being members of U)

$$x \equiv ra+ma, \quad y \equiv sa+na,$$

where $r \in A$, $s \in A$; m, n integers. By direct manipulation,

$$\begin{aligned} x-y &= ra+ma-sa-na \\ &= (r-s)a+(m-n)a \\ &= ta+pa \end{aligned}$$

where $t \equiv r-s \in A$, $p \equiv m-n$ is an integer. Hence, by definition of U,

$$x-y \in U.$$

Also, if $z \in A$,

$$\begin{aligned} zx &= z(ra+ma) \\ &= zra+zma \\ &= (zr)a+(mz)a \qquad \text{(Example 3 above)} \\ &= (zr+mz)a \\ &= va, \end{aligned}$$

say, where $v \equiv zr+mz \in A$. Thus

$$zx = va+0.a \in U.$$

The two requirements for an ideal are therefore verified and the result is established.

REMARKS. (i) Informally, the ideal generated by a is the sum of elements of the type λa where, for each element summed, λ is either an element of the ring A or an integer.

(ii) *When the parent ring A has an identity element e*, the elements $ra+na$ of the ideal can be written in the form

$$ra+nea = (r+ne)a = ua,$$

where u is an element of A. In this case the ideal consists of the 'multiples' ua of the given element a by the elements u of the ring.

We conclude this brief summary of the elementary properties of rings and ideals by a definition which emphasizes the connection between this work and the theory of numbers to be developed shortly.

DEFINITION. Let U be an ideal within a ring A. Two elements $a, b \in A$ are said to be *congruent modulo U* if the difference $a-b$ belongs to U.

The relationship is written

$$a \equiv b \pmod U,$$

so that

$$a \in A, \quad b \in A, \quad a-b \in U \Rightarrow a \equiv b \pmod U.$$

Examples

1. Prove that

$$a \equiv b \pmod U, \quad c \in A \Rightarrow a+c \equiv b+c \pmod U, \quad ac \equiv bc \pmod U.$$

2. Prove that

$$a \equiv p \pmod U, \quad b \equiv q \pmod U \Rightarrow a+b \equiv p+q \pmod U, \quad ab \equiv pq \pmod U.$$

3. Given a ring A and two ideals U, V, prove that the sets of elements

$$\{p+q\} \text{ and } \{pq\},$$

$p \in U, q \in V$, are ideals.

4. Given fixed elements a, b, c of a ring A, prove that the set

$$\{xa+yb+zc\},$$

$x \in A, y \in A, z \in A$, is an ideal.

5. Given a field F (p. 80) and an ideal U having a non-zero element, prove that U contains the unity of the field and hence that U consists of the whole field.

6. Show that, when the algebraic operations involved are defined in the usual way, each of the following systems is both a ring and a vector space:

(i) the set of all polynomials in x with real coefficients,

(ii) the set of all continuous real functions of the real variable x,

(iii) the set of all bounded real functions of the real variable x.

Determine whether or not each system is a field when the algebraic operations retain their usual meanings.

Show that the functions of each system which vanish when x takes a certain fixed value a form an ideal in the corresponding ring.

7. Prove that, if U_1, U_2 are ideals in a ring R, then the intersection $U_1 \cap U_2$ is also an ideal.

8

THEORY OF NUMBERS: CONGRUENCES

The theory of numbers is a major branch of mathematics and it is possible to sketch here only a small portion of the theory. For wider reading a textbook devoted to the subject should be consulted.

The present account is directed mainly towards those topics which seem most relevant to the general ideas hitherto studied in this book.

1. The positive integers: primes. The work of this chapter is confined to the set

$$N \equiv \{1, 2, 3, \ldots\}$$

of positive integers, with occasional references to zero. It is understood that, unless otherwise stated, this is what is meant by the symbols $a, b, c, \ldots, x, y, z, \ldots$.

DEFINITIONS. An integer is *composite* when it is the product of two integers other than itself and unity; otherwise it is *prime*.

Thus 4, 12, 35 are composite; 3, 7, 11 are prime.

It is sometimes found convenient *not* to include 1 among the primes.

A table of primes might be constructed by the *Sieve of Aristosthenes*, writing down the integers

$$1, 2, 3, 4, 5, 6, 7, \ldots$$

and then striking out 1, the multiples of 2, the multiples of 3, the multiples of 5,.... The list is endless, however, for, as we now prove, *there is no last prime*:

If, on the contrary, there were a last prime, p, and $2, 3, 5, \ldots, k$ were all the primes less than p, then the integer

$$1+2.3.5....k.p$$

would have unit remainder after division by each of $2, 3, \ldots, p$. It would therefore be *either* prime *or*, if composite, divisible by a prime greater than p. In either case, p would not be the last prime.

2. Three theorems assumed. The reader will doubtless have spent much time on problems dealing with factorization. As a result, he will be prepared to accept as 'obvious' the following three theorems, whose proofs are, in fact, none too easy:

1. QUOTIENT AND REMAINDER. *If a, k are two given positive integers, there are uniquely determined a quotient q and a remainder r, where*

$$0 \leqslant r < k$$

with the property that

$$a = qk + r.$$

2. THE FACTOR THEOREM. *If n is a composite number in the form*

$$n = ab,$$

and if p is a prime number which is a factor of n, then p is a factor of at least one of a, b.

3. PRIME FACTOR DECOMPOSITION. *Any composite number n can be expressed uniquely as a product of prime factors in the form*

$$n = 2^{m_1} 3^{m_2} 5^{m_3} 7^{m_4} 11^{m_5} \ldots,$$

where $m_i = 0$ when the corresponding prime factor is absent. This product of primes certainly terminates, the highest not exceeding n.

3. Congruences modulo p. The work of this paragraph has been anticipated to some extent in one or two Illustrations, but it is now due for study in its own right. We are concerned with the remainders

$$0, 1, 2, \ldots, p-1$$

which arise when any number is divided by a given number p. There are many problems in which the remainder is more significant than the number divided.

DEFINITION. Two numbers a, b are said to be *congruent modulo p*, written

$$a \equiv b \pmod{p},$$

when their remainders are equal after division by p. The number p is called the *modulus* of the congruence.

Thus, assuming $a > b$,

$$a \equiv b \pmod{p}$$

$\Leftrightarrow \exists$ integer m such that

$$a - b = mp.$$

All the numbers congruent (mod p) to a given number a can be obtained from the formula

$$a+kp,$$

where k is an integer (positive or negative) which varies from number to number.

Example

1. Write down all the numbers, if any, which are less than 40 and satisfy the relations:

(i) $x \equiv 2 \pmod 5$, (ii) $x \equiv 1 \pmod 7$,
(iii) $2x \equiv 3 \pmod 4$, (iv) $3x \equiv 1 \pmod{15}$,
(v) $x \equiv 0 \pmod 8$, (vi) $2x \equiv 0 \pmod 8$,
(vii) $x^2 \equiv 1 \pmod 7$, (viii) $x^3 \equiv x \pmod 8$.

4. The algebraic structure of congruences. For modulus p, the remainders belong to the set

$$P \equiv \{0, 1, 2, \ldots, p-1\}.$$

The laws of structure may be considered in turn:

(i) THE LAWS FOR ADDITION:

The inclusive law

$$a \in P, \quad b \in P \Rightarrow \exists \text{ unique } a+b \in P$$

is obeyed, the sum $a+b$ being defined to mean ordinary addition reduced if necessary by a multiple of p;

The associative law

$$(a+b)+c \equiv a+(b+c)$$

is obeyed, each side being the ordinary arithmetical sum $a+b+c$ reduced if necessary by a multiple of p;

The commutative law

$$a+b \equiv b+a$$

is obeyed, by similar argument.

Note that the congruence has a zero (the 'ordinary' zero) and also a negative $-a$ for every element $a \in P$, the latter being the number $p-a$ of elementary arithmetic. Thus the equation

$$a+b \equiv c$$

can be solved within P when any two of a, b, c are given.

(ii) THE LAWS OF MULTIPLICATION:

The inclusive law

$$a \in P, \quad b \in P \Rightarrow \exists \text{ unique } ab \in P$$

is obeyed, the product ab being defined to mean ordinary multiplication reduced if necessary by a multiple of p;

The associative law

$$(ab)c \equiv a(bc)$$

is obeyed, each side being the ordinary arithmetical product abc reduced if necessary by a multiple of p;

The commutative law

$$ab \equiv ba$$

is obeyed, by similar argument.

(iii) THE DISTRIBUTIVE LAWS:

The distributive laws

$$a(b+c) \equiv ab+ac,$$
$$(b+c)a \equiv ba+ca$$

are obeyed, both being the same in virtue of the commutativity of multiplication, since the two sides are equal in their ordinary arithmetical connotation and are therefore equal after reduction by multiples of p.

5. The congruence as a ring. Reference to the definitions given earlier (p. 73) confirms the important result that, with the rules of composition defined in §4, *the elements of the set*

$$P \equiv \{0, 1, 2, \ldots, p-1\}$$

form a ring. They therefore inherit all the properties of a ring, and results true for rings in general hold for P in particular. The set P may sometimes be called a *congruence ring of modulus p*.

This particular ring P is *commutative*.

It is implicit that we are concerned with cases where $p \geqslant 2$, so that *a congruence ring of modulus p always has a unity*, the number 1 of elementary arithmetic. But *the elements of P may, or may not, have inverses* whose products are equal to the unity. For example, if $p = 7$, then the pairs (2,4), (3,5), (6,6) are inverse, since $2\times 4 \equiv 1$, $3\times 5 \equiv 1$, $6\times 6 \equiv 1$. On the other hand, if $p = 6$, the remainder 1 arises only for $5\times 5 \equiv 1$, and no other element has an inverse. Attention has already been drawn (p. 75) to the *divisors of zero*: for example, $3\times 2 \equiv 0 \pmod 6$ although $3 \not\equiv 0$, $2 \not\equiv 0$.

The net result of all this is that congruences may be treated, in so far as addition and multiplication are concerned, like the numbers of ordinary arithmetic. Subtraction, too, presents no problem, since P is an Abelian group for addition. On the other hand, *division is an operation which must not be undertaken within a congruence*, except in special cases after careful examination; for, as has been seen, an element $a \in P$ need not have an inverse a^{-1}, though it can on occasion do so.

Thus, within P,

$$\begin{aligned} ab &\equiv ac \\ \Rightarrow ab-ac &\equiv 0 \\ \Rightarrow a(b-c) &\equiv 0. \end{aligned}$$

But it does *not* follow that either $a \equiv 0$ or $b-c \equiv 0$.

Again,

$$\begin{aligned} a \equiv u, \quad b &\equiv v \\ \Rightarrow a+b \equiv u+v, \qquad ab &\equiv uv, \\ a^2 &\equiv u^2, \\ a^3 &\equiv u^3, \\ &\cdots\cdots \\ a^n &\equiv u^n \end{aligned}$$

for positive integral n.

Examples

1. Find solutions of the equations

$$2x \equiv 1, \quad 3x \equiv 1, \ldots, \quad (p-1)x \equiv 1 \qquad (\operatorname{mod} p)$$

for $p = 3, 4, 5, 6$.

2. Find solutions, if any, of the equations

$$x^2 \equiv 1, \quad x^3 \equiv 1, \quad x^4 \equiv 1, \quad x^5 \equiv 1$$

for modulus $3, 4, 5, 6, 7$.

6. The cancellation rule for prime modulus. To prove that, *if p is a prime number*, then

$$ab \equiv 0 \qquad (\operatorname{mod} p)$$

$\Rightarrow$ *either* $a \equiv 0$ *or* $b \equiv 0$ *or both.*

For $$ab \equiv 0 \qquad (\operatorname{mod} p)$$

$\Rightarrow \exists$ integer n such that, in terms of ordinary multiplication,

$$ab = np.$$

But, by the factor theorem (p. 95), the property that p is a factor of the product ab ensures that it is a factor of at least one of a, b. Hence either

$$a \equiv 0 \quad \text{or} \quad b \equiv 0 \text{ or both.}$$

7. Incongruent residues. The theory of congruences modulo p is essentially that of the set of residues

$$P \equiv \{0, 1, 2, \ldots, p-1\}$$

which must now be studied in greater detail.

DEFINITION. The p residues are called *incongruent*, since, if r, s are any two distinct members of P, the difference $r-s$, being less than p but yet not zero, cannot be a multiple of p. Hence

$$r-s \not\equiv 0.$$

The p numbers

$$0, 1, 2, \ldots, p-1$$

are called a *complete set of incongruent residues.*

THEOREM. *If to each element of a complete set of incongruent residues is added a given number k, then the resulting sums are themselves congruent to a complete set of incongruent residues.*

Text Example

1. Prove the theorem (i) for $p = 6$, (ii) for $p = 7$, (iii) generally.

THEOREM. *For prime modulus p, if each element of a complete set of incongruent residues is multiplied by an integer k ($k \not\equiv 0$), then the resulting products are themselves congruent to a complete set.*

To prove this, consider the p elements

$$0, k, 2k, \ldots, (p-1)k.$$

It is to be shown first that no two of these are congruent: for

$$\begin{aligned} & rk \equiv sk \\ \Rightarrow\ & (r-s)k \equiv 0 \\ \Rightarrow\ & \text{either } r-s \equiv 0 \quad \text{or} \quad k \equiv 0 \qquad (\S 6) \\ \Rightarrow\ & r-s \equiv 0 \qquad (k \not\equiv 0, \text{ given}) \\ \Rightarrow\ & r = s \qquad (\text{since } r < p, s < p). \end{aligned}$$

Hence the p elements

$$0, k, 2k, \ldots, (p-1)k$$

are mutually incongruent, so that they are congruent to a complete set of incongruent residues.

Text Example

1. By taking $p = 6, p = 8$, show that the result just proved need not be true when p is not prime.

8. The inverse of an integer for prime modulus. To prove that, *if p is a prime integer, then each element a of the set*

$$P \equiv \{0, 1, 2, \ldots, p-1\}$$

other than zero has an inverse a^{-1}.

By the preceding paragraph, the numbers

$$0, a, 2a, \ldots, (p-1)a$$

form the set

$$0, 1, 2, \ldots, p-1$$

in some order. In particular, there is one of the numbers

$$a, 2a, \ldots, (p-1)a$$

whose value is 1. Call that number xa. Then

$$xa \equiv 1 \quad (\text{mod}\, p),$$

so that x is the inverse (mod p) of a.

The set P thus has the property that, when p is prime, every non-zero element, has an inverse which is an *integer*.

Text Example

1. Group in inverse pairs the non-zero elements for the cases $p = 3, 5, 7, 11, 13$.

***Illustration* 1.** *Fermat's theorem.* To prove that, *if p is a prime number, and if p is not a factor of a, then*

$$a^{p-1} \equiv 1 \quad (\text{mod}\, p).$$

The numbers

$$a, 2a, \ldots, (p-1)a \quad (a \not\equiv 0)$$

are congruent, in some order, to the numbers

$$1, 2, \ldots, p-1.$$

Hence the product of the first set is congruent to the product of the second set, so that, multiplication being commutative,

$$a.2a.3a.....(p-1)a \equiv 1.2.3.....(p-1)$$
$$\Rightarrow a^{p-1}(p-1)! \equiv (p-1)!.$$

But each factor of $(p-1)!$ is less than p, so that

$$(p-1)! \not\equiv 0.$$

Hence, *since p is prime*,

$$a^{p-1} \equiv 1.$$

COROLLARY. It follows that

$$a^p \equiv a,$$

and this result is true, when p is prime, even when p *is* a factor of a.

Examples

1. Prove that

$$(p-k)^2 \equiv k^2 \quad (\text{mod}\, p)$$

and deduce that the squares of a complete set of incongruent residues cannot form a complete set of incongruent residues if $p > 2$.

2. Prove that, if $k \not\equiv 0$ (mod 5), then

$$k^4 \equiv 1 \quad (\text{mod}\, 5).$$

3. Prove that, for all positive integral numbers n,

$$(p+k)^n \equiv k^n \quad (\text{mod}\, p).$$

4. Prove that, in arithmetic modulo 3, there are 18 quadratic equations

$$ax^2+bx+c \equiv 0 \quad (a \not\equiv 0)$$

and that six of them have no solutions.

5. Prove that, in arithmetic modulo 4, there are 12 linear equations

$$ax+b \equiv 0 \quad (a \not\equiv 0)$$

and that two of them have no solutions.

6. Prove that, if

$$a \equiv b \quad (\text{mod}\, 6),$$

then

$$a \equiv b \quad (\text{mod}\, 3).$$

State and disprove the converse result. Is it true that, if

$$a \equiv b \quad (\text{mod}\, 2)$$

and

$$a \equiv b \quad (\text{mod}\, 3),$$

then

$$a \equiv b \quad (\text{mod}\, 6)?$$

7. Prove that the product of r consecutive positive integers is divisible by $r!$

8. Prove that, if p is prime, then all the binomial coefficients

$$p, \quad \frac{p(p-1)}{1.2}, \quad \frac{p(p-1)(p-2)}{1.2.3}, \ldots$$

are integers divisible by p.

To what extent is this result true if p is not prime?

9. Prove that, if m, n are integers and p is prime, then

$$m^p + n^p \equiv (m+n)^p \pmod{p}.$$

10. Prove that $(pq)!$ is divisible by $(p!)^q q!$.

11. Prove *Wilson's theorem* that, if p is prime, then

$$1+(p-1)!$$

is a multiple of p.

12. Prove that, if $2p+1$ is prime, then $(p!)^2+(-1)^p$ is divisible by it.

13. If p is a prime number, and a a number prime to p, prove that $a^{p-1}-1$ is a multiple of p.

If a^x-1 is a multiple of p, where x is an integer less than $p-1$, prove that x is a factor of $p-1$.

14. Prove that, if d is a factor of 12, the relation

$$x^d \equiv 1 \pmod{13}$$

has exactly d solutions.

15. Prove that, if the polynomial $f(x)$ of degree n with integral coefficients has more than n roots which are incongruent integers modulo p (p prime), then it is identically congruent to zero.

9. Scales. It is familiar that the numbers of arithmetic are, in normal usage, expressed 'in the scale of 10'. That is, the number written

$$2{,}345$$

signifies the expression

$$2 \times 10^3 + 3 \times 10^2 + 4 \times 10 + 5.$$

It would, however, be perfectly possible to use some other number as base, so that for base, say, 7 the same symbol would mean

$$2 \times 7^3 + 3 \times 7^2 + 4 \times 7 + 5.$$

For bases greater than 10 it would be necessary to use new symbols for digits higher than 9; for example, in the scale of 12 a symbol would be needed for the successor of 9 and yet another for the successor of that one.

It may be of interest to spend a few moments in the scale of 3 where there are three symbols: 0, 1, 2. The first few integers, expressed in that scale, are

$$0, 1, 2$$

$$10, 11, 12, 20, 21, 22,$$

$$100, 101, 102, 110, 111, 112, 120, 121, 122, 200, \ldots.$$

The number written

$$abcdef$$

(where a, b, c, d, e, f are the numbers 0, 1, 2 with $a \neq 0$) is

$$a \times 3^5 + b \times 3^4 + c \times 3^3 + d \times 3^2 + e \times 3 + f,$$

or even, using the scale of three throughout,

$$a \times 3^{12} + b \times 3^{11} + c \times 3^{10} + d \times 3^2 + e \times 3 + f.$$

Thus f is the remainder on dividing the given number by 3; e is the remainder on dividing that quotient by 3; d is the remainder on dividing the new quotient by 3; and so on.

For example, to express the number normally written 87,654 in terms of 3, the successive divisions give:

$$\begin{array}{r|r}
3 & 87654 \\ \hline
3 & 29218+0 \\ \hline
3 & 9739+1 \\ \hline
3 & 3246+1 \\ \hline
3 & 1082+0 \\ \hline
3 & 360+2 \\ \hline
3 & 120+0 \\ \hline
3 & 40+0 \\ \hline
3 & 13+1 \\ \hline
3 & 4+1 \\ \hline
 & 1+1
\end{array}$$

and so, in the scale of three, the number is

$$11{,}110{,}020{,}110.$$

Examples

All the following numbers are written in the scale of 3. Perform the stated operations:

1. $120+2102$; $21{,}101-10{,}222$.
2. 201×121; 1111×1111.
10. $\sqrt{(221)}$; $\sqrt[10]{(10{,}000{,}101)}$.
11. 11^{11}; 12^{21}; 10^{10}.
12. $12{,}012\div 2$; $200{,}100\div 212$.
20. $(10+11+12)^{10}$.

The scale of two has acquired great importance in recent years through its use in computing. For example, the instructions, 'Do this', 'Do not do this', can be coded by means of the two numbers 0, 1 respectively. All numbers in the scale of two have digits either 0 or 1, and so the numbers, though lengthy to write, are simple in form. Thus,

1100 pence make one shilling,

10,100 shillings make one pound.

Numbers expressed in the scale of two are also said to be in the *binary system.*

The arithmetic has unexpected features: for example,

$$111{,}111+1 = 1{,}000{,}000.$$

Examples

1. Evaluate $101+1110+10{,}101$.
10. Evaluate $100{,}001-1010$.
11. Evaluate 111×101 and 111×111.
100. Evaluate $101{,}001\div 1001$.
101. Evaluate 11^{11} and 10^{10}.
110. Evaluate the determinant

$$\begin{vmatrix} 10 & 11 & 101 \\ 11 & 110 & 111 \\ 101 & 111 & 10 \end{vmatrix}.$$

A very elementary example of the way in which a binary arithmetic may be introduced is afforded by the symbolism

$p = 0$ means, 'the statement called P is true',

$p = 1$ means, 'the statement called P is false'.

Let P, Q be two given statements, and let R denote the statement, '*one or other of P, Q is true*'. Then

$$\begin{aligned} & p = 0, \quad q = 0 \\ \Rightarrow\ & P \text{ and } Q \text{ are true} \\ \Rightarrow\ & r = 0; \\ & p = 1, \quad q = 0 \\ \Rightarrow\ & Q \text{ is true (though } P \text{ is not)} \\ \Rightarrow\ & r = 0; \\ & p = 0, \quad q = 1 \\ \Rightarrow\ & P \text{ is true (though } Q \text{ is not)} \\ \Rightarrow\ & r = 0; \\ & p = 1, \quad q = 1 \\ \Rightarrow\ & \text{neither } P \text{ nor } Q \text{ is true} \\ \Rightarrow\ & r = 1. \end{aligned}$$

The values of r may be exhibited by the table

$p \backslash q$	0	1
0	0	0
1	0	1

Note that r obeys the formula

$$r = pq.$$

Example

1. Construct the corresponding table for the statement S, that P and Q are true simultaneously, and show that

$$s = p+q-pq.$$

Finally, consider the converse problem of establishing *an interpretation for a statement T for which*

$$t = p+q:$$

Under binary arithmetic,

$$\begin{aligned} & t = 0 \\ \Rightarrow\ & p = 0 \quad \text{and} \quad q = 0 \end{aligned}$$

or

$$p = 1 \quad \text{and} \quad q = 1,$$

whereas

$$t = 1$$

$$\Rightarrow p = 0 \quad \text{and} \quad q = 1$$

or

$$p = 1 \quad \text{and} \quad q = 0.$$

Thus T is true when P, Q are both true or both false; T is false when one of P, Q is true and the other is false.

A typical example of this situation is afforded by the geometrical properties of a triangle ABC in which

P is the statement, '$AB = AC$',
Q is the statement, '$\angle ACB = \angle ABC$'.

Examples

1. Prove that the numbers

$$11, \quad 111, \quad 11111$$

expressed in the binary system are all prime, but that

$$111111$$

is composite.

2. Prove that, in the binary system, a number all of whose k digits are units is necessarily composite, where k is a composite number.

Is this result true for numbers expressed in any scale?

3. Solve for x the eqations (in arithmetic modulo 2)

$$x^2+1 = 0,$$
$$x^3+1 = 0.$$

4. Prove that in arithmetic modulo 2 every polynomial

$$a_0 x^n + a_1 x^{n-1} + \ldots + a_n \qquad (a_r = 0 \text{ or } 1)$$

with an even number of non-zero coefficients can take the value zero, but that a polynomial with an odd number of non-zero coefficients cannot take the value zero unless $a_n = 0$.

9

MODULES OF INTEGERS

1. First properties. Let S be a given set whose elements $a, b, c, \ldots$ are all integers, positive, negative or zero, subject to the rules of addition and multiplication used in elementary arithmetic. The set S is to be chosen so as to satisfy the condition

$$\begin{aligned} m \in S, \quad & n \in S \\ \Rightarrow m-n & \in S. \end{aligned}$$

It follows that *the elements of S form a group (necessarily commutative since elementary addition is commutative) with respect to the operation of addition:*

For the case $m = n$ gives

$$\begin{aligned} n-n &\in S \\ \Rightarrow 0 &\in S \end{aligned}$$

and then

$$\begin{aligned} 0 \in S, \quad & m \in S \\ \Rightarrow 0-m & \in S \\ \Rightarrow -m & \in S. \end{aligned}$$

Finally,

$$\begin{aligned} m \in S, \quad n \in S \Rightarrow m \in S, \quad & (-n) \in S \\ \Rightarrow m-(-n) & \in S \\ \Rightarrow m+n & \in S. \end{aligned}$$

The conditions for a group are therefore all satisfied.

Definition. A set S of integers subject to the condition

$$m \in S, \quad n \in S \Rightarrow m-n \in S$$

is called a *module*.

Examples

1. The multiples of 7 form a module.
2. The powers of 7 do not form a module.
3. The set $\{0\}$ consisting of zero alone forms a module.

NOTE. Compare the definition of an *ideal* given on p. 90. Much of the work that follows is clearly analogous to that given for ideals.

2. The fundamental theorem. Let x, y be any two integers and a, b any two elements of a module S. To prove that *the element*

$$xa+yb$$

is in the module S:

For

$$\begin{aligned} a \in S &\Rightarrow a+a \in S \Rightarrow 2a \in S; \\ a \in S,\ 2a \in S &\Rightarrow a+2a \in S \Rightarrow 3a \in S; \\ a \in S,\ 3a \in S &\Rightarrow a+3a \in S \Rightarrow 4a \in S; \end{aligned}$$

and so, by direct induction, if k is any positive integer, then

$$ka \in S.$$

Hence

$$\begin{aligned} & 0 \in S,\quad ka \in S \\ \Rightarrow\ & 0-ka \in S \\ \Rightarrow\ & -ka \in S. \end{aligned}$$

It follows that, for any integer x, positive, negative or zero,

$$xa \in S.$$

Similarly

$$yb \in S.$$

Finally,

$$\begin{aligned} & xa \in S,\quad yb \in S \\ \Rightarrow\ & xa+yb \in S. \end{aligned}$$

COROLLARY. *Any module* (other than the zero module) *has positive elements.*

THEOREM. *The elements of any module* (other than the zero module) *consist of the multiples of the least of its positive elements:*

Since the module *has* positive elements, it has a least one, say d. Let n be any positive member of S, and suppose that q, r are the quotient and remainder on dividing n by d; thus (p. 95)

$$n = qd+r.$$

Now

$$\begin{aligned} & n \in S,\quad d \in S \\ \Rightarrow\ & n+(-q)d \in S \\ \Rightarrow\ & r \in S. \end{aligned}$$

But

$$0 \leqslant r < d,$$

where d is the *smallest positive element* in S. Hence

$$r = 0,$$

so that
$$n = qd.$$

If n is a negative integer, then

$$-n \in S,$$

and the argument follows as before.

3. Highest common factor. Let a, b be two given positive integers.

DEFINITION. The *highest common factor* d of a and b is the largest integer which divides both a and b.

To prove that *the module*

$$\{xa+yb\}$$

consists of the multiples of the highest common factor d of a and b:

As above (§2), the module consists of the multiples of some number d'. In particular, a and b are both multiples of d', so that d' is a common factor of a and b. Hence

$$d' \leqslant d.$$

Also d, being a factor of a and b, is a factor of every element $xa+yb$ and, in particular, of d'. Hence

$$d \leqslant d'.$$

Thus $d' = d$, so that the module

$$\{xa+yb\}$$

consists of the multiples of the highest common factor d of a and b.

4. Euclid's algorithm: particular example. It is one thing to imply that, given two numbers a, b with highest common factor d, two numbers x, y can be found so that $xa+yb = d$. It is quite a different matter to find them. This is the problem that must now be tackled. The rule known as *Euclid's algorithm* leads infallibly [subject only to numerical accuracy] to the answer.

Consider, in illustration, the two given numbers 7293 and 798.

The normal process of division of the larger by the smaller is equivalent to the relation

$$7293 = 9 \times 798 + 111.$$

Any factor common to 7293 and 798 (in particular, the highest common factor) must 'run across' this relation and be a factor of 111. The process may therefore be repeated for 798 and 111, giving

$$798 = 7\times 111+21.$$

The common factor, by similar argument, divides 21 also, so the process is repeated for 111 and 21:

$$111 = 5\times 21+6.$$

Similarly

$$21 = 3\times 6 \ +3,$$

$$6 = 2\times 3 \ +0.$$

The number 3 is thus a factor of 6 and hence, 'rising' successively through the equations, of 21, 111, 798 and 7293; further, every factor common to the two given numbers penetrates 'down' the equations. Hence *the highest common factor is* 3.

Before generalizing the procedure, we show how to express the highest common factor 3, in terms of the two given numbers 7293 and 798, so as to exhibit it as a member of the module

$$\{7293x+798y\}.$$

Starting from the second last equation and 'rising' successively through them,

$$\begin{aligned} 3 &= 21-3\times 6 \\ &= 21-3\{111-5\times 21\} \\ &= 16\times 21-3\times 111 \\ &= 16\{798-7\times 111\}-3\times 111 \\ &= 16\times 798-115\times 111 \\ &= 16\times 798-115\{7293-9\times 798\} \\ &= 1051\times 798-115\times 7293. \end{aligned}$$

Thus 3, expressed as a member of the module, is

$$(-115)\times 7293+1051\times 798.$$

5. Euclid's algorithm: general case. Let m,n be two given integers ($m>n$). By successive divisions, in each case until the remainder is less than the divisor, a sequence of relations is obtained in the form:

$$\begin{aligned} m &= a_1 n+r_1, \\ n &= a_2 r_1+r_2, \\ r_1 &= a_3 r_2+r_3, \end{aligned}$$

$$r_2 = a_4 r_3 + r_4,$$

$$\dots\dots\dots\dots\dots\dots$$

$$r_{k-2} = a_k r_{k-1} + r_k,$$

$$r_{k-1} = a_{k+1} r_k + 0.$$

The number r_k, by the last equation, is a factor of r_{k-1}. Hence, 'rising up' the equations, it is a factor of r_{k-2}, of r_{k-3}, ..., of r_1, of n, of m. That is, r_k is a factor common to m and n.

Moreover, *every* factor common to m and n is, by the first equation, a factor of r_1. Hence, 'dropping down' the equations, it is a factor of r_2, of r_3, ..., of r_{k-1}, of r_k. Hence *r_k is the highest common factor of m and n.*

COROLLARY. *If $r_k = 1$, then m and n are mutually prime*, having no factor other than unity in common.

6. The $xm + yn = d$ theorem. To prove that, *if m, n are two given integers whose highest common factor is d, then integers x, y can always be found (not both positive) so that*

$$xm + yn = d.$$

The equations of § 5, with r_k replaced by d, may be written in ascending order in the form

$$d = r_{k-2} - a_k r_{k-1},$$

$$r_{k-1} = r_{k-3} - a_{k-1} r_{k-2},$$

$$r_{k-2} = r_{k-4} - a_{k-2} r_{k-3},$$

$$\dots\dots\dots\dots\dots\dots$$

$$r_2 = n - a_2 r_1,$$

$$r_1 = m - a_1 n.$$

Replace r_{k-1} in the first equation by its expression in terms of r_{k-2}, r_{k-3} from the second; this gives d as a *linear* expression in r_{k-2}, r_{k-3}. Replace r_{k-2} in this new equation by its expression in terms of r_{k-3}, r_{k-4} from the third; this gives d as a *linear* expression in r_{k-3}, r_{k-4}. Continue in this way down the equations. Eventually, replace r_2 by $n - a_2 r_1$ and then r_1 by $m - a_1 n$. In this way d is expressed *linearly* in terms of m, n, the coefficients being somewhat complicated polynomials in the integers a_1, a_2, ..., a_k. Calling these coefficients, themselves integers, x and y, we have the relation

$$d = xm + yn.$$

***Illustration* 1.** *Euclid's factor theorem* (compare p. 95). To prove that, *if a,b have no common factor other than unity, and if a is a factor of the product bc, then a is a factor of c.*

The highest common factor of a,b belongs to the module $\{xa+yb\}$, and so integers x,y can be found so that

$$xa+yb = 1.$$

Hence

$$xac+ybc = c.$$

Now a is a factor of the term xac and also, by the hypothesis, of the term ybc. It is therefore a factor of their sum, which is c.

***Illustration* 2.** *To find integers x,y so that*

$$77x+43y = 1.$$

Euclid's algorithm gives

$$\begin{aligned} 77 &= 1\times 43+34,\\ 43 &= 1\times 34+9,\\ 34 &= 3\times 9+7,\\ 9 &= 1\times 7+2,\\ 7 &= 3\times 2+1. \end{aligned}$$

Using these equations in reverse order,

$$\begin{aligned} 1 &= 7-3\times 2\\ &= 7-3(9-1\times 7) = 4\times 7-3\times 9\\ &= 4(34-3\times 9)-3\times 9 = 4\times 34-15\times 9\\ &= 4\times 34-15(43-1\times 34) = 19\times 34-15\times 43\\ &= 19(77-1\times 43)-15\times 43\\ &= 19\times 77-34\times 43. \end{aligned}$$

Thus

$$x = 19, \quad y = -34.$$

COROLLARY. A *solution in integers* of the equation

$$77x+43y = 1$$

is

$$x = 19, \quad y = -34.$$

This solution is not unique, since

$$\begin{aligned} &19\times 77-34\times 43\\ &\quad = (19+43\lambda)\times 77-(34+77\lambda)\times 43 \end{aligned}$$

for all values of λ. Hence a more general solution is

$$x = 19+43\lambda, \quad y = -34-77\lambda.$$

Examples

Find the highest common factors d of the following pairs of integers m,n (using Euclid's algorithm) and then express d in the form $xm+yn$:

1. $m = 1092, n = 1386.$ **2.** $m = 1001, n = 784.$

3. $m = 1230$, $n = 234$.

Find solutions in integers of the following equations:

1. $17x+13y = 1$. 2. $53x+25y = 1$.

3. $47x+30y = 1$.

7. Solution in integers: Diophantine equations. It is of interest *to solve in integers the equation*

$$ax+by = c,$$

where a, b, c are given integers with no factor in common.

If a, b have a common factor, then that factor is also a factor of $ax+by$ and therefore of c (provided that the equation *is* soluble in integers). But this contradicts the hypothesis that a, b, c have no common factor, and so we assume that the coefficients a, b have no factor in common.

Since a, b have no common factor,

$$\exists\, p,q \quad \text{so that} \quad ap+bq = 1.$$

Hence

$$a(pc)+b(qc) = c.$$

There is therefore a solution

$$x = pc, \quad y = qc.$$

This solution is *not unique*. A more general solution is

$$x = pc+\lambda b, \quad y = qc-\lambda a \qquad (\lambda \text{ arbitrary}),$$

since

$$a(pc+\lambda b)+b(qc-\lambda a) = c.$$

8. Some typical problems in number theory. There are many fascinating properties of integers which lie outside our present purpose, but a few typical examples are given, with fairly detailed explanations, to indicate what can be done.

1. *To prove that* 42 *is a factor of* n^7-n.

By straightforward algebra,

$$\begin{aligned} n^7-n &= n(n^6-1) \\ &= n(n^3-1)(n^3+1) \\ &= n(n-1)(n^2+n+1)(n+1)(n^2-n+1) \\ &= n(n-1)(n+1)(n^2-n+1)(n^2+n+1). \end{aligned}$$

The integers $n-1, n, n+1$ are consecutive, so at least one is even and one is a multiple of 3. Hence the given number is divisible by 6.

Also, by Fermat's theorem (p. 100),

$$n^7 \equiv n \pmod{7},$$

so that 7 is a factor of $n^7 - n$.

Thus 6 and 7 are both factors, so that the number is divisible by 42.

2. *To prove that, if p, q are positive integers such that $4q - p^2$ is a perfect square, then p is even, and, for any other positive integer n,*

$$n^2 + pn + q$$

can be expressed as the sum of two squares.

Let

$$4q - p^2 = k^2,$$

where k is an integer. Then

$$p^2 + k^2 = 4q.$$

Since $4q$ is even, p and k are both *odd* or both *even*. If they are odd, write $p = 2\alpha + 1$, $k = 2\beta + 1$. Then

$$p^2 + k^2 = 4(\alpha^2 + \alpha + \beta^2 + \beta) + 2,$$

which cannot be a multiple of 4. Hence p is even.

Since p, k are both *even*, write

$$p = 2u, \quad k = 2v.$$

Then

$$q = \tfrac{1}{4}(p^2 + q^2) = u^2 + v^2,$$

so that

$$\begin{aligned} n^2 + pn + q &= n^2 + 2un + u^2 + v^2 \\ &= (n+u)^2 + v^2 \\ &= \text{sum of two squares.} \end{aligned}$$

3. *To prove that, if n is any positive integer, then n consecutive odd integers can be found not one of which is a prime.*

The solution is by construction. Let

$$3, 5, 7, 11, 13, \ldots$$

be the first n odd primes and let their product be p. Then the n numbers

$$2p + 3, 2p + 5, 2p + 7, 2p + 9, \ldots, 2p + (2n + 1)$$

are all odd and all composite.

[For example, if $n = 5$,

$$\begin{aligned} p &= 3 \times 5 \times 7 \times 11 \times 13 = 15{,}015, \\ 2p &= 30{,}030 \end{aligned}$$

and the numbers are

$$30{,}033, 30{,}035, 30{,}037, 30{,}039, 30{,}041.]$$

Revision Examples

1. Prove, using induction if you wish, that

(i) $n^4-4n^3+5n^2-2n = m(12)$,
(ii) $(2n+1)^5-2n-1 = m(120)$,
(iii) $2^{2n}+15n-1 = m(9)$,
(iv) $3^{2n+2}-8n-9 = m(64)$,
(v) $3^{2n+3}+40n-27 = m(64)$,
(vi) $3^{2n+5}+160n^2-56n-243 = m(512)$,
(vii) $3.4^{n+1}+10^{n-1}-4 = m(9)$,
(viii) $3^{2n+1}+2^{n+2} = m(7)$,
(ix) $2^{2n}-3n-1 = m(9)$.

where $m(k)$ means *a multiple of k*.

2. Prove that n^2-n+1 $(n > 1)$ is not a perfect square.

3. Prove that $n(n+1)(n+2)(n+3)$ is not a perfect square.

4. Find n consecutive integers none of which are primes.

5. The integers a, b have no factor in common and the product ab is a perfect square. Prove that a and b are both perfect squares.

6. Prove that, if $n-1$, $n+1$ are both prime (> 5), then n must be of one of the forms $30t$, $30t \pm 12$; and that, if $n-2$, $n+2$ are both prime (> 5), then n must be of one of the forms $30t+15$, $30t \pm 9$.

7. Prove that, if n is a positive integer, the number of solutions of the equation $x+2y+3z = 6n$, for which x, y, z are positive integers or zero, is $3n^2+3n+1$.

8. Find a general formula for all the positive integers which, when divided by 5, 6, 7, leave remainders 1, 2, 3 respectively; and show that the least of them is 206.

9. Prove that, if $f(x)$ is a polynomial such that ab is a factor both of $f(a)$ and of $f(b)$, then ab is a factor of $f(a+b)$, where a, b are relatively prime.

[Consider $f(x)-f(a) = (x-a)g(x)$, $f(x)-f(b) = (x-b)h(x)$.]

10. Prove that, if $m-p$ is a factor of $mn+pq$, where $m > p$, then it is also a factor of $mq+np$.

11. Find solutions in integers of the following equations:

(i) $2x+3y = 5$, (ii) $5x+3y = 7$,
(iii) $6x+5y = 13$, (iv) $8x+7y = 5$.

12. By considering the units digits, determine all the members of the sequence

$$1,\ 1+2!,\ 1+2!+3!,\ 1+2!+3!+4!, \ldots,\ 1+2!+\ldots+n!, \ldots$$

which are squares of integers.

Prove also that no member of the sequence other than the first is the cube of an integer.

13. Show that if an integer of the form $4n+3$ is expressed as a product of integers, then one at least of these is of the form $4m+3$.

A sequence of integers is defined by the relations $x_1 = 1$,

$$x_{n+1} = 4x_1 x_2 \ldots x_n + 3 \qquad (n \geqslant 1).$$

Prove that, if any pair of integers x_p, x_q is taken, there is no factor in common.

Deduce that there is an infinity of prime numbers of the form $4n+3$.

14. Prove that, if m, n, p are positive integers, all greater than 1, such that

$$(m^n)^p = m^{(n^p)},$$

then, necessarily, $n = p = 2$.

15. The integers $u_0, u_1, \ldots, u_n, \ldots$ are defined by the relations

$$u_0 = 0, \quad u_1 = 1, \quad u_{r+1} = u_r + u_{r-1} \qquad (r \geqslant 1).$$

Prove that, if r, s are positive integers,

$$u_{r+s} = u_r u_{s+1} + u_s u_{r+1}$$

and that, if $r > s$,

$$u_{r-s} = (-1)^s (u_r u_{s+1} - u_s u_{r+1}).$$

Deduce that the greatest common factor of u_{207} and u_{345} is u_{69}.

16. Find the smallest integer which, when divided by 28, leaves a remainder 21, and, when divided by 19, leaves a remainder 17.

10

POLYNOMIAL FORMS

In Part 1, polynomials in a variable x were considered as *functions* of that variable, so that, for example,

$$f(x) \equiv x^2+x+1$$
$$\Rightarrow f(0) = 1, \quad f(2) = 7, \quad f(-5) = 21.$$

The emphasis was on the *values* $f(0), f(2), f(-5), \ldots$.

There are, however, many aspects of polynomial theory in which the important thing is *form* rather than value, and this is what is now to be studied.

1. The polynomial form. An expression such as

$$a_0+a_1x+a_2x^2+\ldots+a_nx^n$$

is called a *polynomial form* with *coefficients* a_0, a_1, a_2, ..., a_n in the *indeterminate*, or *variable*, x.

It is essential that the set from which the coefficients come, and the laws of structure for operations within that set, should be clearly understood. The usual examples are the integers, the rational numbers and the complex numbers. In actual practice, at the present level of work, the implication is usually that the coefficients are drawn from the complex numbers (often, in fact, very much restricted to integers), but there are many advantages now in explicit definition. For example, if all coefficients are to be real numbers, the polynomial x^2+4 cannot be factorized; if they are complex numbers, then $x+2i$ and $x-2i$ are factors.

The important things in the theory of polynomial forms are the coefficients $a_0, a_1, \ldots, a_n$ drawn, in an agreed order, from the elements of a basic set S whose structure for addition and multiplication is known. Adjoined to these coefficients is an indeterminate x whose nature need not be further specified; there is, indeed, a sense in which its primary duty is simply to attach itself to the coefficients so that they are kept separate and identifiable by means of its powers. It is an 'indicator' specifying the coefficient to which it is attached.

Some of the coefficients may be zero. The *degree* of a given polynomial form is the highest value of r, for which a_r is not zero while all coefficients are zero for greater values of r.

DEFINITION. Two polynomial forms

$$a_0+a_1x+a_2x^2+\ldots+a_nx^n,$$
$$b_0+b_1x+b_2x^2+\ldots+b_mx^m,$$

each with coefficients in S, are said to be *equal* when corresponding coefficients a_r, b_r are equal for all values of r. This implies, in particular, that

$$m = n.$$

In this context the symbol = for equality of the polynomials is used rather than the symbol ≡ for identity as in Part 1. Thus the relation

$$a_0+a_1x+a_2x^2+a_3x^3 = 3-2x^2$$

implies

$$a_0 = 3, \quad a_1 = 0, \quad a_2 = -2, \quad a_3 = 0.$$

2. The sum of two polynomial forms. Given two polynomial forms

$$a_0+a_1x+a_2x^2+\ldots+a_nx^n,$$
$$b_0+b_1x+b_2x^2+\ldots+b_mx^m,$$

their *sum* is defined to be the polynomial form

$$c_0+c_1x+c_2x^2+\ldots+c_kx^k,$$

where

$$c_0 = a_0+b_0,\ c_1 = a_1+b_1,\ c_2 = a_2+b_2, \ldots,\ c_k = a_k+b_k,$$

and where it is assumed, by convention, that, if (say) $m > n$, then terms

$$a_{n+1}x^{n+1}+\ldots+a_mx^m$$

are added to the first polynomial form, the new coefficients $a_{n+1}, \ldots, a_m$ all being zero. Then, automatically, $k = m$.

The *difference* of two polynomial forms is defined similarly.

3. The product of two polynomial forms. Take, as before, two polynomial forms

$$a_0+a_1x+\ldots+a_nx^n,$$
$$b_0+b_1x+\ldots+b_mx^m.$$

Their *product* is defined (see also p. 76) to be the polynomial form

$$c_0+c_1x+c_2x^2+\ldots+c_kx^k,$$

where

$$c_0 = a_0 b_0,$$
$$c_1 = a_0 b_1 + a_1 b_0,$$
$$c_2 = a_0 b_2 + a_1 b_1 + a_2 b_0,$$
$$\dots\dots\dots\dots\dots$$
$$c_r = a_0 b_r + a_1 b_{r-1} + a_2 b_{r-2} + \dots + a_r b_0,$$
$$\dots\dots\dots\dots\dots\dots\dots\dots$$

If necessary, any of the polynomial forms may be 'extended' by introducing terms with zero coefficients to make the formulae for $c_0, c_1, c_2, \dots$ perfectly general.

The degree of the product is $m+n$, the last coefficient being $a_n b_m$.

Illustration 1. The product of

$$1+2x+3x^2, \quad 5+6x+7x^2+8x^5$$

is

$$c_0+c_1x+\dots+c_7x^7,$$

where

$$c_0 = 1.5 = 5,$$
$$c_1 = 1.6+2.5 = 16,$$
$$c_2 = 1.7+2.6+3.5 = 34,$$
$$c_3 = 1.0+2.7+3.6 = 32,$$
$$c_4 = 1.0+2.0+3.7 = 21,$$
$$c_5 = 1.8+2.0+3.0 = 8,$$
$$c_6 = 1.0+2.8+3.0 = 16,$$
$$c_7 = 1.0+2.0+3.8 = 24.$$

The product is thus

$$5+16x+34x^2+32x^3+21x^4+8x^5+16x^6+24x^7.$$

Examples

1. Write down the sum, difference and product of the polynomial forms

$$1-2x+3x^2-4x^3, \quad x+x^2.$$

2. Write down the products

$$(1-2x)(1+2x+4x^2+8x^3),$$
$$(1-2x+x^2)(1+2x+x^2),$$
$$(1+2x+3x^2+4x^3)(1+2x+3x^2+4x^3).$$

4. Factorization. There is a close analogy between the work which follows and that previously given for integers. The treatments will be kept reasonably parallel.

DEFINITION. A polynomial form, with coefficients in a set S, is said to be *composite in* S when it can be expressed as the product of two polynomial forms (each, by implication, of degree greater than zero) with coefficients in S. Otherwise the form is *prime in* S.

As we have seen, a form like

$$x^2+4$$

is prime when the coefficients are real numbers, but composite, in the form

$$(x+2i)(x-2i)$$

when the coefficients are complex.

It is convenient now to deal with *monic* polynomial forms, that is, with polynomial forms having unit coefficient for the highest power of x. With a slight change of notation, a typical polynomial form is then

$$x^n+p_1x^{n-1}+p_2x^{n-2}+\ldots+p_{n-1}x+p_n.$$

It may be denoted briefly by the symbol

$$P_n(x)$$

or even

$$P_n,$$

where the letter P carries the name of the coefficients $p_1, p_2, \ldots$ and the suffix n the degree of the polynomial form.

5. Quotient and remainder. Let

$$U_n \equiv x^n+u_1x^{n-1}+\ldots+u_{n-1}x+u_n,$$
$$V_m \equiv x^m+v_1x^{m-1}+\ldots+v_{m-1}x+v_m$$

be two given polynomial forms, where, for convenience,

$$n \geqslant m.$$

It is familiar, but must now be proved more carefully, that there exist a *quotient polynomial form* Q_{n-m} and a *remainder polynomial form* R, of degree less than m but *not necessarily monic*, such that

$$U_n = V_m Q_{n-m}+R.$$

Consider first a particular example:

$$U_4 = x^4+8x^3+5x+7,$$
$$V_2 = x^2+2x+1.$$

The usual process for division is equivalent to the following sequence:

$$U_4-x^2 V_2 = 6x^3-x^2+5x+7,$$
$$\{U_4-x^2 V_2\}-6xV_2 = -13x^2-x+7,$$
$$\{U_4-x^2 V_2-6xV_2\}+13V_2 = 25x+20,$$

so that

$$U_4 = (x^2+6x-13) V_2+(25x+20).$$

Hence

$$Q_2 = x^2+6x-13, \quad R = 25x+20.$$

The particular example illustrates precisely the procedure to be followed in the general case:

$$U_n-x^{n-m} V_m = (u_1-v_1)x^{n-1}+(u_2-v_2)x^{n-2}+\ldots,$$
$$\{U_n-x^{n-m} V_m\}-(u_1-v_1)x^{n-m-1} V_m$$
$$= \{(u_2-v_2)-(u_1-v_1)\}x^{n-2}+\{(u_3-v_3)-(u_1-v_1)\}x^{n-3}+\ldots.$$

The order of the polynomial on the right is successively reduced in degree by subtracting an appropriate multiple of V_m times a power of x. Ultimately the degree on the right becomes m, and a final subtraction then leaves a polynomial of degree $m-1$ (or less, should the top coefficient(s) then happen to be zero). Thus

$$U_n-\{x^{n-m}+(u_1-v_1)x^{n-m-1}+(u_2-v_2-u_1+v_1)x^{n-m-2}+\ldots\} V_m$$
$$= \text{polynomial of degree } m-1,$$

so that

$$U_n-Q_{n-m} V_m = R,$$

or

$$U_n = Q_{n-m} V_m+R.$$

Note that the polynomial Q_{n-m} is monic.

Examples

Find the quotient and remainder in each of the following cases:

1. $x^4+2x^3+3x^2+4x+5$; x^2+x+1.

2. x^3-4x^2+3x+1; x^2-2x.

3. x^5+5x^3+7x+3; x^3+2x+2.

4. $x^5-6x^4+11x^3-6x^2$; x^2-5x+6.

5. x^4+6x^2-7x+2; x^4+2x^3+3.

6. *The uniqueness of quotient and remainder.* Prove that *the quotient and remainder, for given U_n and V_m, are unique*: that is, if

$$U_n = V_m Q_{n-m}+R$$

and

$$U_n = V_m A_{n-m}+B$$

(where R, B are of degrees less than m), then

$$A_{n-m} = Q_{n-m} \quad \text{and} \quad B = R.$$

6. Euclid's algorithm; application to polynomial forms.

Let U_n, V_m $(n \geqslant m)$ be two given polynomial forms. Successive division, as for Euclid's algorithm, gives a sequence of quotients and remainders. For ease of notation, the use of suffixes to denote degrees is discarded and they are used instead merely to distinguish successive polynomials, thus:

$$\text{quotients} \quad q_1, q_2, q_3, \ldots,$$
$$\text{remainders} \quad r_1, r_2, r_3, \ldots.$$

Then, writing U, V for U_n, V_m,

$$\begin{aligned} U &= q_1 V + r_1, \\ V &= q_2 r_1 + r_2, \\ r_1 &= q_3 r_2 + r_3, \\ r_2 &= q_4 r_3 + r_4, \\ &\ldots\ldots\ldots\ldots\ldots\ldots \\ r_{k-2} &= q_k r_{k-1} + r_k, \\ r_{k-1} &= q_{k+1} r_k + 0. \end{aligned}$$

The degrees of the remainders drop steadily, so the process ultimately terminates.

The polynomial form r_k is, by the last equation, a factor of r_{k-1}, and so, 'rising' successively through the equations, r_k is also a factor of $r_{k-2}, r_{k-3}, \ldots, r_1, V, U$. Thus r_k is a factor common to U and V.

Moreover, *every* factor of U and V is, by the first equation, a factor of r_1 and so, 'dropping' down the equations, of $r_2, r_3, \ldots, r_k$. Hence *r_k is the factor of greatest degree common to U and V*, or, in the usual phrase, r_k is the *highest common factor* of U and V.

If r_k is a constant other than zero (that is, if r_k is a polynomial form of degree zero) then U, V have no proper common factor.

Note that *the polynomial r_k obtained by this process is, in general, not monic.*

***Illustration* 2.** *Highest common factors.* The two examples which follow are designed to emphasize the essential identity of the processes for numbers and for polynomials.

(i) *To find the highest common factor of the numbers* 14,772 *and* 1464.

The algorithm gives

$$\begin{aligned} 14{,}772 &= 10 \times 1464 + 132, \\ 1464 &= 11 \times 132 + \underline{12}, \\ 132 &= 11 \times \underline{12} + 0. \end{aligned}$$

Hence the highest common factor is 12.

(ii) *To find the highest common factor of the polynomials* $x^4+4x^3+7x^2+7x+2$ *and* x^3+4x^2+6x+4.

The algorithm gives

$$\begin{aligned} x^4+4x^3+7x^2+7x+2 &= x(x^3+4x^2+6x+4)+x^2+3x+2, \\ x^3+4x^2+6x+4 &= (x+1)(x^2+3x+2)+\underline{x+2}, \\ x^2+3x+2 &= (x+1)(\underline{x+2})+0. \end{aligned}$$

Hence the highest common factor is $x+2$.

REMARK. Case (i) is, at all stages, a particular example of case (ii) for the specialization $x = 10$; but see Example 5 below.

Examples

Use the method of the text to find the highest common factors of the following pairs of polynomials:

1. $x^3+6x^2+11x+6$, x^2+4x+3.
2. x^6-1, x^3+x^2+x.
3. $x^4+6x^3+10x^2+9x+4$, x^3+3x^2-6x-8.
4. x^4+2x^2+5x+3, x^2-7x+2.
5. Prove that the highest common factor of

$$x^4+4x^3+6x^2+5x+2, \quad 2x^3+7x^2+7x+2$$

is x^2+3x+2.

Is it true that the highest common factor of 14,652 and 2772 is 132?

7. The '$AU+BV = H$' theorem. Let U, V be two given polynomial forms, of degrees n, m $(n \geqslant m)$, having a highest common factor H. It is required *to establish the existence of polynomial forms* A, B *such that*

$$H = AU+BV.$$

The equations of §6, with H written for r_k, give†

$$\begin{aligned} H &= r_{k-2}-q_k r_{k-1}, \\ r_{k-1} &= r_{k-3}-q_{k-1} r_{k-2}, \\ &\cdots\cdots\cdots\cdots\cdots\cdots \\ r_2 &= V-q_2 r_1, \\ r_1 &= U-q_1 V. \end{aligned}$$

Successive substitution gives H *linearly* in terms of r_{k-2}, r_{k-1}, then in terms of r_{k-3}, r_{k-2}, then in terms of $r_{k-4}, r_{k-3}, \ldots$, then in terms of V and

† The calculations are a little more awkward than those given in §6, since the remainders are not monic. See the following Illustration.

r_1; and, finally, in terms of U and V. The coefficients A, B of U, V in this final expression for H are polynomials which are *linear* in $r_1, r_2, \ldots, r_{k-1}$ —though they are complicated expressions if written in terms of $q_1, q_2, \ldots, q_k$.

COROLLARY. If U, V have no common factor, then polynomials A, B can be found so that

$$AU+BV = 1.$$

***Illustration* 3.** Let

$$U \equiv x^4-x^3-x^2-x-2,$$
$$V \equiv x^3-3x^2+3x-2.$$

Then (i)

$$U-Vx = 2x^3-4x^2+x-2,$$
$$(U-Vx)-2V = 2x^2-5x+2,$$

so that

$$U = (x+2)\,V+(2x^2-5x+2)$$
$$= (x+2)\,V+r_1.$$

(ii)

$$V \equiv x^3-3x^2+3x-2,$$
$$r_1 \equiv 2x^2-5x+2.$$

Then

$$2V-r_1x = -x^2+4x-4,$$
$$2(2V-r_1x)+r_1 = 3x-6,$$

so that

$$4V = (2x-1)\,r_1+(3x-6)$$
$$= (2x-1)\,r_1+r_2.$$

(iii)

$$r_1 = 2x^2-5x+2,$$
$$r_2 = 3x-6.$$

Then

$$3r_1-2r_2x = -3x+6,$$
$$(3r_1-2r_2x)+r_2 = 0,$$

so that

$$3r_1 = (2x-1)\,r_2.$$

There is no further remainder, and the process stops. The greatest common factor is r_2 which, taken in its monic form, is $x-2$.

Further, it follows from the relations (ii), (i) successively that

$$3(x-2) = 4V-(2x-1)\,r_1$$
$$= 4V-(2x-1)\{U-(x+2)\,V\}$$
$$= -(2x-1)\,U+\{4+(2x-1)(x+2)\}\,V.$$

Hence *the highest common factor $x-2$ is expressed linearly in terms of U, V in the form*

$$x-2 = (-\tfrac{2}{3}x+\tfrac{1}{3})\,U+(\tfrac{2}{3}x^2+x+\tfrac{2}{3})\,V.$$

Revision Examples

1. Express the highest common factor linearly in terms of U, V in the cases:

(i) $U \equiv x^2+x+1$, $V \equiv x+1$.

(ii) $U \equiv x^3-6x^2+11x-6$, $V \equiv x^2-4x+3$.

(iii) $U \equiv x^3-6x^2+12x-8$, $V \equiv x^2-4x+4$.

(iv) $U \equiv x^5+3x^3+1$, $V \equiv x^3+2x+2$.

2. Obtain an identity of the form

$$A(x)p(x)+B(x)q(x) = h(x),$$

where
$$p(x) = x^{10}-1, \quad q(x) = x^6-1$$

and $h(x)$ is the highest common factor of $p(x)$, $q(x)$.

3. Explain briefly how to find the highest common factor of two integers or two polynomials.

If m and n are positive integers whose H.C.F. is k, prove that the H.C.F. of the integers 2^m-1 and 2^n-1 is 2^k-1; and that the H.C.F. of the polynomials $x^{2^m}-x$ and $x^{2^n}-x$ is $x^{2^k}-x$.

4. Find polynomials $A(x)$, $B(x)$ such that

$$(x^2+1)^2 A(x)+(x-1)^2 B(x) = 1.$$

5. If P, Q are polynomials, the degree of Q being less than that of P, prove that polynomials P_0, P_1, P_2, ... all of degree less than Q can be found such that

$$P \equiv P_0+P_1 Q+P_2 Q^2+P_3 Q^3+\ldots.$$

Prove that the polynomials P_0, P_1, P_2, ... are unique.

6. Prove that a given polynomial $f(x)$ can be expressed in the form

$$f(x) \equiv xg(x^2+1)+h(x^2+1),$$

where $g(x^2+1)$, $h(x^2+1)$ are polynomials in x^2+1.

7. Find the highest common factor of the two polynomials

$$f(x) \equiv x^4-13x^3+58x^2-96x+36,$$
$$g(x) \equiv x^4-11x^3+36x^2-24x-36.$$

Find all the roots of the equation

$$\{f(x)\}^2 = \{g(x)\}^2.$$

8. If $d(f)$ is written for 'the degree of $f(x)$', show that *necessary* and *sufficient* conditions for the polynomials $f(x)$, $g(x)$ to have a factor in common are that polynomials $A(x)$, $B(x)$ exist such that

$$d(A) \leqslant d(g)-1, \quad d(B) \leqslant d(f)-1,$$
$$A(x)f(x) \equiv B(x)g(x).$$

9. Find the highest common factor of

$$x^4+3x^3+9x^2+8x+6,$$
$$2x^4+5x^3-x^2-3x-6.$$

10. Find two polynomials P, Q such that

$$(x^2-3x+2)P+(x^2+x+1)Q = 1.$$

11. Find the highest common factor of

$$30x^5+37x^4+110x^3+110x^2+103x+30,$$
$$30x^3+17x^2+77x+30.$$

12. Show that $(x+\alpha)^3+(x+\beta)^3$ cannot be equal to $2(x+\gamma)^3$ for all values of x unless $\alpha = \beta = \gamma$.

13. If

$$F \equiv 3x^3-8x^2+19x-10,$$
$$G \equiv 3x^4+4x^3-22x^2-9x+14,$$

express the highest common factor of F and G in the form

$$AF+BG.$$

14. Repeat Question 13 when
(i) $F \equiv x^7+1, \quad G \equiv x^5+1$;
(ii) $F \equiv nx^{n+1}-(n+1)x^n+1, \quad G \equiv x^n-nx+(n-1)$.

15. The polynomials $f(x), g(x)$ have no common factor. Prove that, if $f(x), g(x)$ both divide a polynomial $k(x)$, then $f(x)g(x)$ divides $k(x)$.

16. The polynomials $f(x), g(x)$ have no common factor. Prove that, if $f(x)g(x) = \{\phi(x)\}^2$, then the polynomial $\phi(x)$ has a factorization $\phi(x) = \phi_1(x)\phi_2(x)$ such that

$$f(x) = \{\phi_1(x)\}^2, \quad g(x) = \{\phi_2(x)\}^2.$$

SECTION II

MATRICES

11

MATRICES AND MAPPINGS

Before starting the formal theory of matrices, a brief account may be given of some of the reasons which enforce their introduction. Special emphasis is laid on the idea of a *mapping* or *transformation*.

1. Rotations. Suppose that a configuration is given in a plane and that, referred to a fixed set of orthogonal Cartesian axes, the coordinates of a typical point P are (x, y).

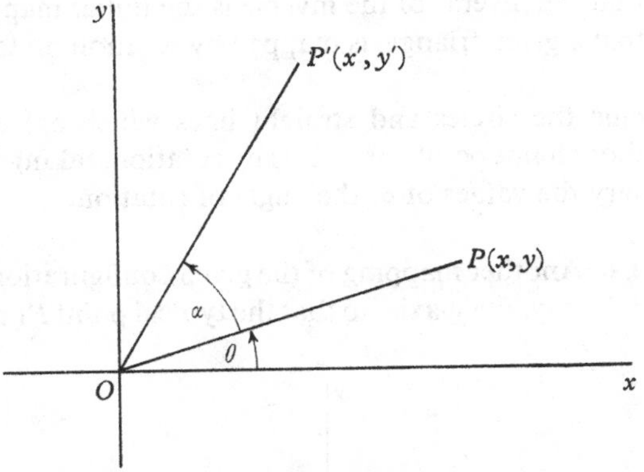

Fig. 8. Rotation

Rotate the configuration as a whole about the origin O in the counter-clockwise sense through an angle α, so that P moves to the position $P'(x', y')$. If $OP = r$ and if the angle between the x-axis and OP, measured in the counter-clockwise sense, is θ, then $OP' = r$ and the angle between the x-axis and OP' is $\theta + \alpha$. Thus

$$x = r\cos\theta, \qquad y = r\sin\theta,$$
$$x' = r\cos(\theta+\alpha), \quad y' = r\sin(\theta+\alpha),$$

so that, on expanding the trigonometric functions, the relations expressing x' and y' in terms of x and y are

$$x' = x\cos\alpha - y\sin\alpha,$$
$$y' = x\sin\alpha + y\cos\alpha.$$

These relations give a *transformation* or *mapping* from x,y to x',y' corresponding geometrically to a *rotation* of the configuration about O through an angle α.

The mapping from a given P' back to P is of similar type, found by solving the above equations so as to express x,y in terms of x',y':

$$x = x'\cos\alpha + y'\sin\alpha,$$
$$y = -x'\sin\alpha + y'\cos\alpha.$$

This mapping is called the *inverse* of the given mapping.

Text Examples

1. Prove that the inverse of the inverse is the initial mapping.

2. Prove that a given triangle is mapped by rotation on to a *congruent* triangle.

3. Determine the circles and straight lines which are mapped into themselves (though not point-for-point) by rotation, taking into account where necessary the values of α, the angle of rotation.

2. Reflexions. Another mapping of the given configuration is obtained by *reflecting* it in, say, the y-axis, so that the typical point $P(x,y)$ is moved

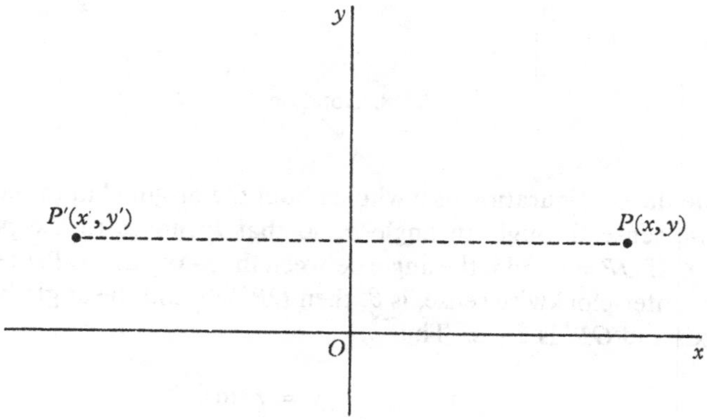

Fig. 9. Reflexion in Oy

to the position $P'(x',y')$, where

$$x' = -x, \quad y' = y.$$

In the same way, reflexion in the x-axis is governed by the mapping

$$x' = x, \quad y' = -y.$$

A similar effect is obtained by *reflexion with respect to the origin O*, whereby P becomes that point P' for which O is the middle point of PP'.

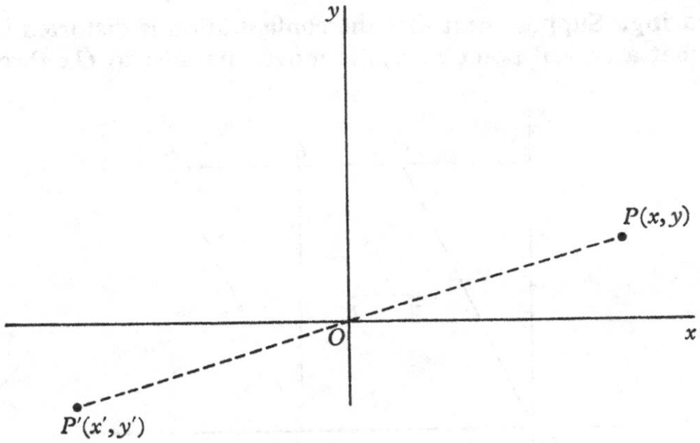

Fig. 10. Reflexion about O

The mapping is

$$x' = -x, \quad y' = -y.$$

Text Examples

1. Prove that the *inverse* of reflexion in the y-axis is the mapping

$$x = -x', \quad y = y'$$

and that the *inverse* of reflexion with respect to the origin is

$$x = -x', \quad y = -y'.$$

2. Prove that a reflexion with respect to the origin is the same as a rotation about the origin through an angle π, and deduce the equations of the former from those of the latter.

3. Prove that a given triangle is mapped into a congruent triangle by reflexion.

A triangle ABC is reflected into a triangle $A'B'C'$. A point moves round the sides of ABC in the sense $A \to B \to C \to A$, and this is found to be clockwise. Prove that, for reflexion in a line, the motion

$$A' \to B' \to C' \to A'$$

is counter-clockwise; and that, for reflexion in a point, the motion $A' \to B' \to C' \to A'$ is clockwise.

4. Determine the straight lines and circles which are mapped into themselves (though not point-for-point) by reflexion.

3. Shearing. Suppose next that the configuration is distorted in such a way that a typical point $P(x,y)$ is moved parallel to Ox through a

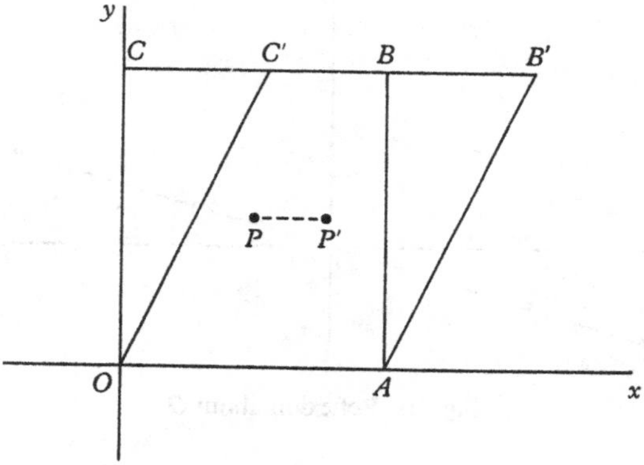

Fig. 11. Shearing

distance ky, where k is constant. The diagram (Fig. 11) shows, in illustration, the effect of shearing a rectangle $OABC$ to the position $OAB'C'$. By definition, the new position of P is given by the mapping

$$x' = x + ky, \quad y' = y.$$

Text Examples

1. Prove that the *inverse* of this mapping is

$$x = x' - ky', \quad y = y'.$$

2. Prove that, in general, the distance between two points is different from the distance between their maps.

3. Determine the circles and straight lines which are mapped into themselves (though not point-for-point) by shearing.

4. Prove that straight lines are mapped into straight lines by shearing.

4. The mappings consolidated. The particular mappings derived in §§1, 2, 3 are all special cases of a more general mapping of the type

$$x' = ax+by,$$
$$y' = cx+dy,$$

where a, b, c, d are constant. These mappings are of great importance and are called *linear mappings* (*transformations*).

NOTATION. The symbolism

$$P' = TP$$

is used to express the fact that P' is the point derived from P by means of a mapping called T.

Suppose, for example, that

$$R \equiv \text{Rotation},$$
$$U \equiv \text{Reflexion about } Ox,$$
$$V \equiv \text{Reflexion about } Oy,$$
$$A \equiv \text{Reflexion about the origin } O,$$
$$S \equiv \text{Shearing}.$$

Then the mappings illustrated in the earlier paragraphs of this chapter are

$$P' = RP, \quad P' = VP, \quad P' = AP, \quad P' = SP.$$

It is natural to express the *inverse mappings* in the forms

$$P = R^{-1}P', \quad P = V^{-1}P', \quad P = A^{-1}P', \quad P = S^{-1}P'.$$

A reader who enjoys notation may like to accustom himself from the start to the standard notation

$$P \xrightarrow{R} P'$$

for 'P is mapped by R to P'' and

$$P' \xrightarrow{R^{-1}} P$$

for 'P' is mapped by R^{-1} to P'.

Text Examples

1. Prove that the mapping

$$x' = ax+by,$$
$$y' = cx+dy$$

maps a straight line $lx+my+n = 0$ to another straight line.

2. Discuss the possibility of the mapping of a straight line to itself.
3. Discuss the possibility of the mapping of a point to itself.

5. The product of two mappings. The notation of §4 may be extended to cover two or more successive mappings. Suppose, for example, that a given configuration is rotated through an angle α, so

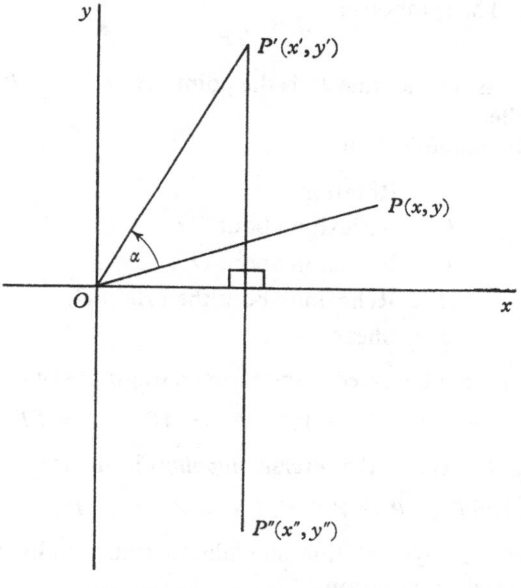

Fig. 12. The transformation *UR*

that a typical point $P(x,y)$ goes to $P'(x',y')$; and that the resulting configuration is then reflected in the x-axis so that P' goes to $P''(x'',y'')$. Then

$$P'' = UP',$$

$$P' = RP,$$

and, in an obvious sense of the symbolism,

$$P'' = URP.$$

DEFINITION. The mapping of P into P'' is called the *product* (in that order) of R followed by U.

In the notation of §4,

$$P \overset{UR}{\rightarrow} P''$$

is the same as

$$P \overset{R}{\rightarrow} P' \overset{U}{\rightarrow} P''.$$

The product mapping RU (in that order) is very different from UR, as the diagrams (Figs. 12, 13) clearly show. In order to obtain RU, the

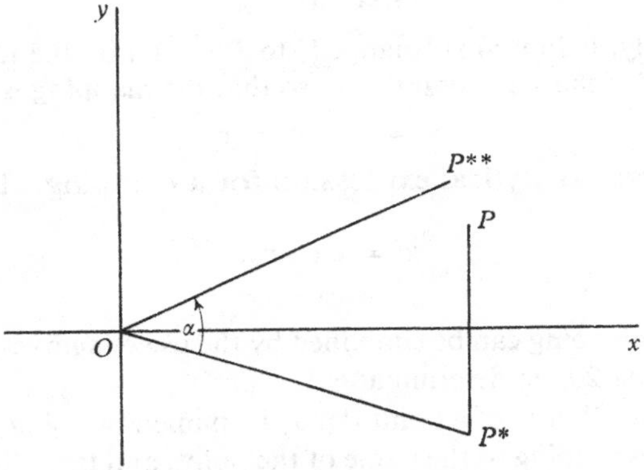

Fig. 13. The transformation *RU*

point P is *first* reflected in the x-axis to P^* and *then* the radius OP^* is rotated through an angle α to OP^{**}. The primitive mappings are

$$P^{**} = RP^*, \quad P^* = UP.$$

Analytically, if $P \equiv (x,y)$, then

$$P' \equiv RP = (xc-ys,\ xs+yc),$$

where c, s are written for $\cos\alpha$, $\sin\alpha$. Hence

$$P'' \equiv UP' = (xc-ys,\ -xs-yc).$$

On the other hand,

$$P^* \equiv UP = (x, -y),$$
$$P^{**} \equiv RP^* = (xc+ys,\ xs-yc).$$

Suppose next that any three of the mappings (not necessarily all different) are denoted by general symbols X, Y, Z. Then *these mappings have 'products' which obey the associative law*

$$X(YZ) = (XY)Z,$$

each product denoting the result of operating first by Z and then by Y and then by X. This mapping is conveniently denoted by the symbol XYZ.

When a mapping is repeated, the product XX is denoted by X^2; similarly $X^3, X^4, \ldots$.

The 'mapping' which leaves every configuration entirely unaltered is called the *identity mapping* and is denoted by the symbol I. It recalls unity in that, for any mapping X,

$$XI = IX = X.$$

Analytically, if $P(x,y)$ is 'mapped' to $P'(x',y')$ by the identity mapping, then P' is the same point as P, so that the mapping is

$$x' = x, \quad y' = y.$$

6. The general analytical expression for a mapping. The notation (§4)

$$x' = ax+by,$$
$$y' = cx+dy$$

for a typical mapping can be simplified by the use of suffixes. (Compare Part 1, Chapter 20, on determinants.)

The two coordinates of a point P may be named by using for each the letter p, corresponding to the name of the point, and then distinguishing between them by the help of *suffixes* 1, 2, so that P is the point (p_1, p_2). Similarly a point Q would be (q_1, q_2).

The mapping T is then, in the first instance,

$$p_1' = ap_1+bp_2,$$
$$p_2' = cp_1+dp_2;$$

but this, in its turn, can be written more economically by replacing the coefficients a, b, c, d, which define the mapping T, by the single letter t and distinguishing individuals by a *double suffix* notation:

$$p_1' = t_{11}p_1+t_{12}p_2,$$
$$p_2' = t_{21}p_1+t_{22}p_2.$$

Note at once that, in general,

$$t_{12} \neq t_{21}.$$

The four coefficients corresponding to the mapping T appear in a *rectangular array*

$$\begin{matrix} t_{11} & t_{12} \\ t_{21} & t_{22} \end{matrix}$$

whose shape gives point to the notation, for *the first suffix indicates the row and the second the column in which the element appears.*

[More generally, we shall later meet arrays like

$$\begin{matrix} t_{11} & t_{12} & t_{13} & t_{14} \\ t_{21} & t_{22} & t_{23} & t_{24} \\ t_{31} & t_{32} & t_{33} & t_{34} \end{matrix}$$

of three rows and four columns. The element t_{23} appears in the second row and the third column.]

The given mapping can be written more simply in the form

$$p'_1 = \sum_{\lambda=1}^{2} t_{1\lambda} p_\lambda, \quad p'_2 = \sum_{\lambda=1}^{2} t_{2\lambda} p_\lambda.$$

Even greater compactness is obtained by the notation

$$p'_i = \sum_{\lambda=1}^{2} t_{i\lambda} p_\lambda \qquad (i = 1,2),$$

the parenthesis on the right indicating that i can take either of the values 1,2.

Finally, for this stage, we introduce the *double suffix convention*, whereby a repeated *Greek* suffix is summed over all relevant values—here, 1 and 2. Thus

$$p'_i = t_{i\lambda} p_\lambda.$$

Note that λ is a *dummy suffix*, which can be replaced by any other Greek letter, so that

$$p'_i = t_{i\mu} p_\mu = t_{i\alpha} p_\alpha.$$

The same Greek letter must not be used simultaneously for two distinct summations in a single composite expression.

7. Analytical expression for a product. Suppose that a mapping A maps a point P to the point $P' \equiv AP$, and that a second mapping B then maps P' to the point $P'' \equiv BP'$, so that

$$P'' = BP' = BAP.$$

[The alternative notation of p. 133 is $P \xrightarrow{BA} P''$.]

If the corresponding equations are

$$p'_i = a_{i\lambda} p_\lambda \qquad (i = 1,2) \text{ for } P \xrightarrow{A} P',$$

$$p''_i = b_{i\lambda} p'_\lambda \qquad (i = 1,2) \text{ for } P' \xrightarrow{B} P'',$$

both summed for $\lambda = 1,2$, then

$$p_i'' = b_{i\lambda}\{a_{\lambda\mu}p_\mu\}$$
$$= b_{i\lambda}a_{\lambda\mu}p_\mu \qquad \text{for } P \xrightarrow{BA} P'',$$

summed for $\lambda, \mu = 1,2$ independently. The mapping

$$P'' = BAP, \quad \text{or} \quad P \xrightarrow{BA} P'',$$

is thus of the form

$$p_i'' = c_{i\mu}p_\mu,$$

summed for $\mu = 1,2$, where

$$c_{i\mu} = b_{i\lambda}a_{\lambda\mu}.$$

Replacing μ (which is not summed in *this* relation) by the Roman letter j, the 'product' mapping has coefficients

$$c_{ij} \equiv b_{i\lambda}a_{\lambda j}.$$

In detail, the coefficients are $c_{11}, c_{12}, c_{21}, c_{22}$, where

$$c_{11} = b_{11}a_{11}+b_{12}a_{21}, \qquad c_{12} = b_{11}a_{12}+b_{12}a_{22},$$
$$c_{21} = b_{21}a_{11}+b_{22}a_{21}, \qquad c_{22} = b_{21}a_{12}+b_{22}a_{22}.$$

Note the rule:
In order to obtain the coefficient c_{ij}, write down the product

$$b_{i\cdot}a_{\cdot j},$$

with b, a in correct order, and then insert λ, to give

$$b_{i\lambda}a_{\lambda j}$$

or, in detail,

$$b_{i1}a_{1j}+b_{i2}a_{2j}.$$

The ideas sketched here reach their fulfilment when they lead to the theory of matrices in the next chapter (see p. 148.)

Examples

1. Verify that the product of two rotations about a given point is a rotation and confirm this result analytically.

2. Prove, in the notation of §4, that, if I is the identity,

$$U^2 = I, \quad V^2 = I, \quad A^2 = I,$$
$$UV = VU = A.$$

3. Prove that reflexion W in the line $y = x \tan \alpha$ is given by the relations

$$x' = x \cos 2\alpha + y \sin 2\alpha, \quad y' = x \sin 2\alpha - y \cos 2\alpha,$$

and verify from these relations that

$$W^2 = I.$$

4. Prove that, if P, Q are the shears

$$x' = x + yp, \quad y' = y$$

and

$$x' = x + yq, \quad y' = y$$

respectively, then their product is also a shear.

Prove that

$$PQ = QP.$$

5. Prove that, if P is the shear

$$x' = x + py, \quad y' = y$$

and U the reflexion

$$x' = x, \quad y' = -y,$$

then

$$PU \neq UP.$$

6. Verify *by analytical calculations* that, in the notation of §4,

$$U(VA) = (UV)A,$$
$$R(UV) = (RU)V,$$
$$S(AU) = (SA)U,$$
$$R(AV) = (RA)V.$$

7. Prove, by solving the equations, that, if T is the mapping

$$p_1' = a_{11}p_1 + a_{12}p_2, \quad p_2' = a_{21}p_1 + a_{22}p_2,$$

then the inverse T^{-1} is the mapping

$$p_1 = b_{11}p_1' + b_{12}p_2', \quad p_2 = b_{21}p_1' + b_{22}p_2',$$

where, if $\Delta \equiv a_{11}a_{22} - a_{12}a_{21}$ (assumed not to be zero),

$$b_{11} = a_{22}/\Delta, \quad b_{22} = a_{11}/\Delta, \quad b_{12} = -a_{12}/\Delta, \quad b_{21} = -a_{21}/\Delta.$$

Verify the relations

$$T(T^{-1}) = (T^{-1}T) = I$$

by considering the products $a_{i\lambda}b_{\lambda j}$ and $b_{i\lambda}a_{\lambda j}$.

12

THE ELEMENTARY MANIPULATION OF MATRICES

1. Introductory. In many branches of mathematics the quantities to be investigated are enumerated by means of symbols written in an assigned order. A typical example is provided by the Cartesian coordinates of a point in a plane, where two numbers, x and y, are written in the form

$$(x, y)$$

signifying that the first denotes the distance of the point from one line while the second denotes its distance from another. The array

$$\begin{pmatrix} t_{11} & t_{12} \\ t_{21} & t_{22} \end{pmatrix}$$

of the preceding chapter supplies another more elaborate example.

***Illustration* 1.** *The multiplication table.* The array

1	2	3	4	5	6
2	4	6	8	10	12
3	6	9	12	15	18
4	8	12	16	20	24

has four *rows* and six *columns*. A typical *element* lies in, say, the ith row and the jth column; for example, 15 lies in the third row and the fifth column, while 16 lies in the fourth row and the fourth column. The element in the ith row and jth column is evaluated by the simple rule that it is equal to the product ij.

The immediate point is that the elements are exhibited in an assigned order whereby each is readily located.

Examples

1. Complete the following *addition table* in which the element in the ith row and jth column has the value $i+j$. (This array has 7 rows and 4 columns.)

$$\begin{matrix} 2 & 3 & 4 & 5 \\ 3 & 4 & 5 & 6 \\ 4 & 5 & . & . \\ 5 & . & . & . \\ 6 & . & . & . \\ 7 & . & . & . \\ 8 & . & . & . \end{matrix}$$

2. Construct a table of three rows and four columns in which the element in the ith row and jth column has the value

(i) i/j, (ii) $2i+j$, (iii) j^2.

2. Definitions. A rectangular array, when organized for rules of addition and multiplication to be defined later, is called a *matrix*. Its *elements* are exhibited in *rows* and *columns*. These elements are usually real or complex numbers according to the nature of the problem being studied, and should be specified precisely when necessary.

The element in the ith row and jth column is often named by means of a *double-suffix notation*

$$a_{ij},$$

where the first suffix, i, identifies the row and the second, j, the column.

A matrix may be named within brackets so that, for m rows and n columns, a typical matrix is

$$\begin{pmatrix} a_{11} & a_{12} & a_{13} & \cdots & a_{1j} & \cdots & a_{1n} \\ a_{21} & a_{22} & a_{23} & \cdots & a_{2j} & \cdots & a_{2n} \\ \cdots & \cdots & \cdots & \cdots & \cdots & \cdots & \cdots \\ a_{i1} & a_{i2} & a_{i3} & \cdots & a_{ij} & \cdots & a_{in} \\ \cdots & \cdots & \cdots & \cdots & \cdots & \cdots & \cdots \\ a_{m1} & a_{m2} & a_{m3} & \cdots & a_{mj} & \cdots & a_{mn} \end{pmatrix}.$$

For brevity, this matrix is often denoted by the symbol

$$(a_{ij})$$

in which a typical element is enclosed within brackets, or, even more compactly, by the single symbol

$$\mathbf{a}$$

in **bold** type. In either case, the number of rows and columns is implicit in the context, or may be stated explicitly if desired. It is always understood that the element a_{ij} stands in the ith row and jth column. The two elements a_{ij}, a_{ji} are, in general, quite distinct.

A matrix of m rows and n columns is said to be *of type* $m \times n$ ('of type m by n').

When the number of rows is equal to the number of columns, the matrix is called *square*. For example, a square matrix of type 2 is

$$\mathbf{t} \equiv \begin{pmatrix} t_{11} & t_{12} \\ t_{21} & t_{22} \end{pmatrix}.$$

A matrix consisting of a single column is called a *column vector*; for example,

$$\begin{pmatrix} u_{11} \\ u_{21} \\ u_{31} \end{pmatrix}.$$

The second suffix, however, is unnecessary and may be dropped, giving the notation

$$\begin{pmatrix} u_1 \\ u_2 \\ u_3 \end{pmatrix}.$$

For convenience of printing, a column vector may be written 'horizontally', using the notation

$$\{u_1, u_2, u_3\}.$$

A matrix consisting of a single row is called a *row vector*. A typical one, of type 1×4, is

$$(v_1, v_2, v_3, v_4).$$

A matrix of type 1×1, consisting of one row and one column, is called a *scalar*. The brackets surrounding the scalar (a) are often omitted.

The present use of the words *vector* and *scalar* is closely allied to, but not identical with, that familiar in physics. A physical vector must transform from one set of coordinate axes to another in accordance with the transformation law obeyed by displacements in space.

A *zero matrix* $\mathbf{0}$ is one whose elements are all zero; it may be of any type $m \times n$ and, on occasion, it may be written $\mathbf{0}_{mn}$.

A *unit matrix* is a *square* one whose elements are *unity* down the *leading diagonal*, and otherwise *zero*. Typical examples, of *orders* 2, 3, 4 are

$$\mathbf{I}_2 \equiv \begin{pmatrix} 1 & 0 \\ 0 & 1 \end{pmatrix}, \quad \mathbf{I}_3 \equiv \begin{pmatrix} 1 & 0 & 0 \\ 0 & 1 & 0 \\ 0 & 0 & 1 \end{pmatrix}, \quad \mathbf{I}_4 \equiv \begin{pmatrix} 1 & 0 & 0 & 0 \\ 0 & 1 & 0 & 0 \\ 0 & 0 & 1 & 0 \\ 0 & 0 & 0 & 1 \end{pmatrix}.$$

The elements of $\mathbf{I}_n$ may be named by means of the *Kroeneker Delta Symbol* δ_{ij}, which is defined by the relations

$$\delta_{ij} = 1 \qquad (i = j),$$

$$\delta_{ij} = 0 \qquad (i \neq j).$$

Thus

$$\mathbf{I}_n \equiv (\delta_{ij}).$$

When the order of a unit matrix is clear from the context, the suffix may be omitted and the unit matrix written simply as $\mathbf{I}$.

Examples

Write down the square matrices of four rows and columns defined by the formulae:

1. $a_{ij} = \begin{cases} -1 & i = j, \\ 0 & i \neq j. \end{cases}$ **2.** $a_{ij} = \begin{cases} 1 & i = j+1, \\ 0 & \text{otherwise.} \end{cases}$

3. $a_{ij} = \begin{cases} 1 & i+j = 3, \\ -1 & \text{otherwise.} \end{cases}$ **4.** $a_{ij} = (-1)^{i+j}$.

5. $a_{ij} = i^j$. **6.** $a_{ij} = i-j$.

3. Addition. Let

$$\mathbf{a} \equiv (a_{ij}), \quad \mathbf{b} \equiv (b_{ij})$$

be two matrices *each of m rows and n columns*. Their *sum* is defined to be the matrix

$$\mathbf{c} \equiv (c_{ij}),$$

where

$$c_{ij} = a_{ij} + b_{ij}$$

for all pairs of values of i, j. The sum is written in the notation

$$\mathbf{a} + \mathbf{b}.$$

For example,

$$\begin{pmatrix} 1 & 3 & 5 \\ 4 & -2 & 0 \end{pmatrix} + \begin{pmatrix} 3 & 2 & 1 \\ -1 & -1 & -1 \end{pmatrix}$$
$$= \begin{pmatrix} 1+3 & 3+2 & 5+1 \\ 4-1 & -2-1 & 0-1 \end{pmatrix}$$
$$= \begin{pmatrix} 4 & 5 & 6 \\ 3 & -3 & -1 \end{pmatrix}.$$

Examples

1. If

$$\mathbf{a} = \begin{pmatrix} 1 & 2 & 3 \\ 4 & 5 & 6 \end{pmatrix}, \quad \mathbf{b} = \begin{pmatrix} 3 & 2 \\ 1 & 0 \end{pmatrix}, \quad \mathbf{c} = \begin{pmatrix} -1 & -2 \\ -3 & 0 \end{pmatrix},$$
$$\mathbf{d} = \begin{pmatrix} 3 & 0 & -3 \\ 0 & -2 & 0 \end{pmatrix}, \quad \mathbf{e} = \begin{pmatrix} 0 & 0 & 1 \\ 2 & -1 & 3 \end{pmatrix}, \quad \mathbf{f} = \begin{pmatrix} 1 & 0 \\ 0 & 1 \end{pmatrix},$$

find the sum of all pairs of matrices to which the word *sum* can be applied.

2. Prove that (conformable) matrices whose elements are complex numbers obey the *associative law*

$$(\mathbf{a}+\mathbf{b})+\mathbf{c} = \mathbf{a}+(\mathbf{b}+\mathbf{c})$$

and the *commutative law*

$$\mathbf{a}+\mathbf{b} = \mathbf{b}+\mathbf{a}.$$

3. The rectangular Cartesian coordinates of two points P_1, P_2 are written in matrix form

$$\mathbf{r}_1 \equiv (x_1, y_1), \quad \mathbf{r}_2 \equiv (x_2, y_2).$$

Prove that the point whose coordinates are given by the matrix $\mathbf{r}_1+\mathbf{r}_2$ is the fourth vertex of the parallelogram of which OP_1, OP_2 are adjacent sides, where O is the origin of coordinates.

4. Find the relation between a_{ij} and b_{ij} if the sum of the matrices $\mathbf{a}$ and $\mathbf{b}$ is the zero matrix.

4. Scalar multiplication. Let

$$\mathbf{a} \equiv (a_{ij})$$

be a matrix of type $m \times n$ and k a given number. The *product of* $\mathbf{a}$ *by the scalar* k is defined to be the matrix

$$\mathbf{c} \equiv (c_{ij}),$$

where
$$c_{ij} = ka_{ij}$$

for all pairs of values of i,j. For example,

$$3\begin{pmatrix}1 & 2 & 3\\ 2 & 1 & 0\end{pmatrix} = \begin{pmatrix}3 & 6 & 9\\ 6 & 3 & 0\end{pmatrix}$$

and
$$-2\begin{pmatrix}0 & 1\\ 2 & 3\\ 4 & 5\end{pmatrix} = \begin{pmatrix}0 & -2\\ -4 & -6\\ -8 & -10\end{pmatrix}.$$

Examples

1. Prove that

$$k\mathbf{I}_3 \equiv \begin{pmatrix}k & 0 & 0\\ 0 & k & 0\\ 0 & 0 & k\end{pmatrix}.$$

2. Evaluate

$$2\begin{pmatrix}1 & 2\\ 3 & 4\end{pmatrix}+5\begin{pmatrix}6 & 7\\ 8 & 9\end{pmatrix}.$$

3. Evaluate

$$\cos\theta\begin{pmatrix}\cos\theta & \sin\theta\\ -\sin\theta & \cos\theta\end{pmatrix}+\sin\theta\begin{pmatrix}\sin\theta & -\cos\theta\\ \cos\theta & \sin\theta\end{pmatrix}.$$

5. Linear combinations. Let

$$\mathbf{a} \equiv (a_{ij}),\quad \mathbf{b} \equiv (b_{ij}),\quad \mathbf{c} \equiv (c_{ij}), \ldots$$

be given matrices, each of type $m \times n$, and let $p, q, r, \ldots$ be given numbers. The *linear combination*

$$p\mathbf{a}+q\mathbf{b}+r\mathbf{c}+\ldots$$

is defined to be the matrix

$$\mathbf{u} \equiv (u_{ij}),$$

where
$$u_{ij} = pa_{ij}+qb_{ij}+rc_{ij}+\ldots$$

for all pairs of values of i,j. For example, if

$$\mathbf{a} = \begin{pmatrix}1 & 2 & 3\\ 2 & 3 & 4\end{pmatrix},\quad \mathbf{b} = \begin{pmatrix}1 & 0 & 0\\ 0 & 1 & 2\end{pmatrix},\quad \mathbf{c} = \begin{pmatrix}-1 & -2 & -3\\ 2 & 2 & 2\end{pmatrix},$$

then

$$2\mathbf{a}+\mathbf{b}+3\mathbf{c} = \begin{pmatrix}2+1-3 & 4+0-6 & 6+0-9\\ 4+0+6 & 6+1+6 & 8+2+6\end{pmatrix}$$

$$= \begin{pmatrix}0 & -2 & -3\\ 10 & 13 & 16\end{pmatrix}.$$

The rules for addition and scalar multiplication given in §§2, 3 are merely particular cases of this more general rule. The *difference* $\mathbf{a}-\mathbf{b}$ is the matrix $\mathbf{c}$, where

$$c_{ij} = a_{ij}-b_{ij}.$$

Examples

With the matrices $\mathbf{a}, \mathbf{b}, \ldots$ defined in §3 (p. 144), form the linear combinations:

1. $\mathbf{a}+\mathbf{d}+\mathbf{e}$, **2.** $2\mathbf{a}-3\mathbf{d}+4\mathbf{e}$, **3.** $5\mathbf{a}+3\mathbf{d}-\mathbf{e}$,
4. $\mathbf{b}+\mathbf{c}+2\mathbf{f}$, **5.** $2\mathbf{b}-3\mathbf{c}$, **6.** $-\mathbf{b}+3\mathbf{c}+4\mathbf{f}$,
7. $3\mathbf{a}-2\mathbf{e}$, **8.** $\mathbf{a}-\mathbf{d}-\mathbf{e}$, **9.** $4\mathbf{b}-3\mathbf{c}+2\mathbf{f}$.

10. Write down the matrices of four rows and columns defined by the formulae:

$$\begin{aligned} \mathbf{a} &\equiv (a_{ij}), \quad \text{where } a_{ij} = i+j, \\ \mathbf{b} &\equiv (b_{ij}), \quad \text{where } b_{ij} = i-j, \\ \mathbf{c} &\equiv (c_{ij}), \quad \text{where } c_{ij} = 4i+3j, \\ \mathbf{d} &\equiv (d_{ij}), \quad \text{where } d_{ij} = 2i-3j. \end{aligned}$$

Verify for these four matrices the relations

$$\begin{aligned} 7\mathbf{a}+\mathbf{b} &= 2\mathbf{c}, \\ \mathbf{a}-5\mathbf{b} &= -2\mathbf{d}, \\ 18\mathbf{a} &= 5\mathbf{c}-\mathbf{d}, \end{aligned}$$

and form the linear combinations

$$\begin{aligned} &2\mathbf{a}+\mathbf{b}-\mathbf{c}+3\mathbf{d}, \\ &3\mathbf{a}-2\mathbf{b}+2\mathbf{c}-\mathbf{d}, \\ &\mathbf{a}+3\mathbf{b}-\mathbf{c}+2\mathbf{d}. \end{aligned}$$

11. Exhibit the matrices of given type $m \times n$ as the vectors of a vector space.

6. The transpose of a matrix. Given a matrix

$$\mathbf{a} \equiv (a_{ij})$$

of type $m \times n$, a second matrix can be derived from it by interchanging rows and columns. The result is called the *transpose* of $\mathbf{a}$ and is denoted by the symbol

$$\mathbf{a}',$$

where $\mathbf{a}'$ is a matrix of type $n \times m$. A typical element of $\mathbf{a}'$ is a'_{ij}, where

$$a'_{ij} = a_{ji}.$$

For example, if

$$\mathbf{a} \equiv \begin{pmatrix} 1 & 2 & 3 \\ 4 & 5 & 6 \end{pmatrix},$$

then

$$\mathbf{a}' \equiv \begin{pmatrix} 1 & 4 \\ 2 & 5 \\ 3 & 6 \end{pmatrix}.$$

The transpose of the transpose is the given matrix itself; that is,

$$(\mathbf{a}')' = \mathbf{a}.$$

DEFINITION. A matrix which is its own transpose is called *symmetrical*. Such a matrix is necessarily square. The condition of symmetry is

$$\mathbf{a}' = \mathbf{a}$$

or

$$a_{ij} = a_{ji} \qquad \text{(all } i,j).$$

Typical symmetrical matrices are

$$\begin{pmatrix} 1 & 2 & 3 \\ 2 & 3 & 4 \\ 3 & 4 & 5 \end{pmatrix}, \quad \begin{pmatrix} a & h & g \\ h & b & f \\ g & f & c \end{pmatrix}, \quad \begin{pmatrix} 1 & 0 \\ 0 & 1 \end{pmatrix}.$$

DEFINITION. A matrix which is the *negative* of its own transpose is called *skew-symmetrical*. Such a matrix is necessarily square. The condition is

$$\mathbf{a}' = -\mathbf{a},$$

or

$$a'_{ij} = -a_{ji} \qquad \text{(all } i,j),$$

so that, in particular,

$$a_{ii} = 0$$

for all values of i.

Typical skew-symmetrical matrices are

$$\begin{pmatrix} 0 & 1 \\ -1 & 0 \end{pmatrix}, \quad \begin{pmatrix} 0 & -h & g \\ h & 0 & -f \\ -g & f & 0 \end{pmatrix}, \quad \begin{pmatrix} 0 & 1 & 2 & 3 \\ -1 & 0 & 4 & 5 \\ -2 & -4 & 0 & 6 \\ -3 & -5 & -6 & 0 \end{pmatrix}.$$

TWO IMPORTANT THEOREMS:

1. *The transpose of a linear combination is the corresponding linear combination of the individual transposes; that is, if*

$$\mathbf{u} \equiv p\mathbf{a} + q\mathbf{b} + r\mathbf{c} + \ldots,$$

then

$$\mathbf{u}' \equiv p\mathbf{a}' + q\mathbf{b}' + r\mathbf{c}' + \ldots.$$

For

$$u_{ij} = pa_{ij}+qb_{ij}+rc_{ij}+\ldots,$$

and
$$u'_{ji} = u_{ij}; \quad a'_{ji} = a_{ij}, \quad b'_{ji} = b_{ij}, \ldots,$$

so that

$$u'_{ji} = pa'_{ji}+qb'_{ji}+rc'_{ji}+\ldots$$

for *all pairs of values of* i, j. Hence

$$\mathbf{u}' = p\mathbf{a}'+q\mathbf{b}'+r\mathbf{c}'+\ldots.$$

2. *Any square matrix can be expressed as the sum of two matrices, of which one is symmetrical and one skew-symmetrical.*

Let **a** be a given square matrix and **a**′ its transpose. Then

$$\mathbf{a} = \tfrac{1}{2}(\mathbf{a}+\mathbf{a}')+\tfrac{1}{2}(\mathbf{a}-\mathbf{a}').$$

Write
$$\mathbf{b} = \tfrac{1}{2}(\mathbf{a}+\mathbf{a}'), \quad \mathbf{c} = \tfrac{1}{2}(\mathbf{a}-\mathbf{a}').$$

Thus, by the preceding theorem,

$$\mathbf{b}' = \tfrac{1}{2}(\mathbf{a}'+\mathbf{a}) = \mathbf{b},$$
$$\mathbf{c}' = \tfrac{1}{2}(\mathbf{a}'-\mathbf{a}) = -\mathbf{c},$$

as required.

7. The product of two matrices. Let

$$\mathbf{a} \equiv (a_{ij}), \quad \mathbf{b} \equiv (b_{ij})$$

be two given matrices. Their product follows the rule derived earlier (p. 138) for mappings.

DEFINITION. The *product* **ab** of two matrices **a**, **b** is defined to be the matrix **c** for which

$$c_{ij} = \sum_{\lambda} a_{i\lambda} b_{\lambda j},$$

summed for all relevant values of λ.

Since the suffix λ refers to columns of **a** and rows of **b**, *the product* **ab** *cannot exist unless the number of columns of* **a** *is equal to the number of rows of* **b**. In that case, the matrices are said to be *conformable* for the product **ab**.

The summations

$$\sum_{\lambda} a_{i\lambda} b_{\lambda}$$

are taken for all values of i corresponding to rows of **a** and for all values of j corresponding to columns of **b**. If, then, **a** is of type $m \times n$ and **b** is of type $n \times p$, the product **ab** is of type $m \times p$; that is, *the types conform to the 'domino' rule*

$$(m \times n)(n \times p) \text{ gives } (m \times p).$$

***Illustration* 2.** If

$$\mathbf{a} \equiv \begin{pmatrix} 1 & 2 & 3 \\ 4 & 5 & 6 \\ 7 & 8 & 9 \end{pmatrix}, \quad \mathbf{b} \equiv \begin{pmatrix} \alpha & \beta & \gamma & \delta \\ p & q & r & s \\ \lambda & \mu & \nu & \rho \end{pmatrix},$$

the element c_{21} is formed by taking from **a** the second row

$$\begin{pmatrix} \cdot & \cdot & \cdot \\ 4 & 5 & 6 \\ \cdot & \cdot & \cdot \end{pmatrix}$$

and from **b** the first column

$$\begin{pmatrix} \alpha & \cdot & \cdot & \cdot \\ p & \cdot & \cdot & \cdot \\ \lambda & \cdot & \cdot & \cdot \end{pmatrix}$$

and multiplying corresponding elements, so that

$$c_{21} = 4\alpha + 5p + 6\lambda.$$

Similarly c_{14} arises from the elements

$$\begin{pmatrix} 1 & 2 & 3 \\ \cdot & \cdot & \cdot \\ \cdot & \cdot & \cdot \end{pmatrix} \begin{pmatrix} \cdot & \cdot & \cdot & \delta \\ \cdot & \cdot & \cdot & s \\ \cdot & \cdot & \cdot & \rho \end{pmatrix},$$

so that

$$c_{14} = \delta + 2s + 3\rho.$$

Informally, the element c_{ij} is obtained by allowing the left eye to pass from left to right along the ith row of **a** while the right eye drops at the same rate down the jth column of **b**; corresponding terms $a_{i\lambda}$, $b_{\lambda j}$ are multiplied as the eyes traverse them.

***Illustration* 3.** *The product* $\mathbf{x}'(\mathbf{ax})$. Let **a** be the 3×3 matrix

$$\mathbf{a} \equiv \begin{pmatrix} a & h & g \\ h & b & f \\ g & f & c \end{pmatrix},$$

and **x** the column vector

$$\mathbf{x} \equiv \{x, y, z\}.$$

Then **ax** is the *column vector*

$$\mathbf{ax} = \begin{pmatrix} a & h & g \\ h & b & f \\ g & f & c \end{pmatrix} \begin{pmatrix} x \\ y \\ z \end{pmatrix}$$

$$= \begin{pmatrix} ax + hy + gz \\ hx + by + fz \\ gx + fy + cz \end{pmatrix},$$

and $\mathbf{x}'(\mathbf{ax})$, where $\mathbf{x}'$ is the transpose of $\mathbf{x}$, is the *scalar*

$$\mathbf{x}'(\mathbf{ax}) = (x, y, z)\begin{pmatrix} ax+hy+gz \\ hx+by+fz \\ gx+fy+cz \end{pmatrix}$$

$$= x(ax+hy+gz)+y(hx+by+fz)+z(gx+fy+cz)$$
$$= ax^2+by^2+cz^2+2fyz+2gzx+2hxy,$$

a *quadratic form* familiar in coordinate geometry.

Once the associative law $(\mathbf{x}'\mathbf{a})\mathbf{x} = \mathbf{x}'(\mathbf{ax})$ has been established, this product may be written without ambiguity in the form $\mathbf{x}'\mathbf{ax}$. (See p. 153.)

Example

1. Form the product **ab** for each of the following pairs of matrices:

(i) $$\mathbf{a} \equiv \begin{pmatrix} 1 & 2 \\ 3 & 4 \end{pmatrix}, \quad \mathbf{b} \equiv \begin{pmatrix} p & q & r \\ u & v & w \end{pmatrix};$$

(ii) $$\mathbf{a} \equiv \begin{pmatrix} 1 & 0 & -1 \\ 2 & 4 & 6 \end{pmatrix}, \quad \mathbf{b} \equiv \begin{pmatrix} 5 & -1 & 3 & 4 \\ 0 & 2 & 1 & -2 \\ 3 & 1 & 0 & 1 \end{pmatrix};$$

(iii) $$\mathbf{a} \equiv \begin{pmatrix} 7 & -9 \\ 2 & 4 \end{pmatrix}, \quad \mathbf{b} \equiv \begin{pmatrix} 1 \\ 2 \end{pmatrix};$$

(iv) $$\mathbf{a} \equiv \begin{pmatrix} 1 \\ 2 \\ 3 \\ 4 \end{pmatrix}, \quad \mathbf{b} \equiv (\alpha \quad \beta \quad \gamma \quad \delta);$$

(v) $$\mathbf{a} \equiv (1 \quad 2 \quad 3 \quad 4), \quad \mathbf{b} \equiv \begin{pmatrix} \alpha \\ \beta \\ \gamma \\ \delta \end{pmatrix};$$

(vi) $$\mathbf{a} \equiv \begin{pmatrix} 1 & 3 & -2 & 4 \\ 3 & -2 & 1 & 4 \\ 4 & 3 & 1 & -2 \\ -2 & 4 & 3 & 1 \end{pmatrix}, \quad \mathbf{b} \equiv \begin{pmatrix} 3 & 0 \\ 4 & 2 \\ 6 & 8 \\ 1 & 3 \end{pmatrix}.$$

In general, *the product* **ba** *is quite different from* **ab**. Indeed, the existence of **ab** does not even imply that of **ba**: if **a**,**b** are of types $m \times n$ and

$n \times p$, then **ba** cannot exist unless also $m = p$. The product **ba** is given by the definition

$$\mathbf{ba} = \mathbf{d},$$

where

$$d_{ij} = \sum_{\mu} b_{i\mu} a_{\mu j},$$

summed for all relevant values of μ.

***Illustration* 4.** Let

$$\mathbf{a} \equiv \begin{pmatrix} 1 & 2 & 3 \\ 4 & 5 & 6 \end{pmatrix}, \quad \mathbf{b} \equiv \begin{pmatrix} p & x \\ q & y \\ r & z \end{pmatrix},$$

then

$$\mathbf{ab} = \begin{pmatrix} p+2q+3r, & x+2y+3z \\ 4p+5q+6r, & 4x+5y+6z \end{pmatrix},$$

$$\mathbf{ba} = \begin{pmatrix} p+4x, & 2p+5x, & 3p+6x \\ q+4y, & 2q+5y, & 3q+6y \\ r+4z, & 2r+5z, & 3r+6z \end{pmatrix}.$$

8. The rule (ab)′ = b′a′. Let **a**, **b** be two given matrices of types $m \times n$ and $n \times p$, and let **c** be their product, so that

$$\mathbf{c} = \mathbf{ab}.$$

To prove that, *if dashes denote transpositions, then*

$$\mathbf{c}' = \mathbf{b}'\mathbf{a}':$$

Note first that **b′**, **a′** are of types $p \times n$ and $n \times m$, so that their product exists and is of type $p \times m$.

Since

$$c_{pq} = \sum_{\lambda} a_{p\lambda} b_{\lambda q}$$

for all pairs of values of p, q, it follows that c'_{ij}, defined by the relation $c'_{ij} = c_{ji}$, is obtained from the formula

$$c'_{ij} = \sum_{\mu} a_{j\mu} b_{\mu i}.$$

But

$$b_{\mu i} = b'_{i\mu}, \quad a_{j\mu} = a'_{\mu j},$$

so that

$$c'_{ij} = \sum_{\mu} a'_{\mu j} b'_{i\mu}$$

$$= \sum_{\mu} b'_{i\mu} a'_{\mu j}.$$

Hence

$$\mathbf{c}' = \mathbf{b}'\mathbf{a}',$$

by definition of the product.

Illustration 5. Let

$$\mathbf{a} \equiv \begin{pmatrix} 1 & 2 \\ 3 & 4 \end{pmatrix}, \quad \mathbf{b} \equiv \begin{pmatrix} u \\ v \end{pmatrix}.$$

Then

$$\mathbf{ab} = \begin{pmatrix} u+2v \\ 3u+4v \end{pmatrix},$$

so that

$$(\mathbf{ab})' = (u+2v, \quad 3u+4v).$$

Also

$$\mathbf{b}' \equiv (u, v), \quad a' \equiv \begin{pmatrix} 1 & 3 \\ 2 & 4 \end{pmatrix},$$

so that

$$\begin{aligned} \mathbf{b}'\mathbf{a}' &= (u, v)\begin{pmatrix} 1 & 3 \\ 2 & 4 \end{pmatrix} \\ &= (u+2v, \quad 3u+4v) \\ &= (\mathbf{ab})'. \end{aligned}$$

9. Multiplication by the unit matrix. To prove that, *if* $\mathbf{a}$ *is a matrix of type* $m \times n$ *and* $\mathbf{I}_m$, $\mathbf{I}_n$ *unit matrices of orders* m, n, *then*

$$\mathbf{I}_m \mathbf{a} = \mathbf{a}, \quad \mathbf{a}\mathbf{I}_n = \mathbf{a}.$$

This is the property that justifies the word *unit*.

For convenience, write $\mathbf{I}$ for $\mathbf{I}_m$ and let $\mathbf{I}_m \mathbf{a} = \mathbf{c}$. Then

$$c_{ij} = \sum_{\lambda} I_{i\lambda} a_{\lambda j}.$$

But $I_{i\lambda} = 1$ when λ is i and $I_{i\lambda} = 0$ otherwise. Hence

$$c_{ij} = I_{ii} a_{ij} = a_{ij},$$

so that

$$\mathbf{c} = \mathbf{a}.$$

The proof of the second equality is similar.

10. The associative law. Once the product $\mathbf{ab}$ has been defined, the extension to products $\mathbf{abc}$, $\mathbf{abcd}$, ... is immediate, subject to conformability at each stage. For example, $\mathbf{abc}$ is formed by the sequence

$$\mathbf{ab}$$

$$(\mathbf{ab})\mathbf{c}$$

and $\mathbf{abcd}$ then follows as the product

$$(\mathbf{abc})\mathbf{d}.$$

One point, however, must be confirmed before the notation $\mathbf{abc}$ can be used without ambiguity.

THEOREM. *The products, when conformable, obey the associative law*

$$(\mathbf{ab})\mathbf{c} = \mathbf{a}(\mathbf{bc}):$$

Writing $[(\mathbf{ab})\mathbf{c}]_{ij}$ for the element in the ith row and jth column of the matrix on the left, with similar notation as it arises elsewhere,

$$\begin{aligned}[(\mathbf{ab})\mathbf{c}]_{ij} &= \sum_{\lambda} (\mathbf{ab})_{i\lambda} c_{\lambda j} \\ &= \sum_{\lambda} \left\{ \sum_{\mu} a_{i\mu} b_{\mu\lambda} \right\} c_{\lambda j}.\end{aligned}$$

The right-hand side of the summation is the sum of terms of the form

$$a_{i\mu} b_{\mu\lambda} c_{\lambda j}$$

where μ ranges over the columns of **a** (or rows of **b**) while λ ranges *independently* over the columns of **b** (or rows of **c**).

Similarly

$$\begin{aligned}[\mathbf{a}(\mathbf{bc})]_{ij} &= \sum_{\mu} a_{i\mu} (\mathbf{bc})_{\mu j} \\ &= \sum_{\mu} a_{i\mu} \left\{ \sum_{\lambda} b_{\mu\lambda} c_{\lambda j} \right\},\end{aligned}$$

where, again, the right-hand side is the sum of terms of the form

$$a_{i\mu} b_{\mu\lambda} c_{\lambda j}$$

over the same ranges as before for λ and μ. This establishes the result.

COROLLARY. The rule for *the transpose of a product* **abcd**.

Since

$$\begin{aligned}(\mathbf{abcd})' &= \{(\mathbf{abc})\,\mathbf{d}\}' \\ &= \mathbf{d}'(\mathbf{abc})' \\ &= \mathbf{d}'\{(\mathbf{ab})\,\mathbf{c}\}' \\ &= \mathbf{d}'\mathbf{c}'(\mathbf{ab})' \\ &= \mathbf{d}'\mathbf{c}'\mathbf{b}'\mathbf{a}',\end{aligned}$$

it follows that *the transpose of a product* **abcd**... *is the product in reverse order of the transposes*... **d′c′b′a′**.

Examples

1. Form the products **abc, bcd, abcd** for the matrices:

(i)
$$\mathbf{a} \equiv \begin{pmatrix} 1 & 2 \\ 3 & 4 \end{pmatrix}, \quad \mathbf{b} \equiv \begin{pmatrix} 9 & 8 & 7 \\ 6 & 5 & 4 \end{pmatrix}, \quad \mathbf{c} \equiv \begin{pmatrix} 7 & -2 \\ 5 & -3 \\ 4 & -4 \end{pmatrix}, \quad \mathbf{d} \equiv \begin{pmatrix} 3 \\ 1 \end{pmatrix};$$

(ii)
$$\mathbf{a} \equiv \begin{pmatrix} 3 \\ -5 \end{pmatrix}, \quad \mathbf{b} \equiv (1, 2, 5), \quad \mathbf{c} \equiv \begin{pmatrix} 0 & 5 \\ 3 & 6 \\ 2 & 1 \end{pmatrix}, \quad \mathbf{d} \equiv \begin{pmatrix} 2 & 3 & 1 \\ 5 & 0 & 4 \end{pmatrix}.$$

2. Form the product **abc** for the matrices:

(i)
$$\mathbf{a} \equiv (x, y), \quad \mathbf{b} \equiv \begin{pmatrix} a & h \\ h & b \end{pmatrix}, \quad \mathbf{c} \equiv \begin{pmatrix} x \\ y \end{pmatrix};$$

(ii)
$$\mathbf{a} \equiv (\alpha, \beta, \gamma), \quad \mathbf{b} \equiv \begin{pmatrix} a & h & g \\ h & b & f \\ g & f & c \end{pmatrix}, \quad \mathbf{c} \equiv \begin{pmatrix} x \\ y \\ z \end{pmatrix};$$

(iii)
$$\mathbf{a} \equiv (x, y, z), \quad \mathbf{b} \equiv \begin{pmatrix} 0 & h & -g \\ -h & 0 & f \\ g & -f & 0 \end{pmatrix}, \quad \mathbf{c} \equiv \begin{pmatrix} x \\ y \\ z \end{pmatrix}.$$

11. The distributive laws for matrices.

THEOREM. *The multiplication of matrices obeys* (subject to the conformity of the operations) *the distributive laws*

$$\mathbf{a}(\mathbf{b}+\mathbf{c}) = \mathbf{ab}+\mathbf{ac},$$
$$(\mathbf{b}+\mathbf{c})\,\mathbf{a} = \mathbf{ba}+\mathbf{ca},$$

The proof is immediate, but should be written out carefully.

12. The 'Pythagoras' formula.

THEOREM. *To establish the formula*

$$\mathbf{x}'\mathbf{x} = x_1^2+x_2^2+\ldots+x_n^2,$$

where $\mathbf{x}$ *is the column vector* $\{x_1, x_2, \ldots, x_n\}$.

By definition,

$$\mathbf{x}'\mathbf{x} = (x_1, x_2, \ldots, x_n)\begin{pmatrix} x_1 \\ x_2 \\ \vdots \\ x_n \end{pmatrix},$$

which, by direct application of the rule for matrix multiplication, is the scalar

$$x_1^2+x_2^2+\ldots+x_n^2.$$

13. The powers of a square matrix. If $\mathbf{a}$ is a square matrix of order n, then the product

$$\mathbf{aa}$$

exists, and is written

$$\mathbf{a}^2.$$

The *powers* $\mathbf{a}^3, \mathbf{a}^4, \mathbf{a}^5, \ldots$ are then defined by the inductive rule

$$\mathbf{a}^{n+1} = \mathbf{a}^n\mathbf{a}.$$

Revision Examples

1. Prove from the definition that, if m, n are positive integers, then

$$\mathbf{a}^m\mathbf{a}^n = \mathbf{a}^n\mathbf{a}^m = \mathbf{a}^{n+m}.$$

2. Prove that, if $\mathbf{x} = \{x_1, x_2, x_3\}$, then

$$\mathbf{xx}' \equiv \begin{pmatrix} x_1^2 & x_1x_2 & x_1x_3 \\ x_2x_1 & x_2^2 & x_2x_3 \\ x_3x_1 & x_3x_2 & x_3^2 \end{pmatrix}.$$

3. Prove that, if

$$\mathbf{a} \equiv \begin{pmatrix} 2 & -1 \\ 1 & 0 \end{pmatrix},$$

then

$$\mathbf{a}^n \equiv \begin{pmatrix} n+1 & -n \\ n & -n+1 \end{pmatrix}.$$

4. Prove that, if the matrices $\mathbf{a}, \mathbf{b}$ are symmetric, so that $\mathbf{a}' = \mathbf{a}$, $\mathbf{b}' = \mathbf{b}$, it is *not* necessary that $\mathbf{ab}$ is also symmetric. Give examples.

5. Prove that, if

$$\mathbf{x} \equiv \{x_1, x_2, x_3\}, \quad \mathbf{y} \equiv \{y_1, y_2, y_3\},$$

then

$$\mathbf{xy}' + \mathbf{yx}' \equiv \begin{pmatrix} 0 & x_1 & y_1 \\ 0 & x_2 & y_2 \\ 0 & x_3 & y_3 \end{pmatrix} \times \begin{pmatrix} 0 & 0 & 0 \\ y_1 & y_2 & y_3 \\ x_1 & x_2 & x_3 \end{pmatrix}.$$

6. Prove that, if $\mathbf{a}$ is an $m \times n$ matrix, then each of the products $\mathbf{aa}'$ and $\mathbf{a}'\mathbf{a}$ exist and are square matrices.

Prove that the matrix $(\mathbf{aa}')^k$ is symmetric for all positive integers k.

7. Prove that, if

$$\mathbf{x} \equiv \{x_1, x_2, x_3, x_4\}, \quad \mathbf{y} \equiv \{y_1, y_2, y_3, y_4\},$$

then $$\mathbf{xy}' + \mathbf{yx}'$$
is a symmetric matrix and
$$\mathbf{xy}' - \mathbf{yx}'$$
is a skew-symmetric matrix.

8. Prove that, if **a** is a square matrix of order n, then
$$\mathbf{a}^2 - \mathbf{I}^2 = (\mathbf{a}+\mathbf{I})(\mathbf{a}-\mathbf{I}) = (\mathbf{a}-\mathbf{I})(\mathbf{a}+\mathbf{I}),$$
$$\mathbf{a}^3 + \mathbf{I}^3 = (\mathbf{a}+\mathbf{I})(\mathbf{a}^2 - \mathbf{a} + \mathbf{I}).$$

9. Prove that the square of the distance between the points $X(x_1, x_2, x_3)$ $Y(y_1, y_2, y_3)$ is
$$(\mathbf{x}-\mathbf{y})'(\mathbf{x}-\mathbf{y}),$$
where **x**, **y** are the column matrices $\{x_1, x_2, x_3\}, \{y_1, y_2, y_3\}$.

10. Prove that the equations
$$x^2 + y^2 = a^2$$
and $$x_1 x + y_1 y = a^2$$
for a circle and the tangent at a point (x_1, y_1) can be expressed in matrix form (with suitable adjustment of notation)
$$\mathbf{x}'\mathbf{x} = a^2,$$
$$\mathbf{x}_1'\mathbf{x} = a^2.$$

11. If **u**, **v** are column vectors of the same length with $\mathbf{u} \neq k\mathbf{v}$ for any scalar k, and if
$$\mathbf{a} = \mathbf{uv}', \quad \mathbf{b} = \mathbf{uv}' - \mathbf{vu}',$$
prove that
$$\mathbf{a}^2 = \alpha\mathbf{a},$$
$$\mathbf{b}^3 = (\alpha^2 - \beta\gamma)\,\mathbf{b},$$
where $\alpha = \mathbf{u}'\mathbf{v}$, $\beta = \mathbf{u}'\mathbf{u}$, $\gamma = \mathbf{v}'\mathbf{v}$.

Prove that a vector **x**, such that $\mathbf{bx} = 0$, satisfies the relations $\mathbf{u}'\mathbf{x} = 0$, $\mathbf{v}'\mathbf{x} = 0$.

12. The matrix **x** satisfies the relation
$$\mathbf{ax} + \mathbf{x}'\mathbf{a} = \mathbf{0}.$$
Prove that **a**, **x** must both be square matrices.

Show that $$\mathbf{ax}^2 = \mathbf{x}'^2\mathbf{a},$$
$$\mathbf{ax}^3 = -\mathbf{x}'^3\mathbf{a}.$$

13. Three matrices **a**, **b**, **c** are conformable for the product **abc**. Establish the relation $(\mathbf{abc})' = \mathbf{c}'\mathbf{b}'\mathbf{a}'$.

Prove that, if **b** and **abc** are both square, then each of the products **aba′** and **c′b′a′** exists.

If, in addition, **b** is skew-symmetric, prove that **aba′** is skew-symmetric and **abc** − **c′ba′** is symmetric.

14. If **a** is the matrix

$$\mathbf{a} \equiv \begin{pmatrix} 1 & 1 & 1 & 1 \\ a_1 & a_2 & a_3 & a_4 \\ a_1^2 & a_2^2 & a_3^2 & a_4^2 \\ a_1^3 & a_2^3 & a_3^3 & a_4^3 \end{pmatrix}$$

and if **b** is the matrix such that

$$b_{ij} = (u+a_j)^{4-i}(v+a_j)^{i-1},$$

find a matrix **x** such that

$$\mathbf{xa} = \mathbf{b}.$$

13

MATRICES OF ORDERS 2 AND 3

The definitions and basic properties of Chapter 12 were given for general matrices, since particular cases are little, if any, easier and since the pattern tends to be obscured by too great simplicity. For the more detailed examination which follows, however, attention is confined to matrices of small orders, beginning with order 3 as the extreme simplicity of 2 makes it somewhat abnormal. Most of the work of this chapter can be generalized readily.

1. The determinant of a square matrix. Let

$$\mathbf{a} \equiv \begin{pmatrix} a_{11} & a_{12} & a_{13} \\ a_{21} & a_{22} & a_{23} \\ a_{31} & a_{32} & a_{33} \end{pmatrix}$$

be a matrix of three rows and columns. The determinant

$$\begin{vmatrix} a_{11} & a_{12} & a_{13} \\ a_{21} & a_{22} & a_{23} \\ a_{31} & a_{32} & a_{33} \end{vmatrix}$$

is called the *determinant of the matrix* and is denoted by the alternative notations

$$|\mathbf{a}|, \quad \det \mathbf{a}.$$

For the properties of determinants, see Part 1, Chapter 20.

THEOREM. If k is a scalar and $\mathbf{a}$ a square matrix of order 3 (more generally, n), then

$$\det(k\mathbf{a}) = k^3 \det \mathbf{a}$$

(more generally, $k^n \det \mathbf{a}$). The factor k appears 3 (or n) times since k is a factor of each of the 3 (or n) rows of $\det(k\mathbf{a})$.

Text Example

1. Prove that $\det(k\mathbf{I}_3) = k^3$.

2. The adjoint (adjugate) of a square matrix. In Part 1, p. 253, the *cofactors* of the elements a_{ij} of a determinant $|a_{ij}|$ were defined:

DEFINITION. The *cofactor* of the element a_{ij} in the determinant of a *square* matrix $\mathbf{a}$ is the determinant of the matrix formed from $\mathbf{a}$ by deleting all the elements in the ith row and the jth column and then taking the appropriate sign (according to the rule given in Part 1). It is convenient to use the temporary notation A^*_{ij} for the cofactor, since the symbol A_{ij} used earlier is now reserved for a slightly different purpose: in fact, we write A_{ij} for what we are now calling A^*_{ji}, so that A_{ij} *is the cofactor of* a_{ij}, *with transposition of suffixes.*

The elements A_{ij} can be written to form a new matrix

$$\begin{pmatrix} A_{11} & A_{12} & A_{13} \\ A_{21} & A_{22} & A_{23} \\ A_{31} & A_{32} & A_{33} \end{pmatrix},$$

called the *adjoint*, or *adjugate*, of $\mathbf{a}$ and denoted by the symbol

$$\mathbf{A} \quad \text{or} \quad \operatorname{adj} \mathbf{a}.$$

3. The products aA, Aa. Consider first the product

$$\mathbf{aA} \equiv \begin{pmatrix} a_{11} & a_{12} & a_{13} \\ a_{21} & a_{22} & a_{23} \\ a_{31} & a_{32} & a_{33} \end{pmatrix} \begin{pmatrix} A_{11} & A_{12} & A_{13} \\ A_{21} & A_{22} & A_{23} \\ A_{31} & A_{32} & A_{33} \end{pmatrix}.$$

The element in the ith row and jth column is

$$a_{i1}A_{1j} + a_{i2}A_{2j} + a_{i3}A_{3j},$$

or (§2)

$$a_{i1}A^*_{j1} + a_{i2}A^*_{j2} + a_{i3}A^*_{j3}.$$

But, from Part 1, p. 254, with a mere change of notation, the value of this expression is $\det \mathbf{a}$ when $i = j$ and zero when $i \neq j$. If we therefore write

$$\det \mathbf{a} = \Delta,$$

it follows that

$$\mathbf{aA} = \begin{pmatrix} \Delta & 0 & 0 \\ 0 & \Delta & 0 \\ 0 & 0 & \Delta \end{pmatrix},$$

so that
$$\mathbf{aA} = \Delta\begin{pmatrix}1 & 0 & 0\\ 0 & 1 & 0\\ 0 & 0 & 1\end{pmatrix}$$
$$= \Delta\mathbf{I},$$

where $\mathbf{I}$ is the unit matrix of order 3.

Similarly
$$\mathbf{Aa} = \Delta\mathbf{I}.$$

Hence
$$\mathbf{aA} = \mathbf{Aa} = \Delta\mathbf{I},$$

where $\Delta = \det \mathbf{a}$. In other notation, this is
$$\mathbf{a}(\operatorname{adj}\mathbf{a}) = (\operatorname{adj}\mathbf{a})\mathbf{a} = \mathbf{I}\det\mathbf{a}.$$

Examples

1. Prove that $\det(\mathbf{aA}) = \Delta^3$.
2. Evaluate $\mathbf{A}$ when $\mathbf{a}$ is $\mathbf{I}_3$.
3. Evaluate $\mathbf{A}$ when

(i) $$\mathbf{a} \equiv \begin{pmatrix}0 & 0 & 1\\ 0 & 1 & 0\\ 1 & 0 & 0\end{pmatrix},$$ (ii) $$\mathbf{a} \equiv \begin{pmatrix}0 & 1 & 0\\ 1 & 0 & 0\\ 0 & 0 & 1\end{pmatrix},$$

(iii) $$\mathbf{a} \equiv \begin{pmatrix}1 & 1 & 1\\ 2 & 2 & 2\\ 3 & 3 & 3\end{pmatrix},$$ (iv) $$\mathbf{a} \equiv \begin{pmatrix}1 & 2 & 3\\ 2 & 3 & 4\\ 3 & 4 & 5\end{pmatrix}.$$

Verify in each case the relation
$$\mathbf{a}\operatorname{adj}\mathbf{a} = \mathbf{I}\det\mathbf{a}.$$

4. The determinant of a product ab. The product of two determinants was obtained in Part 1, p. 268, as a single determinant in the form
$$\begin{vmatrix}a_1 & b_1 & c_1\\ a_2 & b_2 & c_2\\ a_3 & b_3 & c_3\end{vmatrix} \times \begin{vmatrix}p_1 & p_2 & p_3\\ q_1 & q_2 & q_3\\ r_1 & r_2 & r_3\end{vmatrix} = \begin{vmatrix}(11) & (12) & (13)\\ (21) & (22) & (23)\\ (31) & (32) & (33)\end{vmatrix},$$

where
$$(ij) \equiv a_i p_j + b_i q_j + c_i r_j.$$

In double suffix notation, this is equivalent to
$$\begin{vmatrix}a_{11} & a_{12} & a_{13}\\ a_{21} & a_{22} & a_{23}\\ a_{31} & a_{32} & a_{33}\end{vmatrix} \times \begin{vmatrix}b_{11} & b_{12} & b_{13}\\ b_{21} & b_{22} & b_{23}\\ b_{31} & b_{32} & b_{33}\end{vmatrix} = \begin{vmatrix}(11) & (12) & (13)\\ (21) & (22) & (23)\\ (31) & (32) & (33)\end{vmatrix},$$

where, now,

$$(ij) = a_{i1}b_{1j}+a_{i2}b_{2j}+a_{i3}b_{3j},$$

so that (ij) is the element in the ith row and jth column of the product matrix **ab**. Hence

$$(\det \mathbf{a}) \times (\det \mathbf{b}) = \det(\mathbf{ab}),$$

and so *the determinant of the product* **ab** *of two matrices* **a**, **b** *is equal to the product of the individual determinants.*

COROLLARY.

$$\det(\mathbf{abc}) = \det \mathbf{a} \det \mathbf{b} \det \mathbf{c},$$
$$\det(\mathbf{abcd}) = \det \mathbf{a} \det \mathbf{b} \det \mathbf{c} \det \mathbf{d}.$$

Examples

1. Evaluate the determinants of the matrices

$$\begin{pmatrix} 1 & 2 & 3 \\ 4 & 3 & 2 \\ 1 & -1 & 1 \end{pmatrix}, \quad \begin{pmatrix} 2 & 3 & -1 \\ -1 & 2 & 3 \\ 2 & -4 & 1 \end{pmatrix}, \quad \begin{pmatrix} -1 & 0 & 1 \\ 2 & 1 & -3 \\ 0 & 2 & -1 \end{pmatrix}.$$

Write down the matrices which are the products of these matrices taken in pairs (for both orders of products in each case) and verify in all cases the theorem $\det(\mathbf{ab}) = \det \mathbf{a} \det \mathbf{b}$.

2. Find the adjoint of each of the three matrices given in Question 1 and, in each case, verify the relations

$$\text{(i) } \mathbf{aA} = \mathbf{Aa} = \Delta\mathbf{I}, \qquad \text{(ii) } \det \mathbf{A} = \{\det \mathbf{a}\}^2.$$

3. Given that

$$\mathbf{a} \equiv \begin{pmatrix} 1 & 2 \\ -2 & 1 \\ 3 & -1 \end{pmatrix},$$

find the matrices $\mathbf{aa}'$ and $\mathbf{a}'\mathbf{a}$, where $\mathbf{a}'$ is the transpose of **a**. Evaluate $\det(\mathbf{aa}')$ and $\det(\mathbf{a}'\mathbf{a})$.

5. The inverse $\mathbf{a}^{-1}$ of a matrix a. Let **a** be a given *square* matrix of order 3. It is natural to use the name *inverse matrix* and the notation $\mathbf{a}^{-1}$ for a matrix **b**, if such a matrix exists, having the property

$$\mathbf{ab} = \mathbf{ba} = \mathbf{I}.$$

Note first that *a matrix* **a** *cannot have an inverse if* $\det \mathbf{a} = 0$. For

$$\mathbf{ab} = \mathbf{I}$$
$$\Rightarrow \det \mathbf{a} . \det \mathbf{b} = \det \mathbf{I} = 1,$$

which is not possible if $\det \mathbf{a} = 0$.

DEFINITION. A matrix **a** whose determinant is zero is called *singular*. *Such a matrix cannot have an inverse.*

Suppose now that **a** is non-singular, so that

$$\Delta \equiv \det \mathbf{a} \neq 0.$$

It has been proved (§3) that

$$\mathbf{aA} = \mathbf{Aa} = \Delta\mathbf{I},$$

and, since $\Delta \neq 0$, this relation may be divided by it to give

$$\mathbf{a}(\Delta^{-1}\mathbf{A}) = (\Delta^{-1}\mathbf{A})\mathbf{a} = \mathbf{I}.$$

Hence *a non-singular matrix* **a** *has an inverse* $\mathbf{a}^{-1}$ *given by the relation*

$$\mathbf{a}^{-1} = \Delta^{-1}\mathbf{A},$$

where $\Delta \equiv \det \mathbf{a}$ *and* $\mathbf{A} \equiv \operatorname{adj} \mathbf{a}$.

Finally *there is only one such inverse*; for the relations

$$\mathbf{ab} = \mathbf{ba} = \mathbf{I},$$
$$\mathbf{ac} = \mathbf{ca} = \mathbf{I}$$

give

$$\begin{aligned} & \mathbf{ab} = \mathbf{ac} \\ \Rightarrow\ & \mathbf{Aab} = \mathbf{Aac} \\ \Rightarrow\ & (\Delta\mathbf{I})\mathbf{b} = (\Delta\mathbf{I})\mathbf{c} \\ \Rightarrow\ & \Delta\mathbf{b} = \Delta\mathbf{c} \\ \Rightarrow\ & \mathbf{b} = \mathbf{c} \qquad (\Delta \neq 0). \end{aligned}$$

(For a remark on the numerical calculation of inverses, see pp. 164–6.)

Examples

For each of the following matrices, determine whether an inverse exists and find the inverses that do.

1. $\begin{pmatrix} 1 & 3 & 2 \\ 2 & -1 & 1 \\ 3 & -2 & -1 \end{pmatrix}.$

2. $\begin{pmatrix} 2 & 1 & -1 \\ -1 & 2 & 1 \\ 1 & -1 & 2 \end{pmatrix}.$

3. $\begin{pmatrix} 1 & 0 & 0 \\ 0 & \cos\theta & \sin\theta \\ 0 & -\sin\theta & \cos\theta \end{pmatrix}.$

4. $\begin{pmatrix} 3 & 2 & 4 \\ 1 & 2 & 1 \\ 1 & -2 & 2 \end{pmatrix}.$

5. $\begin{pmatrix} a & h & g \\ h & b & f \\ g & f & c \end{pmatrix}.$

6. $\begin{pmatrix} 0 & h & -g \\ -h & 0 & f \\ g & -f & 0 \end{pmatrix}.$

7. $\begin{pmatrix} 5 & 0 & 0 \\ 0 & 5 & 0 \\ 0 & 0 & 5 \end{pmatrix}$.

8. $\begin{pmatrix} 5 & 0 & 0 \\ 0 & -4 & 0 \\ 0 & 0 & 3 \end{pmatrix}$.

9. Find the values of λ for which the following matrices are singular:

(i) $\begin{pmatrix} 1+\lambda & 2 & 3 \\ -1 & 1 & 0 \\ 2 & -3 & 1 \end{pmatrix}$, (ii) $\begin{pmatrix} 1 & 2 & 3 \\ 2 & \lambda+2 & 6 \\ 1 & 2 & \lambda \end{pmatrix}$,

(iii) $\begin{pmatrix} 1-\lambda & 2 & 3 \\ 1 & 1-\lambda & 1 \\ 2 & 2 & 2-\lambda \end{pmatrix}$.

10. Prove that if **a** is singular, so also is the product **ab**.

6. The inverse of a product.

THEOREM. *The inverse* $(\mathbf{ab})^{-1}$ *of the product* **ab** *of two non-singular square matrices is the product* $\mathbf{b}^{-1}\mathbf{a}^{-1}$ *of their inverses in reverse order.*

By definitions of $\mathbf{a}^{-1}$, $\mathbf{b}^{-1}$ and the associative law of multiplication, it follows that

$$\begin{aligned}&(\mathbf{b}^{-1}\mathbf{a}^{-1})(\mathbf{ab})\\ &= \mathbf{b}^{-1}(\mathbf{a}^{-1}\mathbf{a})\mathbf{b} = \mathbf{b}^{-1}\mathbf{I}\mathbf{b}\\ &= \mathbf{b}^{-1}\mathbf{b}\\ &= \mathbf{I},\end{aligned}$$

so that $\mathbf{b}^{-1}\mathbf{a}^{-1}$ is the inverse of **ab**.

COROLLARY.

$$(\mathbf{abc})^{-1} = \mathbf{c}^{-1}\mathbf{b}^{-1}\mathbf{a}^{-1},$$
$$(\mathbf{abcd})^{-1} = \mathbf{d}^{-1}\mathbf{c}^{-1}\mathbf{b}^{-1}\mathbf{a}^{-1}.$$

7. The inverse of the transpose a′ of a matrix a.

THEOREM. *The inverse* $(\mathbf{a}')^{-1}$ *of* $\mathbf{a}'$ *is equal to the transpose* $(\mathbf{a}^{-1})'$ *of the inverse of* **a**, *assumed non-singular.*

It has been proved (p. 151) that the transpose of the product **ab** is the product $\mathbf{b}'\mathbf{a}'$. In particular, if $\mathbf{b} = \mathbf{a}^{-1}$,

$$(\mathbf{aa}^{-1})' = (\mathbf{a}^{-1})'\mathbf{a}',$$

so that $$(\mathbf{a}^{-1})'\mathbf{a}' = \mathbf{I}.$$

Also $$\det \mathbf{a}' = \det \mathbf{a} \neq 0,$$

so that $(\mathbf{a}')^{-1}$ exists. Hence, immediately,

$$(\mathbf{a}')^{-1} = (\mathbf{a}^{-1})'.$$

8. The equations ax = b when a is non-singular.

Three linear equations in three variables x_1, x_2, x_3 can be written in the form

$$\begin{aligned} a_{11}x_1+a_{12}x_2+a_{13}x_3 &= b_1, \\ a_{21}x_1+a_{22}x_2+a_{23}x_3 &= b_2, \\ a_{31}x_1+a_{32}x_2+a_{33}x_3 &= b_3, \end{aligned}$$

where a_{ij}, b_k are constants. These three equations may be written more concisely in the matrix form

$$\mathbf{ax} = \mathbf{b},$$

where

$$\mathbf{a} \equiv \begin{pmatrix} a_{11} & a_{12} & a_{13} \\ a_{21} & a_{22} & a_{23} \\ a_{31} & a_{32} & a_{33} \end{pmatrix}, \quad \mathbf{x} \equiv \begin{pmatrix} x_1 \\ x_2 \\ x_3 \end{pmatrix}, \quad \mathbf{b} \equiv \begin{pmatrix} b_1 \\ b_2 \\ b_3 \end{pmatrix}.$$

DEFINITIONS. The matrix **a** is called the *matrix of the coefficients* and its determinant det **a** the *determinant of the coefficients.*

The variables x_1, x_2, x_3 and the constants b_1, b_2, b_3 appear as the elements of *column vectors.*

If **a** is non-singular, then $\mathbf{a}^{-1}$ exists, so that

$$\begin{aligned} \mathbf{ax} &= \mathbf{b} \\ \Rightarrow \mathbf{a}^{-1}\mathbf{ax} &= \mathbf{a}^{-1}\mathbf{b} \\ \Rightarrow \qquad \mathbf{x} &= \mathbf{a}^{-1}\mathbf{b}, \end{aligned}$$

so that the only possible solution is $\mathbf{a}^{-1}\mathbf{b}$. Further, this value of **x** is readily seen to satisfy the equation $\mathbf{ax} = \mathbf{b}$. Hence *the given equations have the unique solution given in matrix form by the relation*

$$\mathbf{x} = \mathbf{a}^{-1}\mathbf{b}.$$

NOTE. This solution is very important in theoretical work. For numerical examples in which explicit values are required for x_1, x_2, x_3, it may be easier to use the more elementary methods of Part 1.

In this connexion, however, it is worthy of note that, in actual numerical practice, the evaluation of the inverse $\mathbf{a}^{-1}$ of a given matrix **a** is often effected most easily (and, for many people, most surely) by the direct solution of equations:

Suppose, for example, that it is required *to find the inverse of the matrix*

$$\mathbf{a} \equiv \begin{pmatrix} 5 & 3 & 2 \\ 3 & -4 & 6 \\ 1 & 2 & 4 \end{pmatrix}.$$

Denote by **x** the column vector

$$\mathbf{x} \equiv \{x, y, z\},$$

and consider the equation

$$\mathbf{ax} = \mathbf{u},$$

where

$$\mathbf{u} \equiv \{u, v, w\},$$

with solution

$$\mathbf{x} = \mathbf{a}^{-1}\mathbf{u}.$$

The equation $\mathbf{ax} = \mathbf{u}$ is

$$\begin{aligned} 5x+3y+2z &= u, \\ 3x-4y+6z &= v, \\ x+2y+4z &= w. \end{aligned}$$

Solve these equations for x, y, z, first eliminating x between them:

$$\begin{aligned} 5w-u &= 7y+18z, \\ 3w-v &= 10y+\ 6z, \end{aligned}$$

so that, eliminating z,

$$5w-u-3(3w-v) = -23y,$$

or

$$y = \tfrac{1}{23}u-\tfrac{3}{23}v+\tfrac{4}{23}w.$$

It follows that

$$6z = 3w-v-(\tfrac{10}{23}u-\tfrac{30}{23}v+\tfrac{40}{23}w),$$

or

$$z = -\tfrac{10}{138}u+\tfrac{7}{138}v+\tfrac{29}{138}w;$$

and

$$\begin{aligned} x &= w-2y-4z \\ &= \tfrac{28}{138}u+\tfrac{8}{138}v-\tfrac{26}{138}w, \end{aligned}$$

after reduction.

Hence the solution, in matrix form, is

$$138\mathbf{x} = \begin{pmatrix} 28 & 8 & -26 \\ 6 & -18 & 24 \\ -10 & 7 & 29 \end{pmatrix}\mathbf{u},$$

so that

$$\mathbf{a}^{-1} \equiv \begin{pmatrix} \frac{28}{138} & \frac{8}{138} & -\frac{26}{138} \\ \frac{6}{138} & -\frac{18}{138} & \frac{24}{138} \\ -\frac{10}{138} & \frac{7}{138} & \frac{29}{138} \end{pmatrix}.$$

This is a particularly awkward example, and the solution may be compared with the work involved in a direct evaluation of $\mathbf{a}^{-1}$. (It is, of course, in the nature of the case that the three equations in x, y, z must be solved *by elimination* as in the examples: the use of determinants would be contrary to the context.)

For conviction, a simple example is added:

Let
$$\mathbf{a} \equiv \begin{pmatrix} 1 & 1 & 1 \\ 1 & 2 & 3 \\ 1 & 3 & 6 \end{pmatrix}.$$

The equation $\mathbf{ax} = \mathbf{u}$ is
$$\begin{aligned} x+y+z &= u, \\ x+2y+3z &= v, \\ x+3y+6z &= w, \end{aligned}$$

so that
$$\begin{aligned} v-u &= y+2z, \\ w-v &= y+3z \end{aligned}$$

and, subtracting,
$$z = u-2v+w.$$

Hence
$$\begin{aligned} y &= v-u-2z \\ &= -3u+5v-2w, \end{aligned}$$

and
$$\begin{aligned} x &= u-y-z \\ &= 3u-3v+w. \end{aligned}$$

The equations are thus
$$\begin{aligned} 3u-3v+w &= x, \\ -3u+5v-2w &= y, \\ u-2v+w &= z, \end{aligned}$$

so that
$$\mathbf{a}^{-1} \equiv \begin{pmatrix} 3 & -3 & 1 \\ -3 & 5 & -2 \\ 1 & -2 & 1 \end{pmatrix}.$$

Examples

1. Find **x** and **y** from the equations
$$\mathbf{ax} = \mathbf{b}, \quad \mathbf{dy} = \mathbf{x},$$
where **a**, **d** are non-singular square matrices of order 3 and **b**, **x**, **y** are column vectors.

2. Prove that, if **a**, **b** are square matrices of order 3 such that $\mathbf{ab} = \mathbf{0}$, where **0** is the zero matrix, then at least one of **a**, **b** is singular, and **a** (say) can be non-singular only if **b** is the zero matrix.

Give an example in which **a**, **b** are both singular but neither is zero.

3. Establish the relations:
$$\begin{aligned} \det(\mathbf{a}^{-1}) &= \{\det \mathbf{a}\}^{-1}, \\ (\mathbf{a}^{-1})^{-1} &= \mathbf{a}. \end{aligned}$$

4. Prove that if $\mathbf{a}$ is singular so also is $\mathbf{A}$.

5. Find the inverses of the matrices:

(i) $$\mathbf{a} \equiv \begin{pmatrix} 1 & 2 & 3 \\ 2 & 3 & 4 \\ 3 & 4 & 7 \end{pmatrix},$$ (ii) $$\mathbf{a} \equiv \begin{pmatrix} 2 & 0 & 3 \\ -2 & 3 & 0 \\ 1 & -3 & 5 \end{pmatrix}.$$

6. Prove that, if $\mathbf{a}$ is a 3×1 column vector, then the 3×3 matrix $\mathbf{aa}'$ is singular.

7. Find $\det \mathbf{aa}'$, where

$$\mathbf{a} \equiv \begin{pmatrix} 1 & 2 \\ 3 & -1 \\ 5 & 0 \end{pmatrix}.$$

9. Résumé for square matrices of order 2. Let

$$\mathbf{a} \equiv \begin{pmatrix} a_{11} & a_{12} \\ a_{21} & a_{22} \end{pmatrix},$$

so that

$$\det \mathbf{a} \equiv \begin{vmatrix} a_{11} & a_{12} \\ a_{21} & a_{22} \end{vmatrix} = a_{11}a_{22} - a_{12}a_{21}.$$

Then

$$\det (k\mathbf{a}) \equiv \begin{vmatrix} ka_{11} & ka_{12} \\ ka_{21} & ka_{22} \end{vmatrix} = k^2 \det \mathbf{a}.$$

The adjoint $\mathbf{A} \equiv \operatorname{adj} \mathbf{a}$ is found by first taking the matrices of cofactors with proper signs:

$$\begin{pmatrix} a_{22} & -a_{21} \\ -a_{12} & a_{11} \end{pmatrix}$$

and then transposing, so that

$$\mathbf{A} \equiv \operatorname{adj} \mathbf{a} \equiv \begin{pmatrix} a_{22} & -a_{12} \\ -a_{21} & a_{11} \end{pmatrix}.$$

Immediately,

$$\begin{aligned} \mathbf{aA} &= \begin{pmatrix} a_{11} & a_{12} \\ a_{21} & a_{22} \end{pmatrix} \begin{pmatrix} a_{22} & -a_{12} \\ -a_{21} & a_{11} \end{pmatrix} \\ &= \begin{pmatrix} a_{11}a_{22} - a_{12}a_{21}, & -a_{11}a_{12} + a_{12}a_{11} \\ a_{21}a_{22} - a_{22}a_{21}, & -a_{21}a_{12} + a_{22}a_{11} \end{pmatrix} \\ &= \begin{pmatrix} a_{11}a_{22} - a_{12}a_{21} & 0 \\ 0 & a_{11}a_{22} - a_{12}a_{21} \end{pmatrix} \\ &= \det \mathbf{a} \begin{pmatrix} 1 & 0 \\ 0 & 1 \end{pmatrix} \\ &= (\det \mathbf{a}) \mathbf{I}. \end{aligned}$$

The matrix **a** is *non-singular* if

$$\Delta \equiv \det \mathbf{a} \equiv a_{11}a_{22}-a_{12}a_{21} \neq 0.$$

When this inequality holds, the inverse $\mathbf{a}^{-1}$ is given by the formula

$$\mathbf{a}^{-1} = \Delta^{-1}\mathbf{A}$$
$$= \frac{1}{a_{11}a_{22}-a_{12}a_{21}}\begin{pmatrix} a_{22} & -a_{12} \\ -a_{21} & a_{11} \end{pmatrix}.$$

The equations

$$a_{11}x_1+a_{12}x_2 = b_1,$$
$$a_{21}x_1+a_{22}x_2 = b_2,$$

or

$$\mathbf{ax} = \mathbf{b},$$

have, when **a** is non-singular, the unique solution

$$\mathbf{x} = \mathbf{a}^{-1}\mathbf{b}$$
$$= \frac{1}{\Delta}\begin{pmatrix} a_{22} & -a_{12} \\ -a_{21} & a_{11} \end{pmatrix}\begin{pmatrix} b_1 \\ b_2 \end{pmatrix}$$
$$= \frac{1}{\Delta}\begin{pmatrix} a_{22}b_1-a_{12}b_2 \\ -a_{21}b_1+a_{11}b_2 \end{pmatrix},$$

so that

$$x_1 = \frac{a_{22}b_1-a_{12}b_2}{a_{11}a_{22}-a_{12}a_{21}},$$
$$x_2 = \frac{a_{11}b_2-a_{21}b_1}{a_{11}a_{22}-a_{12}a_{21}}.$$

Alternatively, in numerical cases, the equation $\mathbf{ax} = \mathbf{b}$ may be solved directly for **x** to give $\mathbf{a}^{-1}$, as was illustrated in detail (pp. 164–6) for matrices of order 3.

The proof of the formula

$$\det(\mathbf{ab}) = \det\mathbf{a}\,\det\mathbf{b}$$

is worth separate statement for this particular case:

For ease of notation, write

$$\mathbf{a} \equiv \begin{pmatrix} a & b \\ c & d \end{pmatrix}, \quad \mathbf{b} \equiv \begin{pmatrix} p & q \\ r & s \end{pmatrix}.$$

Then

$$\mathbf{ab} = \begin{pmatrix} a & b \\ c & d \end{pmatrix}\begin{pmatrix} p & q \\ r & s \end{pmatrix}$$
$$= \begin{pmatrix} ap+br & aq+bs \\ cp+dr & cq+ds \end{pmatrix},$$

so that

$$\det(\mathbf{ab}) = (ap+br)(cq+ds)-(aq+bs)(cp+dr).$$

Gathering terms in ac, ad, bc, bd, we have

$$\begin{aligned}\det(\mathbf{ab}) &= ac(pq-qp)\\ &\quad +ad(ps-qr)\\ &\quad\quad +bc(rq-sp)\\ &\quad\quad\quad +bd(rs-sr)\\ &= ad(ps-qr)-bc(ps-qr)\\ &= (ad-bc)(ps-qr)\\ &= \det\mathbf{a}\det\mathbf{b}.\end{aligned}$$

10. The expression det A. The adjoint $\mathbf{A}$ of the third order matrix $\mathbf{a}$ has its own determinant $\det\mathbf{A}$. This may be evaluated by combining the formulae obtained in §§3,4 (p. 160):

$$\begin{aligned}\det(\mathbf{aA}) &= \det(\Delta\mathbf{I})\\ &= \Delta^3\det\mathbf{I}\\ &= \Delta^3.\end{aligned}$$

Hence

$$\Delta\det\mathbf{A} = \Delta^3,$$

so that, *if* Δ *is not zero,*

$$\det\mathbf{A} = \{\det\mathbf{a}\}^2.$$

More generally, *if* $\mathbf{a}$ *is of order n and non-singular, then*

$$\det\mathbf{A} = \{\det\mathbf{a}\}^{n-1}.$$

11. The matrix adj A. Apply to $\mathbf{A}$ the rule

$$\text{‘}\mathbf{a}.\operatorname{adj}\mathbf{a} = \det\mathbf{a}.\mathbf{I}\text{’}.$$

Thus

$$\mathbf{A}.\operatorname{adj}\mathbf{A} = (\det\mathbf{A}).\mathbf{I}.$$

Multiply each side on the left by $\mathbf{a}$:

$$\mathbf{aA}\operatorname{adj}\mathbf{A} = (\det\mathbf{A})\mathbf{a},$$

or (if $\mathbf{a}$ is of order n and non-singular)

$$\Delta\operatorname{adj}\mathbf{A} = \Delta^{n-1}\mathbf{a},$$

so that

$$\operatorname{adj}\mathbf{A} = \Delta^{n-2}\mathbf{a}.$$

In particular,

if $\qquad n = 2, \quad \text{then} \quad \operatorname{adj} \mathbf{A} = \mathbf{a},$

if $\qquad n = 3, \quad \text{then} \quad \operatorname{adj} \mathbf{A} = (\det \mathbf{a})\,\mathbf{a}.$

***Illustration* 1.** The matrix

$$\mathbf{a} \equiv \begin{pmatrix} a & h & g \\ h & b & f \\ g & f & c \end{pmatrix}$$

is often used in geometry. By direct computation,

$$\Delta \equiv \det \mathbf{a} \equiv abc + 2fgh - af^2 - bg^2 - ch^2,$$

$$\mathbf{A} \equiv \operatorname{adj} \mathbf{a} \equiv \begin{pmatrix} bc-f^2 & fg-ch & hf-bg \\ fg-ch & ca-g^2 & gh-af \\ hf-bg & gh-af & ab-h^2 \end{pmatrix}.$$

The relation

$$\operatorname{adj} \mathbf{A} = \Delta \mathbf{a}$$

gives six algebraic relations, of which typical instances are

$$(ca-g^2)(ab-h^2)-(gh-af)^2 = (abc+2fgh-af^2-bg^2-ch^2)\,a,$$
$$(hf-bg)(fg-ch)-(bc-f^2)(gh-af) = (abc+2fgh-af^2-bg^2-ch^2)\,f.$$

Revision Examples

1. Given that

$$\mathbf{a} \equiv \begin{pmatrix} 1 & 0 & 0 \\ 0 & 0 & -1 \\ 0 & 1 & 0 \end{pmatrix}, \quad \mathbf{b} \equiv \begin{pmatrix} -1 & 0 & 0 \\ 0 & 1 & 0 \\ 0 & 0 & -1 \end{pmatrix},$$

prove that $\qquad \mathbf{a}^4 = \mathbf{I}, \quad \mathbf{b}^2 = \mathbf{I}, \quad \mathbf{ab} = \mathbf{ba}^3.$

2. By considering the product

$$\begin{pmatrix} x+iy & z+iw \\ -z+iw & x-iy \end{pmatrix} \begin{pmatrix} a+ib & c+id \\ -a+ib & c-id \end{pmatrix},$$

where $i^2 = -1$, prove that

$$(x^2+y^2+z^2+w^2)(a^2+b^2+c^2+d^2)$$

can be expressed as the sum of the squares of four functions each of which is linear in a, b, c, d and also in x, y, z, w.

3. Given that

$$\mathbf{a} \equiv \begin{pmatrix} \frac{1}{2} & 0 \\ 0 & -\frac{1}{2} \end{pmatrix}, \quad \mathbf{b} \equiv \begin{pmatrix} 0 & \frac{1}{2}i \\ -\frac{1}{2}i & 0 \end{pmatrix}, \quad \mathbf{c} \equiv \begin{pmatrix} 0 & \frac{1}{2} \\ \frac{1}{2} & 0 \end{pmatrix}$$

and that, for example, $[\mathbf{a}, \mathbf{b}]$ is written for $\mathbf{ab} - \mathbf{ba}$, prove that

$$[\mathbf{b}, \mathbf{c}] = i\mathbf{a}, \quad [\mathbf{c}, \mathbf{a}] = i\mathbf{b}, \quad [\mathbf{a}, \mathbf{b}] = i\mathbf{c}.$$

Prove also that

$$[\mathbf{a}, \mathbf{a}^2+\mathbf{b}^2+\mathbf{c}^2] = [\mathbf{b}, \mathbf{a}^2+\mathbf{b}^2+\mathbf{c}^2] = [\mathbf{c}, \mathbf{a}^2+\mathbf{b}^2+\mathbf{c}^2] = \mathbf{0}.$$

4. Show that a 3×3 square matrix $\mathbf{a}$ may in general be expressed in the form $\mathbf{lu}$, where $\mathbf{l}$ has all elements on the leading diagonal unity and all elements above the diagonal zero, while $\mathbf{u}$ has all elements below the diagonal zero.

5. If

$$\mathbf{u} \equiv \begin{pmatrix} 0 & 1 \\ 1 & 0 \end{pmatrix}, \quad \mathbf{a} \equiv \begin{pmatrix} a & b \\ c & d \end{pmatrix},$$

where $ad-bc = 1$, write down $\mathbf{ua}$, $\mathbf{au}$, $\mathbf{a}^{-1}$, and verify that

$$(\mathbf{ua})^{-1} = \mathbf{a}^{-1}\mathbf{u}^{-1}.$$

6. Show that the inverse of the matrix

$$\mathbf{F}(\alpha)\,\mathbf{G}(\beta) \equiv \begin{pmatrix} \cos\alpha & -\sin\alpha & 0 \\ \sin\alpha & \cos\alpha & 0 \\ 0 & 0 & 1 \end{pmatrix} \begin{pmatrix} \cos\beta & 0 & \sin\beta \\ 0 & 1 & 0 \\ -\sin\beta & 0 & \cos\beta \end{pmatrix}$$

is $\mathbf{G}(-\beta)\,\mathbf{F}(-\alpha)$.

7. Prove that, if

$$\mathbf{a} \equiv \begin{pmatrix} \cosh a & b\sinh a \\ b^{-1}\sinh a & \cosh a \end{pmatrix},$$

then

$$\mathbf{a}^n = \begin{pmatrix} \cosh na & b\sinh na \\ b^{-1}\sinh na & \cosh na \end{pmatrix},$$

where n is an integer, positive or negative.

8. Obtain the inverse of the matrices

$$\begin{pmatrix} 1 & \lambda & 0 \\ 0 & 1 & \lambda \\ 0 & 0 & 1 \end{pmatrix}, \quad \begin{pmatrix} 1 & 0 & 0 \\ \mu & 1 & 0 \\ 0 & \mu & 1 \end{pmatrix}$$

and hence that of the matrix

$$\begin{pmatrix} 1+\lambda\mu & \lambda & 0 \\ \mu & 1+\lambda\mu & \lambda \\ 0 & \mu & 1 \end{pmatrix}.$$

9. The matrix $\mathbf{a}$ is given by the formula

$$\mathbf{a} \equiv \begin{pmatrix} 2 & 1 \\ 2 & 3 \end{pmatrix}.$$

Verify that

$$\mathbf{a}^2-5\mathbf{a}+4\mathbf{I} = \mathbf{0}.$$

Find all matrices **u** of the form

$$\mathbf{u} \equiv \begin{pmatrix} p & q \\ r & s \end{pmatrix}$$

which satisfy the relation

$$\mathbf{u}^2-5\mathbf{u}+4\mathbf{I} = \mathbf{0},$$

(i) when $q=r=0$, (ii) when q,r are both non-zero.

10. If **a** is a 2×2 matrix and $\mathbf{ax}=\mathbf{xa}$ for every 2×2 matrix **x**, prove that **a** is necessarily of the form $k\mathbf{I}$, where k is a scalar and **I** the unit matrix.

If **a** is not of the form $k\mathbf{I}$ and if

$$\mathbf{ab} = \mathbf{ba},$$

prove that **b** can be expressed in the form

$$\mathbf{b} = p\mathbf{a}+q\mathbf{I},$$

where p,q are scalars.

11. The matrix **a** satisfies the equation $\mathbf{a}^2=\mathbf{a}$ and is not the unit matrix **I**. Prove that $\det\mathbf{a}=0$.

Show that, if n is a positive or negative integer, then

$$(\mathbf{I}+\mathbf{a})^n = \mathbf{I}+(2^n-1)\,\mathbf{a}.$$

12. Prove that, if

$$\mathbf{a} \equiv \begin{pmatrix} a_1 & a_2 & a_3 \\ 1 & 0 & 0 \\ 0 & 1 & 0 \end{pmatrix},$$

then
$$\det(\mathbf{a}-\lambda\mathbf{I}) \equiv a_3+a_2\lambda+a_1\lambda^2-\lambda^3.$$

13. Prove that, if

$$\mathbf{a} \equiv \begin{pmatrix} 0 & -\tan\theta \\ \tan\theta & 0 \end{pmatrix},$$

then

$$\mathbf{I}+\mathbf{a} = \begin{pmatrix} \cos 2\theta & -\sin 2\theta \\ \sin 2\theta & \cos 2\theta \end{pmatrix}(\mathbf{I}-\mathbf{a}).$$

14. Prove that, if n is a positive integer and

$$\mathbf{a} \equiv \begin{pmatrix} a & 1 \\ 0 & a \end{pmatrix},$$

then
$$\mathbf{a}^n \equiv \begin{pmatrix} a^n & na^{n-1} \\ 0 & a^n \end{pmatrix}.$$

Can the result be extended to the case when n is a negative integer?

15. Given that

$$\mathbf{a} \equiv \begin{pmatrix} 1 & -2 \\ -3 & 4 \end{pmatrix},$$

find all the matrices $\mathbf{x}$ such that

$$\mathbf{ax} = \mathbf{xa}.$$

16. Given that

$$\mathbf{a} \equiv \begin{pmatrix} 0 & -1 \\ 1 & 0 \end{pmatrix}, \quad \mathbf{b} \equiv \begin{pmatrix} 0 & -a \\ a & 0 \end{pmatrix}$$

and that a matrix $\mathbf{c}$ exists such that

$$\mathbf{abc} = -\mathbf{I}, \quad \mathbf{ca} = \mathbf{b},$$

find a.

17. Given a 3×3 matrix $\mathbf{a}$ and its adjoint $\mathbf{A}$, prove that, if

$$\det(\mathbf{a}+\lambda\mathbf{I}) \equiv \lambda^3 + p_1\lambda^2 + p_2\lambda + p_3,$$

then

$$\det(\mathbf{A}+\lambda\mathbf{I}) \equiv \lambda^3 + p_2\lambda^2 + p_1 p_3\lambda + p_3^2.$$

18. If $\mathbf{a}$, $\mathbf{b}$ are non-singular 3×3 matrices, and if

$$\operatorname{adj}(\lambda\mathbf{a}+\mathbf{b}) = \lambda^2\mathbf{A} + \lambda\mathbf{C} + \mathbf{B},$$

$$\det(\lambda\mathbf{a}+\mathbf{b}) = \lambda^3\alpha + \lambda^2\gamma + \lambda\delta + \beta,$$

show that $\mathbf{A} = \operatorname{adj}\mathbf{a}$, $\mathbf{B} = \operatorname{adj}\mathbf{b}$, $\alpha = \det\mathbf{a}$, $\beta = \det\mathbf{b}$.

Express $\mathbf{C}$ in terms of α, γ, $\mathbf{b}$ and $\mathbf{A}$, and show that

$$\det\mathbf{C} = \gamma\delta - \alpha\beta.$$

[Note that $(\lambda\mathbf{a}+\mathbf{b})\operatorname{adj}(\lambda\mathbf{a}+\mathbf{b}) = \det(\lambda\mathbf{a}+\mathbf{b}).\mathbf{I}$.]

19. If $\mathbf{a}$ is a non-singular square matrix and $\mathbf{b}$ a square matrix of the same order, prove that

$$\mathbf{ab} = \mathbf{0} \Rightarrow \mathbf{b} = \mathbf{0}.$$

If $\mathbf{c}$ is another square matrix of the same order, prove that there are unique matrices $\mathbf{x}$, $\mathbf{y}$ such that

$$\mathbf{ax} = \mathbf{c}, \quad \mathbf{ya} = \mathbf{c}.$$

Find $\mathbf{x}$ and $\mathbf{y}$ if

$$\mathbf{a} \equiv \begin{pmatrix} 1 & 2 \\ 4 & 5 \end{pmatrix}, \quad \mathbf{c} \equiv \begin{pmatrix} 5 & 17 \\ 14 & 47 \end{pmatrix}.$$

20. Given that

$$\mathbf{a} \equiv \begin{pmatrix} 1 & 1 & 0 \\ -2 & -1 & 0 \\ 0 & 0 & 2 \end{pmatrix},$$

calculate $\mathbf{a}^2$ and $\mathbf{a}^{-1}$.

Verify that

$$(\mathbf{a}^2)^{-1} = (\mathbf{a}^{-1})^2.$$

12. Some metrical applications. Some of the metrical work outlined in Chapter 4 can be modified by using matrix notation, especially with regard to scalar and vector products. This paragraph exhibits the modifications and also one or two examples from mechanics.

Given an origin O, the position of a point P in space can be fixed by means of its *coordinate vector* $\mathbf{p}\ (p_1, p_2, p_3)$ after the manner indicated in Chapter 4, §§4, 5 (pp. 47–50). That vector may now be regarded alternatively as a *matrix*, of three rows and one column: that is, as a *column vector*

$$\mathbf{p} \equiv \{p_1, p_2, p_3\}$$

in the sense in which the notation has been used in matrix theory.

In the same way, the coordinates of the points of a line l through the origin can be given the *column vector* form $r\mathbf{l}$, where $\mathbf{l}$ is a fixed vector defining the line and r a variable scalar defining the various points upon it. For convenience, $\mathbf{l}$ is usually *normalized* so that

$$\mathbf{l}'\mathbf{l} = 1,$$

and r is then the distance (with proper sign attached) of the varying point from the origin. The normalizing relation in expanded form is

$$l_1^2 + l_2^2 + l_3^2 = 1.$$

The *scalar product* of two vectors $\mathbf{l}$, $\mathbf{m}$ is (p. 46)

$$l_1 m_1 + l_2 m_2 + l_3 m_3$$

and this, in matrix notation, is

$$\mathbf{l}'\mathbf{m} \equiv \mathbf{m}'\mathbf{l}.$$

Two lines whose directions are defined by $\mathbf{l}$, $\mathbf{m}$ are thus perpendicular if

$$\mathbf{l}'\mathbf{m} = 0.$$

The *vector product* $\mathbf{l} \wedge \mathbf{m}$ (p. 58) of two vectors $\mathbf{l}$, $\mathbf{m}$ presents more difficulty. It is the vector whose coordinates are

$$l_2 m_3 - l_3 m_2, \quad l_3 m_1 - l_1 m_3, \quad l_1 m_2 - l_2 m_1,$$

and this, by forming the actual product, can be written in either of the forms

$$\mathbf{l} \wedge \mathbf{m} \equiv \begin{pmatrix} 0 & m_3 & -m_2 \\ -m_3 & 0 & m_1 \\ m_2 & -m_1 & 0 \end{pmatrix} \begin{pmatrix} l_1 \\ l_2 \\ l_3 \end{pmatrix},$$

or

$$\mathbf{l} \wedge \mathbf{m} \equiv \begin{pmatrix} 0 & -l_3 & l_2 \\ l_3 & 0 & -l_1 \\ -l_2 & l_1 & 0 \end{pmatrix} \begin{pmatrix} m_1 \\ m_2 \\ m_3 \end{pmatrix},$$

whereby $\mathbf{l} \wedge \mathbf{m}$ appears as a 3×3 matrix multiplying $\mathbf{l}$ or $\mathbf{m}$ according to the needs of a particular problem.

The Illustrations 2 and 3 which follow show how these ideas may be applied in metrical geometry and in dynamics.

Example

1. Given that $\mathbf{x} \equiv \{1,2,3\}$, $\mathbf{y} \equiv \{6,5,4\}$, obtain the vector product $\mathbf{x} \wedge \mathbf{y}$ (i) in the form $\mathbf{ax}$, (ii) in the form $\mathbf{by}$.

Repeat the question for

$$\mathbf{x} \equiv \{0,2,-1\}, \quad \mathbf{y} \equiv \{-3,1,0\},$$

and for

$$\mathbf{x} \equiv \{0,1,1\}, \quad \mathbf{y} \equiv \{1,0,0\}.$$

***Illustration* 2.** *The projection of a point upon a line through the origin.*

Let l be a line through the origin defined in direction by the vector $\mathbf{l}$, normalized so that

$$\mathbf{l}'\mathbf{l} = 1.$$

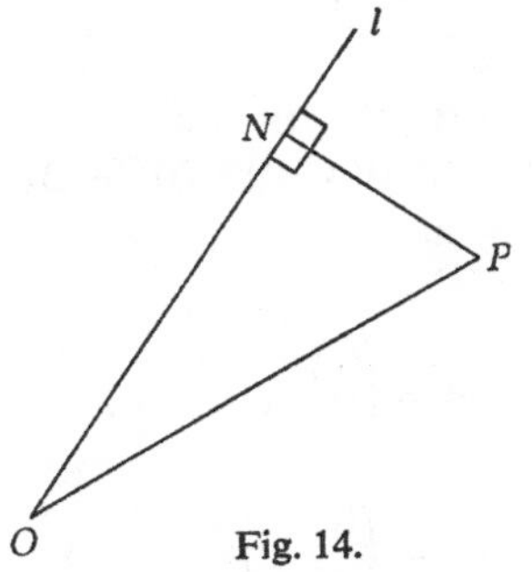

Fig. 14.

If P, with coordinate vector $\mathbf{p}$, is any point of space, the foot of the perpendicular from P to l is a point N, of coordinate vector $\mathbf{n}$, say. Since N is on l, it follows that $\mathbf{n}$ is of the form

$$\mathbf{n} = r\mathbf{l}.$$

Now the vector NP is given by the column matrix

$$\mathbf{p}-\mathbf{n}$$

or

$$\mathbf{p}-r\mathbf{l},$$

and the condition for this to be orthogonal to $\mathbf{l}$ is that

$$\mathbf{l}'(\mathbf{p}-r\mathbf{l}) = 0,$$

or

$$\mathbf{l}'\mathbf{p}-r\mathbf{l}'\mathbf{l} = 0,$$

or, since

$$\mathbf{l}'\mathbf{l} = 1,$$

$$r = \mathbf{l}'\mathbf{p}.$$

Hence

$$\mathbf{n} = (\mathbf{l}'\mathbf{p})\mathbf{l}.$$

The benefits of matrix theory may now be used to express this as the 'multiple' of $\mathbf{p}$ by a 3×3 matrix $\mathbf{a}$ whose elements depend on $\mathbf{l}$ only; for a *given* line l, the matrix $\mathbf{a}$ thus assumes a determinate form. The formula for $\mathbf{n}$ may be re-written

$$\mathbf{n} = \mathbf{l}(\mathbf{l}'\mathbf{p}),$$

and *the three column-vectors* $\mathbf{l}$, $\mathbf{l}'$, $\mathbf{p}$, *taken in this order, are conformable for matrix multiplication.* Hence

$$\mathbf{n} = (\mathbf{l}\mathbf{l}')\mathbf{p}$$

$$= \mathbf{a}\mathbf{p},$$

say, where $\mathbf{a}$ is the 3×3 matrix

$$\mathbf{a} \equiv \mathbf{l}\mathbf{l}'.$$

Thus, *given a line l through the origin, the foot of the perpendicular upon it from a point of coordinate vector* $\mathbf{p}$ *is the point whose coordinate vector is*

$$\mathbf{a}\mathbf{p},$$

where

$$\mathbf{a} \equiv \mathbf{l}\mathbf{l}'.$$

Note that

(i) $$\mathbf{a}' = (\mathbf{l}')'\mathbf{l}' = \mathbf{l}\mathbf{l}' = \mathbf{a},$$

(ii) $$\mathbf{a}^2 = (\mathbf{l}\mathbf{l}')(\mathbf{l}\mathbf{l}') = \mathbf{l}(\mathbf{l}'\mathbf{l})\mathbf{l}' = \mathbf{l}\mathbf{I}\mathbf{l}' = \mathbf{l}\mathbf{l}'$$
$$= \mathbf{a}.$$

(iii) Since

$$\mathbf{a} \equiv \begin{pmatrix} l_1 \\ l_2 \\ l_3 \end{pmatrix} (l_1 \quad l_2 \quad l_3)$$

$$\equiv \begin{pmatrix} l_1^2 & l_1 l_2 & l_1 l_3 \\ l_2 l_1 & l_2^2 & l_2 l_3 \\ l_3 l_1 & l_3 l_2 & l_3^2 \end{pmatrix},$$

all minors of two rows and columns vanish. In particular,

$$\det \mathbf{a} = 0.$$

Examples

1. Find the feet of the perpendiculars from the points with coordinate vectors $\{1,2,3\}$, $\{2,-1,0\}$, $\{5,1,-3\}$, $\{0,2,5\}$ on the line through the origin defined by the direction $\{\frac{3}{13}, \frac{4}{13}, -\frac{12}{13}\}$.

2. If, in the notation of the text, $\mathbf{b}$ is written for $\mathbf{I}-\mathbf{a}$, prove that

$$\mathbf{b}^2 = \mathbf{b}, \quad \mathbf{ab} = \mathbf{ba} = \mathbf{0}.$$

***Illustration* 3.**† *The inertia tensor.* Suppose that a rigid body is rotating with angular velocity ω about a line l. It is a known fact that this angular velocity may be treated as a vector $\boldsymbol{\omega}$, where (with $\mathbf{l'l} = 1$ as on p. 175)

$$\boldsymbol{\omega} = \omega \mathbf{l}.$$

It is also a known fact that the point P of the body located with respect to some suitable fixed origin by the coordinate vector $\mathbf{p}$ then has a (linear) velocity $\mathbf{v}$ given as a vector product in the form

$$\mathbf{v} = \boldsymbol{\omega} \wedge \mathbf{p}.$$

This vector product may be written (p. 174) in the matrix notation

$$\begin{pmatrix} 0 & p_3 & -p_2 \\ -p_3 & 0 & p_1 \\ p_2 & -p_1 & 0 \end{pmatrix} \begin{pmatrix} \omega_1 \\ \omega_2 \\ \omega_3 \end{pmatrix},$$

or

$$\boldsymbol{\pi}\boldsymbol{\omega},$$

say, where $\boldsymbol{\pi}$ is a 3×3 matrix whose elements are independent of the angular velocity. Note that $\boldsymbol{\pi}$ is *skew-symmetric*, having

$$\boldsymbol{\pi}' = -\boldsymbol{\pi}.$$

It follows that, for a particle of mass m situated at P, the (linear) momentum is

$$\begin{aligned} &m\boldsymbol{\pi}\boldsymbol{\omega} \\ &\equiv \mathbf{k}, \end{aligned}$$

say.

A further dynamical fact is that the *angular momentum about O* of the particle at P having momentum $\mathbf{k}$ is the *vector* $\mathbf{h}$ given by the formula

$$\mathbf{h} = \mathbf{p} \wedge \mathbf{k},$$

and this, again, can be written in matrix form

$$\begin{aligned} &\begin{pmatrix} 0 & -p_3 & p_2 \\ p_3 & 0 & -p_1 \\ -p_2 & p_1 & 0 \end{pmatrix} \begin{pmatrix} k_1 \\ k_2 \\ k_3 \end{pmatrix} \\ &\equiv -\boldsymbol{\pi}\mathbf{k}. \end{aligned}$$

Hence

$$\begin{aligned} \mathbf{h} &= -\boldsymbol{\pi}\mathbf{k} \\ &= -\boldsymbol{\pi}(m\boldsymbol{\pi}\boldsymbol{\omega}) \\ &= -m\boldsymbol{\pi}^2\boldsymbol{\omega}, \end{aligned}$$

† To be omitted if the dynamical ideas are found too hard.

where

$$-\pi^2 \equiv \begin{pmatrix} 0 & -p_3 & p_2 \\ p_3 & 0 & -p_1 \\ -p_2 & p_1 & 0 \end{pmatrix} \begin{pmatrix} 0 & p_3 & -p_2 \\ -p_3 & 0 & p_1 \\ p_2 & -p_1 & 0 \end{pmatrix}$$

$$\equiv \begin{pmatrix} p_2^2+p_3^2 & -p_1p_2 & -p_1p_3 \\ -p_2p_1 & p_3^2+p_1^2 & -p_2p_3 \\ -p_3p_1 & -p_3p_2 & p_1^2+p_2^2 \end{pmatrix}.$$

Suppose now that the rigid body consists of a system of particles of typical mass m situated at a typical point $\mathbf{p}$. Summed for all the particles of the body, the *total angular momentum of the rigid body* is the *vector*

$$\mathbf{h} \equiv \mathbf{A}\boldsymbol{\omega},$$

where $\mathbf{A}$, obtained by summation over all the particles of the body, is given by the formula

$$\mathbf{A} \equiv \begin{pmatrix} \Sigma\, m(p_2^2+p_3^2) & -\Sigma\, mp_1p_2 & -\Sigma\, mp_1p_3 \\ -\Sigma\, mp_2p_1 & \Sigma\, m(p_3^2+p_1^2) & -\Sigma\, mp_2p_3 \\ -\Sigma\, mp_3p_1 & -\Sigma\, mp_3p_2 & \Sigma\, m(p_1^2+p_2^2) \end{pmatrix}.$$

DEFINITION. The matrix $\mathbf{A}$ may be called the *inertia matrix* of the rigid body. In the more normal context of rigid dynamics, the name is *inertia tensor.*

Note that *the matrix* $\mathbf{A}$ *is symmetric, having*

$$\mathbf{A}' = \mathbf{A}.$$

Finally, the *kinetic energy* of the rigid body can also be obtained in terms of $\mathbf{A}$. For a single particle, the kinetic energy T is given by the formula

$$2T = mv^2 = m\mathbf{v}'\mathbf{v},$$

so that, here,

$$2T = \Sigma\, m\mathbf{v}'\mathbf{v},$$

summed for all the particles of the system, giving

$$\begin{aligned} 2T &= \Sigma\, m(\boldsymbol{\pi}\boldsymbol{\omega})'(\boldsymbol{\pi}\boldsymbol{\omega}) \\ &= \Sigma\, m\boldsymbol{\omega}'(\boldsymbol{\pi}'\boldsymbol{\pi})\,\boldsymbol{\omega} \\ &= \boldsymbol{\omega}'\{\Sigma\, m(\boldsymbol{\pi}'\boldsymbol{\pi})\}\,\boldsymbol{\omega}, \end{aligned}$$

since the vector $\boldsymbol{\omega}$ does not depend on m or $\boldsymbol{\pi}$. But

$$\begin{aligned} \Sigma\, m\boldsymbol{\pi}'\boldsymbol{\pi} &= -\Sigma\, m\boldsymbol{\pi}^2 \\ &= \mathbf{A}, \end{aligned}$$

so that

$$T = \tfrac{1}{2}\boldsymbol{\omega}'\mathbf{A}\boldsymbol{\omega}.$$

In explanation of the idea behind the concept of $\mathbf{A}$ as a *tensor*, suppose that the coordinate system in terms of which the components of the vectors are calculated is subjected to the 'orthogonal transformation' of matrix $\mathbf{b}$, where

$$\mathbf{b}'\mathbf{b} = \mathbf{b}\mathbf{b}' = \mathbf{I}.$$

Then any vector $\mathbf{x}$ is transformed to $\mathbf{x}^*$ by the rule

$$\mathbf{x}^* = \mathbf{b}\mathbf{x}.$$

For calculations in the *new* system,

$$\boldsymbol{\omega}^* = \omega \mathbf{l}^*$$

(but ω, the scalar angular velocity, is unaltered), and

$$\begin{aligned}\mathbf{v}^* &= \boldsymbol{\omega}^* \wedge \mathbf{p}^*,\\ &= \boldsymbol{\pi}^*\boldsymbol{\omega}^*,\end{aligned}$$

where

$$\boldsymbol{\pi}^* \equiv \begin{pmatrix} 0 & p_3^* & -p_2^* \\ -p_3^* & 0 & p_1^* \\ p_2^* & -p_1^* & 0 \end{pmatrix}.$$

Consider, for example, the formula for the kinetic energy. This is

$$\begin{aligned}2T^* &= \boldsymbol{\omega}^{*\prime}\{\Sigma\, m(\boldsymbol{\pi}^{*\prime}\boldsymbol{\pi}^*)\}\,\boldsymbol{\omega}^*\\ &= (\mathbf{b}\boldsymbol{\omega})'\{\Sigma\, m(\boldsymbol{\pi}^{*\prime}\boldsymbol{\pi}^*)\}(\mathbf{b}\boldsymbol{\omega})\\ &= \boldsymbol{\omega}'\mathbf{b}'\mathbf{A}^*\mathbf{b}\boldsymbol{\omega}.\end{aligned}$$

But it is known that

$$\begin{aligned}2T^* &= 2T\\ &= \boldsymbol{\omega}'\mathbf{A}\boldsymbol{\omega},\end{aligned}$$

and so, since $\mathbf{A}$ and $\mathbf{A}^*$ are symmetric (see later, p. 234),

$$\mathbf{b}'\mathbf{A}^*\mathbf{b} = \mathbf{A},$$

or

$$\mathbf{A}^* = \mathbf{b}\mathbf{A}\mathbf{b}'.$$

It is the fact of transforming under $\mathbf{b}$ according to these rules that characterizes the arrays known as tensors. But the idea is difficult, and a text-book devoted to the subject should be consulted.

Examples

1. Calculate the inertia tensor for a system consisting of 8 particles of mass m situated at the points $(\pm 1, \pm 1, \pm 1)$.

2. Calculate the inertia tensor for a system consisting of 4 particles of mass m situated at the points $(\pm 1, \pm 1, 0)$.

3. Calculate the inertia tensor for a system consisting of particles of mass m at $(1,1,1)$, of mass $2m$ at $(-1,1,1)$, of mass $3m$ at $(1,-1,1)$ and of mass $4m$ at $(1,1,-1)$.

4. Find for each of the above three systems (i) the angular momentum, (ii) the kinetic energy, for rotation with angular velocity ω about the line $\left\{\frac{1}{\sqrt{3}}, \frac{1}{\sqrt{3}}, \frac{1}{\sqrt{3}}\right\}$.

5. Given a vector $\mathbf{u}$, its reflexion in the plane $\mathbf{l}'\mathbf{x} = 0$, where $\mathbf{l}$ is a unit vector, is a vector $\mathbf{v}$ and the *reflexion matrix* $\mathbf{R}$ is then defined by the relation

$$\mathbf{v} = \mathbf{R}\mathbf{u}.$$

Express **R** as a matrix whose elements depend on the components l_1, l_2, l_3 of **l**. Deduce that

$$\mathbf{R}^2 = \mathbf{I},$$
$$\mathbf{R}\mathbf{l} = -\mathbf{l},$$
$$\mathbf{l}'\mathbf{w} = 0 \Leftrightarrow \mathbf{R}\mathbf{w} = \mathbf{w}.$$

14

PARTITIONED MATRICES

This chapter is inserted at this early point in the study of matrices only after a great deal of consideration and reconsideration. The technique is not hard, and it will be used extensively in the sequel, where the benefits should be considerable.

1. The idea of partition. A matrix such as

$$\begin{pmatrix} 1 & 3 & 5 & 7 \\ 2 & 4 & 0 & 1 \\ 3 & 5 & 2 & 6 \end{pmatrix}$$

may be divided arbitrarily by vertical and horizontal lines; say

$$\left(\begin{array}{c|ccc} 1 & 3 & 5 & 7 \\ 2 & 4 & 0 & 1 \\ \hline 3 & 5 & 2 & 6 \end{array}\right).$$

The result is four blocks of numbers each of which may, if desired, be considered as a matrix in its own right:

$$\begin{pmatrix} 1 \\ 2 \end{pmatrix} \quad \begin{pmatrix} 3 & 5 & 7 \\ 4 & 0 & 1 \end{pmatrix}$$

$$(3) \quad (5 \quad 2 \quad 6).$$

More generally, a given matrix **a** may be *partitioned* into a number of submatrices, in a form such as

$$\mathbf{a} \equiv \left(\begin{array}{c|c|c|c} \mathbf{u} & \mathbf{v} & \mathbf{w} & \dots \\ \hline \mathbf{p} & \mathbf{q} & \mathbf{r} & \dots \\ \hline \mathbf{l} & \mathbf{m} & \mathbf{n} & \dots \\ \hline \dots & \dots & \dots & \dots \end{array}\right),$$

where matrices in horizontal layers have the same number of rows and in vertical layers the same number of columns. It is sometimes convenient to omit the horizontal and vertical lines of division in this general case; then **a** looks like a matrix whose elements are themselves matrices.

2. Block addition. Let **a**, **b** be two given matrices which, since they are to be subjected to a process of addition, must be of the same type $m \times n$. When they are partitioned, after the manner proposed in §1, they will be said to be *partitioned conformably for addition* if the pattern is the same for both, so that corresponding submatrices can be added according to the rule typified by:

$$\left(\begin{array}{c|c|c} \mathbf{a} & \mathbf{b} & \mathbf{c} \\ \hline \mathbf{u} & \mathbf{v} & \mathbf{w} \end{array}\right) + \left(\begin{array}{c|c|c} \mathbf{l} & \mathbf{m} & \mathbf{n} \\ \hline \mathbf{p} & \mathbf{q} & \mathbf{r} \end{array}\right) = \left(\begin{array}{c|c|c} \mathbf{a}+\mathbf{l} & \mathbf{b}+\mathbf{m} & \mathbf{c}+\mathbf{n} \\ \hline \mathbf{u}+\mathbf{p} & \mathbf{v}+\mathbf{q} & \mathbf{w}+\mathbf{r} \end{array}\right).$$

For example,

$$\left(\begin{array}{cc|cc} 3 & 1 & 2 & 6 \\ 4 & 2 & 7 & 3 \\ \hline 1 & 4 & 0 & 5 \end{array}\right) + \left(\begin{array}{cc|cc} 2 & 3 & 1 & 7 \\ 2 & 9 & 2 & 3 \\ \hline 3 & 1 & 4 & 2 \end{array}\right) = \left(\begin{array}{cc|cc} 5 & 4 & 3 & 13 \\ 6 & 11 & 9 & 6 \\ \hline 4 & 5 & 4 & 7 \end{array}\right).$$

In practice, there is no great virtue in *block addition*, as the process may be called. The value of partition lies in the much harder case of multiplication (§3).

Examples

1. Prove that a matrix partitioned by p horizontal and q vertical lines is divided into $(p+1)(q+1)$ submatrices.

2. Prove by particular examples that, for square matrices of appropriate order,

$$\det\left(\begin{array}{c|c} \mathbf{a} & \mathbf{b} \\ \hline \mathbf{c} & \mathbf{d} \end{array}\right) + \det\left(\begin{array}{c|c} \mathbf{u} & \mathbf{v} \\ \hline \mathbf{w} & \mathbf{x} \end{array}\right) \neq \det\left(\begin{array}{c|c} \mathbf{a}+\mathbf{u} & \mathbf{b}+\mathbf{v} \\ \hline \mathbf{c}+\mathbf{w} & \mathbf{d}+\mathbf{x} \end{array}\right).$$

3. By considering examples such as

$$\left(\begin{array}{cc|cc} 2 & 1 & 3 & 4 \\ 5 & 3 & 2 & 3 \\ \hline 2 & 1 & 2 & 3 \\ 9 & 5 & 1 & 2 \end{array}\right),$$

prove that, for square matrices,

$$\det\left(\begin{array}{c|c} \mathbf{a} & \mathbf{b} \\ \hline \mathbf{c} & \mathbf{d} \end{array}\right) \neq \det\begin{pmatrix} \det \mathbf{a} & \det \mathbf{b} \\ \det \mathbf{c} & \det \mathbf{d} \end{pmatrix}.$$

3. Block multiplication. Attention here is confined mainly to partitions of the simple type

$$\left(\begin{array}{c|c} \mathbf{a} & \mathbf{b} \\ \hline \mathbf{c} & \mathbf{d} \end{array}\right)$$

which cover most of what is required for subsequent work.

Given two matrices written in partitioned form

$$\left(\begin{array}{c|c} \mathbf{a} & \mathbf{b} \\ \hline \mathbf{c} & \mathbf{d} \end{array}\right), \quad \left(\begin{array}{c|c} \mathbf{p} & \mathbf{q} \\ \hline \mathbf{r} & \mathbf{s} \end{array}\right),$$

there will be obvious advantages if it proves possible to exhibit their product in the partitioned form

$$\left(\begin{array}{c|c} \mathbf{ap}+\mathbf{br} & \mathbf{aq}+\mathbf{bs} \\ \hline \mathbf{cp}+\mathbf{dr} & \mathbf{cq}+\mathbf{ds} \end{array}\right)$$

as if the submatrices $\mathbf{a}, \mathbf{b}, \ldots, \mathbf{r}, \mathbf{s}$ were 'ordinary' elements.

When this can be done, the two given matrices are said to be *partitioned conformably for multiplication*, and the process itself is known as *block multiplication.*

The interesting point is that, *provided the eight products that appear in the answer do exist, then the proposed formula is in fact correct.*

Consider first the types of the matrices. Suppose that the numbers of rows and columns are those indicated by the scheme:

$$\begin{array}{c} \\ m_1 \\ m_2 \end{array}\!\!\begin{array}{c} \begin{array}{cc} n_1 & n_2 \end{array} \\ \left(\begin{array}{c|c} \mathbf{a} & \mathbf{b} \\ \hline \mathbf{c} & \mathbf{d} \end{array}\right) \end{array}, \quad \begin{array}{c} \\ m_1' \\ m_2' \end{array}\!\!\begin{array}{c} \begin{array}{cc} n_1' & n_2' \end{array} \\ \left(\begin{array}{c|c} \mathbf{p} & \mathbf{q} \\ \hline \mathbf{r} & \mathbf{s} \end{array}\right) \end{array}.$$

Thus $\mathbf{b}$, for example, is of type $m_1 \times n_2$ and $\mathbf{r}$ is of type $m_2' \times n_1'$. *The eight products* $\mathbf{ap}, \mathbf{br}, \mathbf{aq}, \mathbf{bs}, \mathbf{cp}, \mathbf{dr}, \mathbf{cq}, \mathbf{ds}$ *then all exist provided only that*

$$n_1 = m_1' \quad \textit{and} \quad n_2 = m_2'.$$

These two relations ensure, incidentally, that the given matrices are themselves conformable for multiplication, since

$$n_1 + n_2 = m_1' + m_2'.$$

***Illustration* 1.** Before proceeding more generally, consider the particular example:

$$\left(\begin{array}{ccc|c} 2 & 3 & 4 & 5 \\ 6 & 7 & 8 & 9 \\ \hline 10 & 11 & 12 & 13 \\ 14 & 15 & 16 & 17 \end{array}\right) \left(\begin{array}{cc} g & h \\ p & q \\ l & m \\ \hline u & v \end{array}\right).$$

Ignoring partition lines, the product is

$$\begin{pmatrix} 2g+\ 3p+\ 4l+\ 5u, & 2h+\ 3q+\ 4m+\ 5v \\ 6g+\ 7p+\ 8l+\ 9u, & 6h+\ 7q+\ 8m+\ 9v \\ 10g+11p+12l+13u, & 10h+11q+12m+13v \\ 14g+15p+16l+17u, & 14h+15q+16m+17v \end{pmatrix}.$$

Now write the matrices in abbreviated form

$$\left(\begin{array}{c|c} \mathbf{a} & \mathbf{b} \\ \hline \mathbf{c} & \mathbf{d} \end{array}\right), \quad \left(\begin{array}{c} \mathbf{x} \\ \hline \mathbf{y} \end{array}\right),$$

with product

$$\left(\begin{array}{c} \mathbf{ax}+\mathbf{by} \\ \hline \mathbf{cx}+\mathbf{dy} \end{array}\right).$$

Since

$$\mathbf{a} \equiv \begin{pmatrix} 2 & 3 & 4 \\ 6 & 7 & 8 \end{pmatrix}, \quad \mathbf{b} \equiv \begin{pmatrix} 5 \\ 9 \end{pmatrix},$$

$$\mathbf{c} \equiv \begin{pmatrix} 10 & 11 & 12 \\ 14 & 15 & 16 \end{pmatrix}, \quad \mathbf{d} \equiv \begin{pmatrix} 13 \\ 17 \end{pmatrix}$$

and

$$\mathbf{x} \equiv \begin{pmatrix} g & h \\ p & q \\ l & m \end{pmatrix},$$

$$\mathbf{y} \equiv (u \quad v),$$

the relevant products are

$$\mathbf{ax} \equiv \begin{pmatrix} 2g+3p+4l, & 2h+3q+4m \\ 6g+7p+8l, & 6h+7q+8m \end{pmatrix}, \quad \mathbf{by} \equiv \begin{pmatrix} 5u & 5v \\ 9u & 9v \end{pmatrix},$$

$$\mathbf{cx} \equiv \begin{pmatrix} 10g+11p+12l, & 10h+11q+12m \\ 14g+15p+16l, & 14h+15q+16m \end{pmatrix}, \quad \mathbf{dy} \equiv \begin{pmatrix} 13u & 13v \\ 17u & 17v \end{pmatrix}.$$

Hence

$$\mathbf{ax}+\mathbf{by} \equiv \begin{pmatrix} 2g+3p+4l+5u, & 2h+3q+4m+5v \\ 6g+7p+8l+9u, & 6h+7q+8m+9v \end{pmatrix},$$

$$\mathbf{cx}+\mathbf{dy} \equiv \begin{pmatrix} 10g+11p+12l+13u, & 10h+11q+12m+13v \\ 14g+15p+16l+17u, & 14h+15q+16m+17v \end{pmatrix},$$

so that the matrix

$$\left(\begin{array}{c} \mathbf{ax}+\mathbf{by} \\ \hline \mathbf{cx}+\mathbf{dy} \end{array}\right)$$

is precisely the product matrix detailed before.

Returning to the general discussion, consider a product which (though not itself general) is adequate for present purposes:

$$\left(\begin{array}{ccc|cc} a_{11} & a_{12} & a_{13} & a_{14} & a_{15} \\ a_{21} & a_{22} & a_{23} & a_{24} & a_{25} \\ \hline a_{31} & a_{32} & a_{33} & a_{34} & a_{35} \end{array}\right) \left(\begin{array}{c|c} b_{11} & b_{12} \\ b_{21} & b_{22} \\ b_{31} & b_{32} \\ \hline b_{41} & b_{42} \\ b_{51} & b_{52} \end{array}\right),$$

partitioned conformably for multiplication. Write it in the form

$$\left(\begin{array}{c|c} \mathbf{a} & \mathbf{b} \\ \hline \mathbf{c} & \mathbf{d} \end{array}\right) \left(\begin{array}{c|c} \mathbf{p} & \mathbf{q} \\ \hline \mathbf{r} & \mathbf{s} \end{array}\right).$$

It is required to prove that *the product is equal to the partitioned matrix*

$$\left(\begin{array}{c|c} \mathbf{ap}+\mathbf{br} & \mathbf{aq}+\mathbf{bs} \\ \hline \mathbf{cp}+\mathbf{dr} & \mathbf{cq}+\mathbf{ds} \end{array}\right).$$

Consider first the term $\mathbf{ap}+\mathbf{br}$. This is

$$\begin{pmatrix} a_{11} & a_{12} & a_{13} \\ a_{21} & a_{22} & a_{23} \end{pmatrix} \begin{pmatrix} b_{11} \\ b_{21} \\ b_{31} \end{pmatrix} + \begin{pmatrix} a_{14} & a_{15} \\ a_{24} & a_{25} \end{pmatrix} \begin{pmatrix} b_{41} \\ b_{51} \end{pmatrix}$$

$$= \begin{pmatrix} a_{11}b_{11}+a_{12}b_{21}+a_{13}b_{31} \\ a_{21}b_{11}+a_{22}b_{21}+a_{23}b_{31} \end{pmatrix} + \begin{pmatrix} a_{14}b_{41}+a_{15}b_{51} \\ a_{24}b_{41}+a_{25}b_{51} \end{pmatrix}$$

$$= \begin{pmatrix} a_{11}b_{11}+a_{12}b_{21}+a_{13}b_{31}+a_{14}b_{41}+a_{15}b_{51} \\ a_{21}b_{11}+a_{22}b_{21}+a_{23}b_{31}+a_{24}b_{41}+a_{25}b_{51} \end{pmatrix}$$

$$= \begin{pmatrix} \sum_1^5 a_{1\lambda}b_{\lambda 1} \\ \sum_1^5 a_{2\lambda}b_{\lambda 1} \end{pmatrix}.$$

By similar argument, which should be verified by explicit calculation,

$$\mathbf{aq}+\mathbf{bs} = \begin{pmatrix} \sum_1^5 a_{1\lambda}b_{\lambda 2} \\ \sum_1^5 a_{2\lambda}b_{\lambda 2} \end{pmatrix},$$

$$\mathbf{cp}+\mathbf{dr} = \left(\sum_1^5 a_{3\lambda}b_{\lambda 1}\right),$$

$$\mathbf{cq}+\mathbf{ds} = \left(\sum_1^5 a_{3\lambda}b_{\lambda 2}\right).$$

Hence

$$\left(\begin{array}{c|c} \mathbf{ap}+\mathbf{br} & \mathbf{aq}+\mathbf{bs} \\ \hline \mathbf{cp}+\mathbf{dr} & \mathbf{cq}+\mathbf{ds} \end{array}\right)$$

$$= \left(\begin{array}{c|c} \sum_1^5 a_{1\lambda}b_{\lambda 1} & \sum_1^5 a_{1\lambda}b_{\lambda 2} \\ \sum_1^5 a_{2\lambda}b_{\lambda 1} & \sum_1^5 a_{2\lambda}b_{\lambda 2} \\ \hline \sum_1^5 a_{3\lambda}b_{\lambda 1} & \sum_1^5 a_{3\lambda}b_{\lambda 2} \end{array}\right),$$

and this, when the partition lines are removed, is precisely the product of the two given matrices.

***Illustration* 2.**

(i)
$$\left(\begin{array}{cc} 1 & 2 \\ 3 & 4 \\ \hline 5 & 6 \end{array}\right) \begin{pmatrix} 9 & 8 \\ 7 & 6 \end{pmatrix}$$

$$= \left(\begin{array}{c} \begin{pmatrix} 1 & 2 \\ 3 & 4 \end{pmatrix} \begin{pmatrix} 9 & 8 \\ 7 & 6 \end{pmatrix} \\ \hline (5 \;\; 6) \begin{pmatrix} 9 & 8 \\ 7 & 6 \end{pmatrix} \end{array}\right) = \left(\begin{array}{cc} 9+14, & 8+12 \\ 27+28, & 24+24 \\ \hline 45+42, & 40+36 \end{array}\right) = \left(\begin{array}{cc} 23, & 20 \\ 55, & 48 \\ \hline 87, & 76 \end{array}\right).$$

(ii)
$$\left(\begin{array}{cc|c} 1 & 0 & 5 \\ 0 & 1 & 6 \\ \hline 1 & 0 & 7 \\ 0 & 1 & 8 \end{array}\right) \left(\begin{array}{cc} 2 & 0 \\ 0 & 2 \\ \hline 9 & 10 \end{array}\right)$$

$$= \left(\begin{array}{c} \begin{pmatrix} 1 & 0 \\ 0 & 1 \end{pmatrix} \begin{pmatrix} 2 & 0 \\ 0 & 2 \end{pmatrix} + \begin{pmatrix} 5 \\ 6 \end{pmatrix} (9 \quad 10) \\ \hline \begin{pmatrix} 1 & 0 \\ 0 & 1 \end{pmatrix} \begin{pmatrix} 2 & 0 \\ 0 & 2 \end{pmatrix} + \begin{pmatrix} 7 \\ 8 \end{pmatrix} (9 \quad 10) \end{array}\right)$$

$$= \left(\begin{array}{c} \begin{pmatrix} 2 & 0 \\ 0 & 2 \end{pmatrix} + \begin{pmatrix} 45 & 50 \\ 54 & 60 \end{pmatrix} \\ \hline \begin{pmatrix} 2 & 0 \\ 0 & 2 \end{pmatrix} + \begin{pmatrix} 63 & 70 \\ 72 & 80 \end{pmatrix} \end{array}\right)$$

$$= \left(\begin{array}{cc} 47 & 50 \\ 54 & 62 \\ \hline 65 & 70 \\ 72 & 82 \end{array}\right).$$

REMARK. In products like these, the technique of block multiplication is of no real help. The examples which follow do, however, illustrate the process and should be worked carefully for practice.

Examples

Evaluate the products of the partitioned matrices:

1.
$$\left(\begin{array}{ccc} 1 & 2 & 3 \\ 4 & 5 & 6 \\ \hline 7 & 8 & 9 \end{array}\right) \begin{pmatrix} 3 \\ 2 \\ 1 \end{pmatrix}.$$

2. $$\left(\begin{array}{ccc|c} 1 & 0 & 0 & 0 \\ 0 & \sin\theta & \cos\theta & -1 \\ \hline -1 & -\cos\theta & \sin\theta & 1 \end{array}\right) \left(\begin{array}{c|c} 1 & \sin\theta \\ \sin\theta & 1 \\ \cos\theta & 1 \\ \hline 1 & \cos\theta \end{array}\right).$$

3. $$\left(\begin{array}{cc|c} a & h & g \\ h & b & f \\ \hline g & f & c \end{array}\right) \left(\begin{array}{c|cc} a & h & g \\ h & b & f \\ \hline g & f & c \end{array}\right).$$

4. $$\left(\begin{array}{cc} a & b \\ \hline c & d \end{array}\right) \left(\begin{array}{c} x \\ \hline y \end{array}\right).$$

5. $$\left(\begin{array}{cc|cc} 1 & 2 & 3 & 4 \\ \hline a & b & c & d \\ p & q & r & s \end{array}\right) \left(\begin{array}{cc|cc} 1 & 0 & 0 & 0 \\ 0 & 1 & 0 & 0 \\ \hline 0 & 0 & 1 & 0 \\ 0 & 0 & 0 & 1 \end{array}\right).$$

6. $$\left(\begin{array}{cc|c} a & h & x \\ h & b & y \\ \hline x & y & 0 \end{array}\right) \left(\begin{array}{ccc} 1 & 0 & 0 \\ 0 & 1 & 0 \\ \hline 1 & 1 & 1 \end{array}\right).$$

7. $$\left(\begin{array}{cc|c} 0 & -h & g \\ h & 0 & -f \\ \hline -g & f & 0 \end{array}\right) \left(\begin{array}{c|cc} 0 & -h & g \\ h & 0 & -f \\ \hline -g & f & 0 \end{array}\right).$$

8. $$\left(\begin{array}{c|c} \mathbf{a} & \mathbf{b} \\ \hline \mathbf{c} & \mathbf{d} \end{array}\right) \left(\begin{array}{c|c} \mathbf{I} & \mathbf{0} \\ \hline \mathbf{0} & \mathbf{I} \end{array}\right), \quad \left(\begin{array}{c|c} \mathbf{a}^{-1} & \mathbf{0} \\ \hline -\mathbf{c}\mathbf{a}^{-1} & \mathbf{I} \end{array}\right) \left(\begin{array}{c|c} \mathbf{a} & \mathbf{b} \\ \hline \mathbf{c} & \mathbf{d} \end{array}\right),$$

$$\left(\begin{array}{c|c} \mathbf{a} & \mathbf{b} \\ \hline \mathbf{c} & \mathbf{d} \end{array}\right) \left(\begin{array}{c|c} \mathbf{I} & \mathbf{0} \\ \hline \mathbf{0} & \mathbf{0} \end{array}\right), \quad \left(\begin{array}{c|c} \mathbf{a} & \mathbf{b} \\ \hline \mathbf{c} & \mathbf{d} \end{array}\right) \left(\begin{array}{c|c} \mathbf{I} & -\mathbf{a}^{-1}\mathbf{b} \\ \hline \mathbf{0} & \mathbf{I} \end{array}\right),$$

where the types of the matrices are conformable for multiplication and $\mathbf{I}$, $\mathbf{0}$ are unit and zero matrices of appropriate orders.

Verify your results independently when

$$\mathbf{a} \equiv \begin{pmatrix} 0 & 1 \\ 1 & 0 \end{pmatrix}, \quad \mathbf{b} \equiv \begin{pmatrix} 1 & 1 \\ 1 & 1 \end{pmatrix}, \quad \mathbf{c} \equiv \begin{pmatrix} 1 & 1 \\ 1 & 0 \end{pmatrix}, \quad \mathbf{d} \equiv \begin{pmatrix} 0 & 1 \\ 1 & 1 \end{pmatrix}.$$

9. Prove that, subject to conformability,

$$\left(\begin{array}{c|c} \mathbf{a} & \mathbf{I} \\ \hline \mathbf{I} & \mathbf{a}^{-1} \end{array}\right) \left(\begin{array}{c|c} \mathbf{a}^{-1} & \mathbf{I} \\ \hline \mathbf{I} & \mathbf{a} \end{array}\right) = 2\left(\begin{array}{c|c} \mathbf{I} & \mathbf{a} \\ \hline \mathbf{a}^{-1} & \mathbf{I} \end{array}\right),$$

$$\left(\begin{array}{c|c} \mathbf{a} & \mathbf{0} \\ \hline \mathbf{I} & \mathbf{a}^{-1} \end{array}\right) \left(\begin{array}{c|c} \mathbf{a}^{-1} & \mathbf{I} \\ \hline \mathbf{0} & \mathbf{a} \end{array}\right) = \left(\begin{array}{c|c} \mathbf{I} & \mathbf{a} \\ \hline \mathbf{a}^{-1} & 2\mathbf{I} \end{array}\right).$$

Some Harder Examples

1. Given four matrices **a**, **b**, **c**, **d**, each of type 2×2, where **a** is non-singular and where **ac** = **ca**, and given also the zero matrix **0** and the unit matrix **I**, prove that

(i) $$\left(\begin{array}{c|c} \mathbf{a} & \mathbf{b} \\ \hline \mathbf{c} & \mathbf{d} \end{array}\right) \left(\begin{array}{c|c} \mathbf{I} & -\mathbf{a}^{-1}\mathbf{b} \\ \hline \mathbf{0} & \mathbf{I} \end{array}\right) = \left(\begin{array}{c|c} \mathbf{a} & \mathbf{0} \\ \hline \mathbf{c} & \mathbf{d}-\mathbf{c}\mathbf{a}^{-1}\mathbf{b} \end{array}\right),$$

(ii) $$\det\left(\begin{array}{c|c} \mathbf{a} & \mathbf{0} \\ \hline \mathbf{c} & \mathbf{d} \end{array}\right) = \det \mathbf{a} \det \mathbf{d},$$

(iii) $$\det\left(\begin{array}{c|c} \mathbf{a} & \mathbf{b} \\ \hline \mathbf{c} & \mathbf{d} \end{array}\right) = \det(\mathbf{ad}-\mathbf{cb}).$$

2. Given a 3×2 matrix **a** and a 2×3 matrix **b**, prove that ($\mathbf{I}_2$, $\mathbf{I}_3$ being unit matrices)

$$\left(\begin{array}{c|c} \mathbf{I}_3 & \mathbf{a} \\ \hline \mathbf{b} & \mathbf{I}_2 \end{array}\right) = \left(\begin{array}{c|c} \mathbf{I}_3-\mathbf{ab} & \mathbf{a} \\ \hline \mathbf{0} & \mathbf{I}_2 \end{array}\right) \left(\begin{array}{c|c} \mathbf{I}_3 & \mathbf{0} \\ \hline \mathbf{b} & \mathbf{I}_2 \end{array}\right)$$

$$= \left(\begin{array}{c|c} \mathbf{I}_3 & \mathbf{0} \\ \hline \mathbf{b} & \mathbf{I}_2 \end{array}\right) \left(\begin{array}{c|c} \mathbf{I}_3 & \mathbf{a} \\ \hline \mathbf{0} & \mathbf{I}_2-\mathbf{ba} \end{array}\right)$$

and deduce that

$$\det(\mathbf{I}_3-\mathbf{ab}) = \det(\mathbf{I}_2-\mathbf{ba}).$$

3. Prove that, if **a**, **b** are square 2×2 matrices and $i^2 = -1$, then

$$\left(\begin{array}{c|c} \mathbf{I} & i\mathbf{I} \\ \hline \mathbf{0} & \mathbf{I} \end{array}\right) \left(\begin{array}{c|c} \mathbf{a} & -\mathbf{b} \\ \hline \mathbf{b} & \mathbf{a} \end{array}\right) \left(\begin{array}{c|c} \mathbf{I} & -i\mathbf{I} \\ \hline \mathbf{0} & \mathbf{I} \end{array}\right) = \left(\begin{array}{c|c} \mathbf{a}+i\mathbf{b} & \mathbf{0} \\ \hline \mathbf{b} & \mathbf{a}-i\mathbf{b} \end{array}\right).$$

Deduce that

$$\det\left(\begin{array}{c|c} \mathbf{a} & -\mathbf{b} \\ \hline \mathbf{b} & \mathbf{a} \end{array}\right) = \det(\mathbf{a}+i\mathbf{b})\det(\mathbf{a}-i\mathbf{b}).$$

4. Column operators: particular example. Suppose that a matrix of type, say, 4×3 is given; for example,

$$\mathbf{a} \equiv \begin{pmatrix} 1 & 2 & 1 \\ 2 & 1 & -1 \\ -1 & 0 & 1 \\ 3 & -1 & -4 \end{pmatrix}.$$

Its three columns may be regarded as three column-vectors:

$$\mathbf{x} \equiv \{1, 2, -1, 3\}, \quad \mathbf{y} \equiv \{2, 1, 0, -1\}, \quad \mathbf{z} \equiv \{1, -1, 1, 4\},$$

and the matrix can be written in partitioned form (with commas instead of vertical lines, for neatness)

$$\mathbf{a} \equiv (\mathbf{x}, \mathbf{y}, \mathbf{z}).$$

Now let **p** be a column vector the number of whose elements is equal to the number of columns of **a**; say

$$\mathbf{p} \equiv \{4, 3, 2\},$$

and form the product **ap**. In partitioned form, this gives a column-vector of four elements,

$$\mathbf{ap} \equiv (\mathbf{x}, \mathbf{y}, \mathbf{z})\begin{pmatrix}4\\3\\2\end{pmatrix}$$

$$\equiv (4\mathbf{x}+3\mathbf{y}+2\mathbf{z}).$$

In other words, *post-multiplication by the vector* **p** *gives a vector which is a linear combination* (p. 145) *of the columns of* **a**.

Any number of such linear combinations may be formed, and they can be 'kept separate' by collecting the *combining vectors* such as **p** into the columns of a single matrix. For example, two such linear combinations, say

$$4\mathbf{x}+3\mathbf{y}+2\mathbf{z} \quad \text{and} \quad 3\mathbf{x}-2\mathbf{y}+\mathbf{z}$$

can be obtained from the product

$$(\mathbf{x}, \mathbf{y}, \mathbf{z})\begin{pmatrix}4 & 3\\3 & -2\\2 & 1\end{pmatrix}$$

which, on multiplying in the partitioned form

$$(\mathbf{x}, \mathbf{y}, \mathbf{z})\left(\begin{array}{c|c}4 & 3\\3 & -2\\2 & 1\end{array}\right),$$

gives the 4×2 matrix

$$(4\mathbf{x}+3\mathbf{y}+2\mathbf{z}, \quad 3\mathbf{x}-2\mathbf{y}+\mathbf{z})$$

whose two columns are precisely the required linear combinations.

Examples

1. Obtain in this way the matrix **q** of type 3×3 whose three columns are the linear combinations

$$3\mathbf{x}-\mathbf{y}+\mathbf{z}, \quad 2\mathbf{x}+\mathbf{y}+4\mathbf{z}, \quad \mathbf{x}-\mathbf{z}.$$

2. Prove that (with **a** as defined), if **r** is the column vector $\{1, -1, 1\}$, then

$$\mathbf{ar} = \mathbf{0},$$

where **0** is a zero column vector.

DEFINITION. A matrix used in this way to form linear combinations of the columns of a given matrix **a** may be called a *column operator*, operating on the columns of **a**.

5. Column operators: more general account.

Let **a** be a given matrix of type $m \times n$ whose n columns are the vectors **x**, **y**, **z**, ... of m elements each. Suppose that **p** is a matrix of n rows and having, say, k columns, so that

$$\mathbf{p} \equiv \begin{pmatrix} p_{11} & p_{21} & \cdots & p_{k1} \\ p_{12} & p_{22} & \cdots & p_{k2} \\ \cdots & \cdots & \cdots & \cdots \\ p_{1n} & p_{2n} & \cdots & p_{kn} \end{pmatrix}.$$

The product **ap** is a matrix whose k columns are k linear combinations of the rows of **a**:

$$\mathbf{ap} \equiv (\mathbf{x}, \mathbf{y}, \mathbf{z}, \ldots) \begin{pmatrix} p_{11} & p_{21} & \cdots & p_{k1} \\ p_{12} & p_{22} & \cdots & p_{k2} \\ p_{13} & p_{23} & \cdots & p_{k3} \\ \cdots & \cdots & \cdots & \cdots \end{pmatrix}$$

$$\equiv (p_{11}\mathbf{x}+p_{12}\mathbf{y}+p_{13}\mathbf{z}+\ldots, \quad \ldots, \quad p_{k1}\mathbf{x}+p_{k2}\mathbf{y}+p_{k3}\mathbf{z}+\ldots).$$

***Illustration* 3.** *Special cases.* For definiteness, let **a** be the 4×3 matrix considered in §4:

$$\mathbf{a} \equiv (\mathbf{x}, \mathbf{y}, \mathbf{z})$$

$$\equiv \begin{pmatrix} 1 & 2 & 1 \\ 2 & 1 & -1 \\ -1 & 0 & 1 \\ 3 & -1 & -4 \end{pmatrix}.$$

(i) Select **p** to be the square 3×3 matrix

$$\mathbf{p} \equiv \begin{pmatrix} 1 & \lambda & 0 \\ 0 & 1 & 0 \\ 0 & \mu & 1 \end{pmatrix}.$$

Then

$$\mathbf{ap} \equiv (\mathbf{x}, \mathbf{y}, \mathbf{z}) \left(\begin{array}{c|c|c} 1 & \lambda & 0 \\ 0 & 1 & 0 \\ 0 & \mu & 1 \end{array}\right)$$

$$\equiv (\mathbf{x}, \quad \lambda\mathbf{x}+\mathbf{y}+\mu\mathbf{z}, \quad \mathbf{z});$$

that is, *the matrix* **ap** *is formed from* **a** *by adding* λ *times the elements of column* 1 *and* μ *times the elements of column* 3 *to the elements of column* 2—a manoeuvre familiar in the evaluation of determinants.

(ii) Select **p** to be the square 3×3 matrix

$$\mathbf{p} \equiv \begin{pmatrix} \lambda & 0 & 0 \\ 0 & 1 & 0 \\ 0 & 0 & 1 \end{pmatrix}.$$

Then

$$\mathbf{ap} \equiv (\mathbf{x}, \mathbf{y}, \mathbf{z}) \left(\begin{array}{c|c|c} \lambda & 0 & 0 \\ 0 & 1 & 0 \\ 0 & 0 & 1 \end{array}\right)$$

$$\equiv (\lambda\mathbf{x}, \mathbf{y}, \mathbf{z});$$

that is, **ap** *is the matrix formed from* **a** *by multiplying the elements of the first column by* λ.

(iii) Select **p** to be the square 3×3 matrix

$$\mathbf{p} \equiv \begin{pmatrix} 0 & 1 & 0 \\ 1 & 0 & 0 \\ 0 & 0 & 1 \end{pmatrix}.$$

Then

$$\mathbf{ap} \equiv (\mathbf{x}, \mathbf{y}, \mathbf{z}) \left(\begin{array}{c|c|c} 0 & 1 & 0 \\ 1 & 0 & 0 \\ 0 & 0 & 1 \end{array}\right)$$

$$\equiv (\mathbf{y}, \quad \mathbf{x}, \quad \mathbf{z});$$

that is, **ap** *is the matrix formed from* **a** *by interchanging the first two columns.*

Revision Examples

1. Given an $m \times n$ matrix **a**, prove that it is possible to

(i) interchange row r and row s,

(ii) multiply row r by λ,

(iii) add $\lambda \times$ row s to row r

by forming the matrix **ba**, where **b** is an $m \times m$ matrix which differs from the unit matrix $\mathbf{I}_m$ in having

(i) $b_{rr} = b_{ss} = 0, b_{rs} = b_{sr} = 1$,

(ii) $b_{rr} = \lambda$,

(iii) $b_{rs} = \lambda$.

2. Given that

$$\mathbf{a} \equiv \left(\begin{array}{c|c} \mathbf{I}_r & \mathbf{p} \\ \hline 0 & \mathbf{I}_s \end{array}\right)$$

obtain (do not merely verify) the inverse of **a** in the form

$$\mathbf{a}^{-1} = \begin{pmatrix} \mathbf{I}_r & -\mathbf{p} \\ 0 & \mathbf{I}_s \end{pmatrix},$$

where $\mathbf{I}_r$, $\mathbf{I}_s$ are unit matrices of orders r, s.

3. Given that

$$\mathbf{a} \equiv \begin{pmatrix} 3 & 5 & 7 \\ 4 & 2 & 6 \\ 1 & 9 & 8 \\ -2 & 3 & -5 \end{pmatrix},$$

find the matrix **p** such that **ap** differs from **a** in having

(i) column 1 replaced by column $1+(\lambda$ times column 3),
(ii) columns 1 and 3 interchanged,
(iii) columns 1, 2, 3 multiplied by 2, -3, 4,
(iv) column 3 replaced by column $3-(7$ times column 1).

4. The *Pauli* matrices are

$$\mathbf{a} \equiv \begin{pmatrix} 0 & 1 \\ 1 & 0 \end{pmatrix}, \quad \mathbf{b} \equiv \begin{pmatrix} 0 & -i \\ i & 0 \end{pmatrix}, \quad \mathbf{c} \equiv \begin{pmatrix} 1 & 0 \\ 0 & -1 \end{pmatrix}.$$

Prove that

(i) $\mathbf{a}^2 = \mathbf{b}^2 = \mathbf{c}^2 = \mathbf{I}$;

(ii) $\mathbf{bc} = -\mathbf{cb} = i\mathbf{a}$,

$\mathbf{ca} = -\mathbf{ac} = i\mathbf{b}$,

$\mathbf{ab} = -\mathbf{ba} = i\mathbf{c}$.

The Eddington-Dirac matrices are

$$\mathbf{E}_1 \equiv \left(\begin{array}{c|c} i\mathbf{a} & \mathbf{0} \\ \hline \mathbf{0} & i\mathbf{a} \end{array}\right), \quad \mathbf{E}_2 \equiv \left(\begin{array}{c|c} i\mathbf{c} & \mathbf{0} \\ \hline \mathbf{0} & i\mathbf{c} \end{array}\right), \quad \mathbf{E}_3 \equiv \left(\begin{array}{c|c} \mathbf{0} & -\mathbf{b} \\ \hline \mathbf{b} & \mathbf{0} \end{array}\right),$$

$$\mathbf{E}_4 \equiv \left(\begin{array}{c|c} -i\mathbf{b} & \mathbf{0} \\ \hline \mathbf{0} & i\mathbf{b} \end{array}\right), \quad \mathbf{E}_5 \equiv \left(\begin{array}{c|c} \mathbf{0} & -i\mathbf{b} \\ \hline -i\mathbf{b} & \mathbf{0} \end{array}\right).$$

Prove that, for all values of p,

$$\mathbf{E}_p^2 = -\mathbf{I},$$

and that, for all distinct values of p and q,

$$\mathbf{E}_p\mathbf{E}_q = -\mathbf{E}_q\mathbf{E}_p.$$

5. Prove that, if

$$\mathbf{a} \equiv \begin{pmatrix} a_{11} & a_{12} & a_{13} & a_{14} \\ a_{21} & a_{22} & a_{23} & a_{24} \\ a_{31} & a_{32} & a_{33} & a_{34} \\ a_{41} & a_{42} & a_{43} & a_{44} \end{pmatrix},$$

then

$$\mathbf{a}\begin{pmatrix} 0 & -1 & 0 & 0 \\ 1 & 0 & 0 & 0 \\ 0 & 0 & 0 & -1 \\ 0 & 0 & 1 & 0 \end{pmatrix} = \begin{pmatrix} a_{12} & -a_{11} & a_{14} & -a_{13} \\ a_{22} & -a_{21} & a_{24} & -a_{23} \\ a_{32} & -a_{31} & a_{34} & -a_{33} \\ a_{42} & -a_{41} & a_{44} & -a_{43} \end{pmatrix}.$$

Denoting the last matrix by the symbol $\mathbf{b}$, prove, by considering the product $\mathbf{b}'\mathbf{a}$, that

$$\{\det \mathbf{a}\}^2$$

can be expressed as the determinant of a skew-symmetric matrix.

6. A square matrix $\mathbf{a}$ has all its elements 0, 1 or -1, and in each row or column there is exactly one element that is not zero. Prove that $\mathbf{a}^2$, $\mathbf{a}^3, \ldots$ are of the same type and hence show that there is a positive integer k such that $\mathbf{a}^k = \mathbf{I}$.

7. Prove that Question 6 remains true if the non-zero elements of $\mathbf{a}$ are replaced by arbitrary pth roots of unity, where p is a given positive integer.

15

MAPPINGS AND VECTOR SPACES

This chapter seeks to explain, from an experimental and even intuitive point of view, some elementary properties about vector spaces, with a particular emphasis on those features which arise in the discussion of systems of linear equations. A proper logical treatment is hard, but it is hoped that this account will serve as a suitable introduction.

1. First ideas. Let $\mathbf{x}$ denote the column vector

$$\mathbf{x} \equiv \{x_1, x_2, \ldots, x_n\}$$

of n elements selected from, say, the field of complex numbers. *The set of all such vectors forms a vector space subject to the rules*

$$\mathbf{x}+\mathbf{y} \equiv \{x_1+y_1, x_2+y_2, \ldots, x_n+y_n\},$$
$$k\mathbf{x} \equiv \{kx_1, kx_2, \ldots, kx_n\}.$$

Text Example

1. Prove the statement just made. (The conditions to be satisfied are given on p. 64.)

DEFINITIONS. The vectors $\mathbf{x}, \mathbf{y}, \mathbf{z}, \ldots$ are said to be *linearly independent* if the relation

$$\lambda\mathbf{x}+\mu\mathbf{y}+\nu\mathbf{z}+\ldots = 0$$

can be satisfied only when

$$\lambda = \mu = \nu = \ldots = 0.$$

Otherwise they are *linearly dependent.*

For example, the three column vectors

$$\mathbf{x} \equiv \{1, 1, 1\}, \quad \mathbf{y} \equiv \{1, 2, 3\}, \quad \mathbf{z} \equiv \{2, 3, 5\}$$

are linearly independent, whereas the vectors

$$\mathbf{x} \equiv \{1, 1, 1\}, \quad \mathbf{y} \equiv \{1, 2, 3\}, \quad \mathbf{z} \equiv \{2, 3, 4\}$$

are dependent, satisfying the relation

$$\mathbf{x}+\mathbf{y}-\mathbf{z} = 0.$$

A standard set of linearly independent vectors is (for $n = 4$)

$$\{1, 0, 0, 0\}, \quad \{0, 1, 0, 0\}, \quad \{0, 0, 1, 0\}, \quad \{0, 0, 0, 1\}.$$

2. The dimension of a vector space. Given any set of vectors **x**, **y**, **z**, ..., **w**, the totality of vectors

$$\lambda\mathbf{x}+\mu\mathbf{y}+\nu\mathbf{z}+\ldots+\rho\mathbf{w},$$

for varying $\lambda, \mu, \nu, \ldots, \rho$, forms a vector space said to be *generated* by them. For example, the three column vectors

$$\mathbf{x} \equiv \{1, 1, 1\}, \quad \mathbf{y} \equiv \{1, 2, 3\}, \quad \mathbf{z} \equiv \{2, 3, 4\}$$

generate a space in which an arbitrary vector can be expressed in the form

$$\{\lambda+\mu+2\nu, \quad \lambda+2\mu+3\nu, \quad \lambda+3\mu+4\nu\}.$$

It becomes an urgent problem *to decide what, starting from given* **x**, **y**, **z**, *is the least number of multipliers* λ, μ, ν *that will 'cover' the whole of the space.*

It was proved, in § 1, that the three particular vectors **x**, **y**, **z** are linearly dependent, being subject to the restriction

$$\mathbf{x}+\mathbf{y}-\mathbf{z} = 0$$

and so a typical vector

$$\lambda\mathbf{x}+\mu\mathbf{y}+\nu\mathbf{z}$$

can be written in the alternative form

$$\lambda\mathbf{x}+\mu\mathbf{y}+\nu(\mathbf{x}+\mathbf{y}),$$

or

$$(\lambda+\nu)\mathbf{x}+(\mu+\nu)\mathbf{y}$$

depending on **x** and **y** only.

This means two things:

(i) a typical vector

$$\lambda\mathbf{x}+\mu\mathbf{y}+\nu\mathbf{z}$$

defined in terms of **x**, **y**, **z** can be written in the form

$$\lambda'\mathbf{x}+\mu'\mathbf{y} \qquad (\lambda' = \lambda+\nu, \mu' = \mu+\nu)$$

depending on **x**, **y** only;

(ii) a typical vector

$$\alpha\mathbf{x}+\beta\mathbf{y}$$

defined in terms of $\mathbf{x}$, $\mathbf{y}$ can be written (in many ways) in the form

$$\alpha'\mathbf{x}+\beta'\mathbf{y}+\gamma'\mathbf{z}$$

depending on all of $\mathbf{x}$, $\mathbf{y}$, $\mathbf{z}$, where the constants α', β', γ' are subject to the two relations

$$\alpha'+\gamma' = \alpha, \quad \beta'+\gamma' = \beta.$$

In other words, *the totality of vectors generated by* $\mathbf{x}$, $\mathbf{y}$, $\mathbf{z}$ *is precisely the same as the totality of vectors generated by* $\mathbf{x}$, $\mathbf{y}$.

Text Example

1. Prove that, in the notation of this paragraph, the totality of vectors generated by $\mathbf{x}$, $\mathbf{z}$ is the same as the totality of vectors generated by $\mathbf{y}$, $\mathbf{z}$.

DEFINITIONS. The least number of vectors (necessarily linearly independent) in terms of which all the vectors of a given vector space can be expressed in the form

$$\lambda\mathbf{x}+\mu\mathbf{y}+\nu\mathbf{z}+\ldots$$

is called the *dimension* of the vector space; the linearly independent vectors used in such an expression are said to form a *basis* for the space.

If all the vectors of a space can be expressed in terms of vectors $\mathbf{x}$, $\mathbf{y}$, $\mathbf{z}, \ldots, \mathbf{w}$ in the form

$$\lambda\mathbf{x}+\mu\mathbf{y}+\nu\mathbf{z}+\ldots+\rho\mathbf{w},$$

then the vectors $\mathbf{x}, \mathbf{y}, \mathbf{z}, \ldots, \mathbf{w}$, *whether linearly independent or not*, are said to *span* the space.

The number of vectors in a system spanning a space cannot be less than the dimension of the space.

Examples

1. Find the dimensions of the spaces spanned by the following sets of column vectors:

(i) $\{1, 0, 0\}, \{0, 1, 0\}, \{0, 0, 1\}$;
(ii) $\{1, 1\}, \{2, 3\}, \{5, 7\}, \{0, 1\}, \{-3, -4\}$;
(iii) $\{1, 0, 0\}, \{0, 1, 0\}, \{1, 1, 1\}, \{1, 3, 5\}, \{1, -2, 6\}$;
(iv) $\{1, 0, 1\}, \{0, 1, 1\}, \{1, 1, 0\}, \{2, 4, 6\}, \{3, 6, 9\}$;
(v) $\{1, 1, 1, 1\}, \{1, 2, 3, 4\}, \{1, 0, 0, 1\}, \{2, 1, 1, 2\}, \{3, 2, 3, 6\}$;
(vi) $\{1, 2, 3, 4\}, \{5, 6, 7, 8\}$;
(vii) $\{1, 2, 3, 4\}, \{2, 4, 6, 8\}$;
(viii) $\{0, 1, -1\}, \{-1, 0, 1\}, \{1, -1, 0\}$.

2. Determine whether the vectors

$$\{1, 3, 5\}, \{2, 4, 6\}$$

belong to the space spanned by the vectors

$$\{1, 0, 0\}, \{0, 1, 0\}, \{0, 3, 5\}, \{4, 3, 2\}.$$

3. Find a vector common to the space spanned by

$$\{1, 2, 3\}, \{3, 2, 1\}$$

and the space spanned by

$$\{1, 0, 1\}, \{3, 4, 3\}.$$

Find also a vector common to the space spanned by

$$\{1, 3, 2\}, \{4, -1, 1\}$$

and the space spanned by

$$\{2, 5, 1\}, \{3, -14, 1\}.$$

3. Subspaces. Given a set of vectors **x**, **y**, **z**, ..., **w** spanning a vector space V, a selection, say **p**, **q**, ..., **t**, taken from them span, in similar manner, a vector space W of which a typical vector is

$$\alpha\mathbf{p}+\beta\mathbf{q}+\ldots+\delta\mathbf{t}.$$

Such a space W is a *subspace* of V.

***Illustration* 1.** The six vectors

$$\begin{pmatrix}1\\0\\0\end{pmatrix}, \begin{pmatrix}0\\1\\0\end{pmatrix}, \begin{pmatrix}0\\0\\1\end{pmatrix}, \begin{pmatrix}1\\1\\1\end{pmatrix}, \begin{pmatrix}1\\2\\3\end{pmatrix}, \begin{pmatrix}2\\4\\0\end{pmatrix}$$

generate a vector space V. The three vectors

$$\begin{pmatrix}1\\0\\0\end{pmatrix}, \begin{pmatrix}0\\1\\0\end{pmatrix}, \begin{pmatrix}2\\4\\0\end{pmatrix}$$

generate a subspace W; the four vectors

$$\begin{pmatrix}1\\0\\0\end{pmatrix}, \begin{pmatrix}0\\1\\0\end{pmatrix}, \begin{pmatrix}0\\0\\1\end{pmatrix}, \begin{pmatrix}1\\1\\1\end{pmatrix}$$

generate a subspace U.

The space V is of dimension 3. The subspace W is of dimension 2, and the subspace U is of dimension 3. The subspace U is, in fact, the same as the space V, a basis in each being the triplet

$$\begin{pmatrix}1\\0\\0\end{pmatrix}, \begin{pmatrix}0\\1\\0\end{pmatrix}, \begin{pmatrix}0\\0\\1\end{pmatrix}.$$

4. The matrix of linear dependence. Let

$$\mathbf{x}, \mathbf{y}, \mathbf{z}, \ldots, \mathbf{w}$$

be p column vectors of n elements each. Form the matrix

$$\mathbf{b} \equiv (\mathbf{x}, \mathbf{y}, \mathbf{z}, \ldots, \mathbf{w}),$$

of type $n \times p$, having them as columns.

A relation of linear dependence

$$\xi_1 \mathbf{x} + \xi_2 \mathbf{y} + \ldots + \xi_p \mathbf{w} = \mathbf{0}$$

can be expressed in the matrix form

$$\mathbf{b}\boldsymbol{\xi} = \mathbf{0},$$

where $\boldsymbol{\xi}$ is the column vector

$$\boldsymbol{\xi} \equiv \{\xi_1, \xi_2, \ldots, \xi_p\}.$$

The existence of k relations

$$\xi_{1i}\mathbf{x} + \xi_{2i}\mathbf{y} + \ldots + \xi_{pi}\mathbf{w} = \mathbf{0} \quad (i = 1, 2, \ldots, k)$$

can be expressed similarly in the matrix form

$$\mathbf{b}\boldsymbol{\xi} = \mathbf{0}$$

where $\boldsymbol{\xi}$ is now the $p \times k$ matrix

$$\boldsymbol{\xi} \equiv (\xi_{ij})$$

and where $\mathbf{0}$ is the zero matrix of n rows and k columns.

The establishment of linear relations connecting $\mathbf{x}, \mathbf{y}, \mathbf{z}, \ldots, \mathbf{w}$ is thus equivalent to finding vectors $\boldsymbol{\xi}$ to satisfy the equation

$$\mathbf{b}\boldsymbol{\xi} = \mathbf{0}.$$

[See Chapter 16, §8 (p. 215).]

***Illustration* 2.** The vectors $\mathbf{x}, \mathbf{y}, \mathbf{z}, \mathbf{u}, \mathbf{v}$ given respectively by

$$\begin{pmatrix}1\\1\\1\end{pmatrix}, \begin{pmatrix}1\\2\\3\end{pmatrix}, \begin{pmatrix}2\\3\\1\end{pmatrix}, \begin{pmatrix}4\\6\\5\end{pmatrix}, \begin{pmatrix}8\\13\\12\end{pmatrix}$$

may be expressed as the columns of the matrix

$$\mathbf{b} \equiv \begin{pmatrix} 1 & 1 & 2 & 4 & 8 \\ 1 & 2 & 3 & 6 & 13 \\ 1 & 3 & 1 & 5 & 12 \end{pmatrix}.$$

They are, in fact, connected (on inspection) by the relations

$$\begin{aligned} \mathbf{x}+\mathbf{y}+\mathbf{z}-\mathbf{u} \quad &= \mathbf{0}, \\ 2\mathbf{y}+\mathbf{z}+\mathbf{u}-\mathbf{v} &= \mathbf{0} \end{aligned}$$

which can be expressed in the form

$$(\mathbf{x}, \mathbf{y}, \mathbf{z}, \mathbf{u}, \mathbf{v}) \begin{pmatrix} 1 & 0 \\ 1 & 2 \\ 1 & 1 \\ -1 & 1 \\ 0 & -1 \end{pmatrix} = \mathbf{0},$$

where $\mathbf{0}$ is the zero 3×2 matrix.

These relations are by no means unique. For example,

$$(\mathbf{x}, \mathbf{y}, \mathbf{z}, \mathbf{u}, \mathbf{v}) \begin{pmatrix} 1 & 0 & 3 \\ 1 & 2 & 5 \\ 1 & 1 & 4 \\ -1 & 1 & -2 \\ 0 & -1 & -1 \end{pmatrix} = \mathbf{0},$$

where $\mathbf{0}$ is now the zero 3×3 matrix. A *systematic search* for relations may be conducted as follows:

Consider the relation

$$\mathbf{b}\boldsymbol{\xi} = \mathbf{0},$$

where $\boldsymbol{\xi}$ is the column vector $\{\xi_1, \xi_2, \xi_3, \xi_4, \xi_5\}$. This gives the linear equations

$$\begin{aligned} \xi_1+\xi_2+2\xi_3+4\xi_4+\ 8\xi_5 &= 0, \\ \xi_1+2\xi_2+3\xi_3+6\xi_4+13\xi_5 &= 0, \\ \xi_1+3\xi_2+\ \xi_3+5\xi_4+12\xi_5 &= 0. \end{aligned}$$

It is reasonable to suppose that these equations may be solved for three of ξ_i when the two others are given arbitrary values. (Compare Section I, p. 66.) If, then, $\xi_4 = \lambda$, $\xi_5 = \mu$, the equations are

$$\begin{aligned} \xi_1+\ \xi_2+2\xi_3 &= -4\lambda-\ 8\mu, \\ \xi_1+2\xi_2+3\xi_3 &= -6\lambda-13\mu, \\ \xi_1+3\xi_2+\ \xi_3 &= -5\lambda-12\mu, \end{aligned}$$

and the solution, after a short calculation, is

$$\begin{aligned} \xi_1 &= -\lambda-\mu, \\ \xi_2 &= -\lambda-3\mu, \\ \xi_3 &= -\lambda-2\mu. \end{aligned}$$

Hence $\boldsymbol{\xi}$ is the column vector

$$\{-\lambda-\mu,\ -\lambda-3\mu,\ -\lambda-2\mu,\ \lambda,\ \mu\}.$$

Note that this vector is a linear combination of the two vectors

$$\mathbf{g} \equiv \{-1, -1, -1, 1, 0\}, \quad \mathbf{h} \equiv \{-1, -3, -2, 0, 1\},$$

being $$\lambda\mathbf{g}+\mu\mathbf{h}.$$

Hence $$\mathbf{b}(\lambda\mathbf{g}+\mu\mathbf{h}) = 0$$

for *arbitrary* values of λ and μ, so that

$$\mathbf{bg} = 0, \quad \mathbf{bh} = 0.$$

The given vectors **x, y, z, u, v** thus satisfy the matrix equation

$$(\mathbf{x}, \mathbf{y}, \mathbf{z}, \mathbf{u}, \mathbf{v})\begin{pmatrix} -1 & -1 \\ -1 & -3 \\ -1 & -2 \\ 1 & 0 \\ 0 & 1 \end{pmatrix} = 0,$$

and so two relations appear naturally in the form

$$\mathbf{x}+\mathbf{y}+\mathbf{z}-\mathbf{u} = 0,$$
$$\mathbf{x}+3\mathbf{y}+2\mathbf{z}-\mathbf{v} = 0.$$

Examples

Find similarly linear relations between the sets of vectors:

1. $\begin{pmatrix} 2 \\ 1 \\ 3 \end{pmatrix}, \begin{pmatrix} 1 \\ 3 \\ 2 \end{pmatrix}, \begin{pmatrix} 3 \\ 2 \\ 1 \end{pmatrix}, \begin{pmatrix} 1 \\ 1 \\ 1 \end{pmatrix}, \begin{pmatrix} 4 \\ 3 \\ 2 \end{pmatrix}.$

2. $\begin{pmatrix} 1 \\ 0 \\ 0 \end{pmatrix}, \begin{pmatrix} 1 \\ -1 \\ 0 \end{pmatrix}, \begin{pmatrix} 3 \\ 2 \\ 1 \end{pmatrix}, \begin{pmatrix} 5 \\ 0 \\ 1 \end{pmatrix}, \begin{pmatrix} 4 \\ 6 \\ 2 \end{pmatrix}.$

3. $\begin{pmatrix} 4 \\ 3 \end{pmatrix}, \begin{pmatrix} 3 \\ 2 \end{pmatrix}, \begin{pmatrix} 2 \\ 1 \end{pmatrix}, \begin{pmatrix} 1 \\ 0 \end{pmatrix}.$

4. $\begin{pmatrix} 1 \\ 2 \\ 3 \\ 4 \end{pmatrix}, \begin{pmatrix} 0 \\ 2 \\ 3 \\ 4 \end{pmatrix}, \begin{pmatrix} 1 \\ 0 \\ 3 \\ 4 \end{pmatrix}, \begin{pmatrix} 1 \\ 2 \\ 0 \\ 4 \end{pmatrix}, \begin{pmatrix} 3 \\ -2 \\ -3 \\ 0 \end{pmatrix}.$

5. The matrix as a mapping. Let V be the vector space containing all column vectors of n elements each, and let **a** be a given matrix of type $m \times n$.

Corresponding to any vector $\mathbf{x} \in V$ may be defined a vector

$$\mathbf{y} \equiv \mathbf{ax}$$

of m elements.

DEFINITION. The matrix $\mathbf{a}$ is said to *map* the vector $\mathbf{x}$ to the vector $\mathbf{ax}$. In the notation of p. 133, the mapping is

$$\mathbf{x} \overset{\mathbf{a}}{\to} \mathbf{ax}.$$

It is important to observe *that every* $\mathbf{x} \in V$ *gives rise to a definite vector* $\mathbf{ax}$ *as its map.*

THEOREM. *The vectors* $\mathbf{y} \equiv \mathbf{ax}$ *obtained by the mapping of all the vectors* $\mathbf{x} \in V$ *form a vector space* W:

To prove this theorem, the vector-space properties (p. 64) must be verified in detail. The most important of these are now given.

Write

$$\mathbf{y}_i \equiv \mathbf{ax}_i \qquad (i = 1, 2, 3, \ldots).$$

Then

$$\begin{aligned}(\mathbf{y}_1+\mathbf{y}_2)+\mathbf{y}_3 &= (\mathbf{ax}_1+\mathbf{ax}_2)+\mathbf{ax}_3\\ &= \mathbf{a}\{(\mathbf{x}_1+\mathbf{x}_2)+\mathbf{x}_3\}\\ &= \mathbf{a}\{\mathbf{x}_1+(\mathbf{x}_2+\mathbf{x}_3)\} \qquad (\mathbf{x}_i \in V)\\ &= \mathbf{ax}_1+\mathbf{a}(\mathbf{x}_2+\mathbf{x}_3)\\ &= \mathbf{y}_1+(\mathbf{y}_2+\mathbf{y}_3).\end{aligned}$$

Similarly,

$$\begin{aligned}\lambda(\mathbf{y}_1+\mathbf{y}_2) &= \lambda(\mathbf{ax}_1+\mathbf{ax}_2)\\ &= \mathbf{a}\{\lambda(\mathbf{x}_1+\mathbf{x}_2)\}\\ &= \mathbf{a}(\lambda\mathbf{x}_1+\lambda\mathbf{x}_2) \qquad (\mathbf{x}_i \in V)\\ &= \mathbf{a}\lambda\mathbf{x}_1+\mathbf{a}\lambda\mathbf{x}_2\\ &= \lambda\mathbf{y}_1+\lambda\mathbf{y}_2;\end{aligned}$$

and, again,

$$\begin{aligned}\lambda(\mu\mathbf{y}) &= \lambda(\mu\mathbf{ax})\\ &= \mathbf{a}\{\lambda(\mu\mathbf{x})\}\\ &= \mathbf{a}\{\lambda\mu\mathbf{x}\} \qquad (\mathbf{x} \in V)\\ &= \lambda\mu\mathbf{ax}\\ &= \lambda\mu\mathbf{y}.\end{aligned}$$

In this way the theorem is proved.

6. The 'linear mapping' property. The work of the preceding paragraph has, by implication, touched on an important class of mappings, known as *linear mappings*. It is not our aim to pursue this subject in great detail, but it may conveniently be enunciated here with reference to the particular mapping

$$\mathbf{y} = \mathbf{ax}.$$

The mapping of **x** upon **y** is said to be linear if

(i) the map of the sum $\mathbf{u}+\mathbf{v}$ is the sum $\mathbf{au}+\mathbf{av}$ of the individual maps,

(ii) the map of the product $\lambda\mathbf{u}$ of **u** by the scalar multiplier λ is equal to $\lambda(\mathbf{au})$, the corresponding multiple of the map of **u**.

The proofs of these properties are immediate.

7. The reverse mapping. In the mapping (§5)

$$\mathbf{y} = \mathbf{ax},$$

each element $\mathbf{x} \in V$ gives rise to precisely one **y**. It is, however, possible for a particular vector **y** to arise from several vectors **x**. Suppose, for example, that a particular **y** arises from two distinct vectors **u**, **v**. There are then two relations

$$\mathbf{y} = \mathbf{au}, \quad \mathbf{y} = \mathbf{av} \qquad (\mathbf{u} \neq \mathbf{v})$$

and so

$$\mathbf{a}(\mathbf{u}-\mathbf{v}) = \mathbf{0}.$$

Hence $\mathbf{u}-\mathbf{v}$ *is a non-zero solution, if such exists, of the equation*

$$\mathbf{ax} = \mathbf{0}.$$

REMARK. If **a** is a non-singular square matrix, the inverse mapping is

$$\mathbf{x} = \mathbf{a}^{-1}\mathbf{y},$$

or (p. 133)

$$\mathbf{y} \xrightarrow{\mathbf{a}^{-1}} \mathbf{x},$$

in which **x** is *uniquely* defined by **y**. The interest arises when **a** is singular, in which case (Part 1, p. 259) non-zero solutions of the equation $\mathbf{ax} = \mathbf{0}$ do exist.

Examples

1. Prove that all the vectors $\mathbf{x} \in V$ which satisfy the relation

$$\mathbf{ax} = \mathbf{0}$$

form a vector space U which is a subspace of V.

2. In the mapping

$$\begin{pmatrix} y_1 \\ y_2 \\ y_3 \end{pmatrix} = \begin{pmatrix} 1 & 2 \\ 2 & 1 \\ 3 & 3 \end{pmatrix} \begin{pmatrix} x_1 \\ x_2 \end{pmatrix},$$

prove that the point $\mathbf{y} \equiv \{3,3,6\}$ arises from the point $\{1,1\}$ only; but that, in the mapping

$$\begin{pmatrix} x_1 \\ x_1 \end{pmatrix} = \begin{pmatrix} 1 & 2 & 3 \\ 2 & 1 & 3 \end{pmatrix} \begin{pmatrix} y_1 \\ y_2 \\ y_3 \end{pmatrix},$$

the point $\{1,1\}$ arises from $\{\frac{1}{3}-\lambda, \frac{1}{3}-\lambda, \lambda\}$ for arbitrary values of λ.

3. In the mapping

$$\begin{pmatrix} y_1 \\ y_2 \\ y_3 \end{pmatrix} = \begin{pmatrix} 1 & 2 & 3 \\ 2 & 3 & 1 \\ 3 & 5 & 4 \end{pmatrix} \begin{pmatrix} x_1 \\ x_2 \\ x_3 \end{pmatrix},$$

prove that all the vectors

$$\mathbf{x} \equiv \{p-7\lambda, q+5\lambda, r-\lambda\}$$

are mapped on the same vector $\mathbf{y}$, whatever the value of λ.

4. In the mapping

$$\begin{pmatrix} y_1 \\ y_2 \\ y_3 \end{pmatrix} = \begin{pmatrix} 1 & 0 & -1 \\ 0 & -1 & 1 \\ 1 & -1 & 0 \end{pmatrix} \begin{pmatrix} x_1 \\ x_2 \\ x_3 \end{pmatrix},$$

find all the vectors $\mathbf{x}$ which give rise to the vector $\mathbf{y} \equiv \{2,-1,1\}$.

DEFINITIONS. The vectors $\mathbf{x}$ being mapped by the matrix $\mathbf{a}$ may not cover the whole vector space generated by all $n \times 1$ column vectors, but only a part of it. In any case, the space V of all the vectors *actually being mapped* is called the *domain* of the mapping.

The space W containing all the vectors $\mathbf{ax}(\mathbf{x} \in V)$ is called the *range* of the mapping.

The space U ($\in V$) of vectors $\mathbf{x}$ for which $\mathbf{ax} = \mathbf{0}$ is called the *null-space* or *kernel* of the mapping.

The dimension of the range W is called the *rank* of the mapping; the dimension of the null-space U is called the *nullity* of the mapping.

[See also Chapter 16, §9 (p. 216).]

***Illustration* 3.** Let V be the vector space of all vectors of 3 elements each and consider the mapping determined by the matrix

$$\mathbf{a} \equiv \begin{pmatrix} 1 & 2 & 3 \\ 2 & 5 & 7 \end{pmatrix}.$$

Then
$$\mathbf{y} = \mathbf{ax},$$
where
$$\mathbf{x} \equiv \{x_1, x_2, x_3\}, \quad \mathbf{y} \equiv \{y_1, y_2\}.$$

The *domain* V of the mapping $\mathbf{a}$ is the vector space of three dimensions whose vectors may be expressed in the form

$$\mathbf{x} \equiv x_1\mathbf{u}+x_2\mathbf{v}+x_3\mathbf{w},$$

where $\mathbf{u} \equiv \{1, 0, 0\}, \quad \mathbf{v} \equiv \{0, 1, 0\}, \quad \mathbf{w} \equiv \{0, 0, 1\}.$

The *null space* U is the set of vectors whose elements satisfy the relation

$$\mathbf{ax} = \mathbf{0}$$

or

$$\begin{pmatrix} 1 & 2 & 3 \\ 2 & 5 & 7 \end{pmatrix} \begin{pmatrix} x_1 \\ x_2 \\ x_3 \end{pmatrix} = 0,$$

or

$$x_1+2x_2+3x_3 = 0,$$
$$2x_1+5x_2+7x_3 = 0.$$

If x_3 has an arbitrary value λ, then

$$x_1+2x_2 = -3\lambda,$$
$$2x_1+5x_2 = -7\lambda,$$

so that $x_1 = -\lambda, \quad x_2 = -\lambda.$

The vectors of the null-space are thus given by the formula

$$\lambda(-\mathbf{u}-\mathbf{v}+\mathbf{w})$$

for arbitrary values of λ. This is of dimension 1 and so the nullity of the mapping is 1.

The *range* W of the mapping is the space defined by all the vectors $\mathbf{ax}$ which are not zero. These vectors are

$$\begin{pmatrix} 1 & 2 & 3 \\ 2 & 5 & 7 \end{pmatrix} \left[x_1\begin{pmatrix} 1 \\ 0 \\ 0 \end{pmatrix} + x_2\begin{pmatrix} 0 \\ 1 \\ 0 \end{pmatrix} + x_3\begin{pmatrix} 0 \\ 0 \\ 1 \end{pmatrix} \right]$$

$$\equiv x_1\begin{pmatrix} 1 \\ 2 \end{pmatrix} + x_2\begin{pmatrix} 2 \\ 5 \end{pmatrix} + x_3\begin{pmatrix} 3 \\ 7 \end{pmatrix},$$

and so W is the space spanned by the vectors $\{1, 2\}, \{2, 5\}, \{3, 7\}$. But these vectors are not independent, in virtue of the identity

$$\begin{pmatrix} 3 \\ 7 \end{pmatrix} \equiv \begin{pmatrix} 1 \\ 2 \end{pmatrix} + \begin{pmatrix} 2 \\ 5 \end{pmatrix}.$$

Hence W is the vector space of dimension 2 having as a basis the two vectors $\{1, 2\}, \{2, 5\}$. The rank of the mapping is thus 2.

Example

1. Find the rank and nullity of the mappings $\mathbf{y} = \mathbf{ax}$ when $\mathbf{a}$ is the matrix

(i) $\begin{pmatrix} 1 & 2 & 3 \\ 1 & 3 & 5 \end{pmatrix}$, (ii) $\begin{pmatrix} 1 & 3 & 7 \\ 2 & 7 & 16 \end{pmatrix}$,

(iii) $\begin{pmatrix} 1 & 2 & 4 & -1 \\ 1 & 3 & 5 & -2 \end{pmatrix}$, (iv) $\begin{pmatrix} 0 & 1 & -1 \\ -1 & 0 & 1 \\ 1 & -1 & 0 \end{pmatrix}$,

(v) $\begin{pmatrix} 0 & 1 & 2 \\ 1 & 2 & 3 \\ 2 & 3 & 4 \end{pmatrix}$, (vi) $\begin{pmatrix} 1 & 1 & 1 & 1 \\ 1 & 2 & 3 & 4 \\ 0 & 1 & 2 & 3 \end{pmatrix}$.

16

LINEAR EQUATIONS: RANK

This chapter seeks to set in more precise arithmetical form the somewhat general ideas studied in the preceding chapter. The work on block multiplication is now important.

There will be a certain amount of repetition of earlier work so that the argument may be seen as a connected whole.

1. Notation and first ideas. The subject-matter of this chapter is the system of linear equations

$$\begin{aligned} a_{11}x_1+a_{12}x_2+\ldots+a_{1n}x_n &= b_1,\\ a_{21}x_1+a_{22}x_2+\ldots+a_{2n}x_n &= b_2,\\ \ldots\ldots\ldots\ldots\ldots\ldots\ldots\ldots\ldots\ldots&\\ a_{m1}x_1+a_{m2}x_2+\ldots+a_{mn}x_n &= b_m, \end{aligned}$$

or

$$\mathbf{ax} = \mathbf{b},$$

where $\mathbf{a}$ is the $m \times n$ matrix

$$\mathbf{a} \equiv (a_{ij}),$$

$\mathbf{x}$ is the $n \times 1$ column vector

$$\mathbf{x} \equiv \{x_1, x_2, \ldots, x_n\}$$

and $\mathbf{b}$ the $m \times 1$ column vector

$$\mathbf{b} \equiv \{b_1, b_2, \ldots, b_m\}.$$

2. The equations when $m = n$ and a is non-singular.

If $m = n$ and $\mathbf{a}$ is non-singular the solution of the equation

$$\mathbf{ax} = \mathbf{b}$$

exists (p. 164) and is unique, given by

$$\mathbf{x} = \mathbf{a}^{-1}\mathbf{b}.$$

COROLLARY. *If $m = n$ and* $\mathbf{a}$ *is non-singular, then the homogeneous system*

$$\mathbf{ax} = \mathbf{0}$$

is satisfied only by the zero vector

$$\mathbf{x} = \mathbf{0}.$$

This solution is called the *trivial* solution of the homogeneous system. It exists whether $\mathbf{a}$ is singular or not, but in the former case there are other, more interesting, solutions.

There is a slightly different way of stating the Corollary:

If $m = n$ and the homogeneous system

$$\mathbf{ax} = \mathbf{0}$$

is satisfied by a non-zero vector $\mathbf{u}$, *then*

$$\det \mathbf{a} = 0.$$

For
$$\det \mathbf{a} \neq 0$$
$$\Rightarrow \exists\, \mathbf{a}^{-1},$$
and so
$$\mathbf{au} = \mathbf{0}$$
$$\Rightarrow \mathbf{a}^{-1}\mathbf{au} = \mathbf{0}$$
$$\Rightarrow \mathbf{u} = \mathbf{0},$$

a contradiction.

3. The complementary system. Given a system of equations

$$\mathbf{ax} = \mathbf{b},$$

the system
$$\mathbf{ax} = \mathbf{0}$$

is a *homogeneous* system said to be *complementary* to the given system.

If $\mathbf{x} = \boldsymbol{\xi}$ is any solution of the complementary system, so that

$$\mathbf{a}\boldsymbol{\xi} = \mathbf{0},$$

and if $\mathbf{x} = \mathbf{u}$ is any one solution whatever of the given system, so that

$$\mathbf{au} = \mathbf{b},$$

then, for all values of λ,

$$\begin{aligned} \mathbf{a}(\lambda\boldsymbol{\xi}+\mathbf{u}) &= \lambda\mathbf{a}\boldsymbol{\xi}+\mathbf{au} \\ &= \lambda.\mathbf{0}+\mathbf{b} \\ &= \mathbf{b}, \end{aligned}$$

so that the vector $\lambda\boldsymbol{\xi}+\mathbf{u}$ satisfies the given equation.

4. The homogeneous equation. Consider the equation

$$\mathbf{ax} = \mathbf{0},$$

where

$$\mathbf{a} \equiv (a_{ij})$$

is a given matrix of type $m \times n$.

If all the elements are zero, the equations have little interest, so it is assumed that there are some non-zero values of the coefficients a_{ij}.

Form now a series of determinants, of orders 1, 2, 3, ..., where a typical determinant of order ρ is formed by choosing elements in an assigned ρ rows and an assigned ρ columns.

When $\rho = 1$, a typical determinant is just $|a_{ij}| \equiv a_{ij}$, and not all of these are zero.

As ρ increases, determinants of order 2, 3, 4, ... are obtained, until ρ reaches the smaller of m, n. It may, however, happen that, before then, ρ reaches a value r *after which all the determinants are zero*: that is, there are non-zero determinants of order r but no non-zero determinants of order greater than r.

DEFINITION. This value of r is called the *rank* of the matrix.

If all the determinants are non-zero, then r is the smaller of m, n.

When $\mathbf{a} \equiv \mathbf{0}$ so that all the elements are zero, then we define r to be zero; otherwise $r \geqslant 1$.

Note that, automatically,

$$r \leqslant m, \quad r \leqslant n.$$

Example

1. By evaluating determinants, find the ranks of the matrices:

(i) $\begin{pmatrix} 1 & 2 & 0 \\ 2 & 4 & 0 \end{pmatrix}$, (ii) $\begin{pmatrix} 1 & 2 & 3 \\ 2 & 4 & 5 \end{pmatrix}$,

(iii) $\begin{pmatrix} 1 & 0 & 0 \\ 0 & 1 & 0 \\ 0 & 0 & 0 \end{pmatrix}$, (iv) $\begin{pmatrix} 1 & 2 & 3 \\ 2 & 3 & 4 \\ 3 & 4 & 0 \end{pmatrix}$,

(v) $\begin{pmatrix} 1 & 2 & 3 \\ 1 & 2 & 3 \\ 1 & 2 & 3 \end{pmatrix}$, (vi) $\begin{pmatrix} -2 & 1 & 1 \\ 1 & -2 & 1 \\ 1 & 1 & -2 \end{pmatrix}$,

(vii) $\begin{pmatrix} 1 & 2 & 3 & 4 \\ 2 & 3 & 3 & 2 \\ 5 & 9 & 12 & 14 \end{pmatrix}$, (viii) $\begin{pmatrix} 2 & 3 & -1 & 4 \\ 1 & 2 & 1 & 1 \\ 3 & 5 & 1 & 6 \end{pmatrix}$.

Illustration 1. Before proceeding further, consider the particular set of equations

$$\begin{aligned} x+\ y-2z+5u-3v &= 0,\\ x+2y+\ z-2u+4v &= 0,\\ 3x+4y-3z+8u-2v &= 0. \end{aligned}$$

It can be verified that all determinants of order 3 in the matrix of coefficients vanish. On the other hand, the 'top left-hand' determinant

$$\begin{vmatrix} 1 & 1 \\ 1 & 2 \end{vmatrix}$$

is not zero. Hence the rank is 2. (If the 'top left-hand' determinant had been zero, the equations and the variables could have been re-arranged to secure a non-zero determinant there.)

The equations may be written in the matrix form

$$\mathbf{ax} = \mathbf{0},$$

where $$\mathbf{a} \equiv \begin{pmatrix} 1 & 1 & -2 & 5 & -3 \\ 1 & 2 & 1 & -2 & 4 \\ 3 & 4 & -3 & 8 & -2 \end{pmatrix},$$

$$\mathbf{x} \equiv \{x, y, z, u, v\}.$$

Partitioned with reference to the non-zero determinant, the matrix equation is

$$\left(\begin{array}{cc|ccc} 1 & 1 & -2 & 5 & -3 \\ 1 & 2 & 1 & -2 & 4 \\ \hline 3 & 4 & -3 & 8 & -2 \end{array}\right) \begin{pmatrix} x \\ y \\ \hline z \\ u \\ v \end{pmatrix} = \mathbf{0} \equiv \begin{pmatrix} 0 \\ 0 \\ \hline 0 \end{pmatrix},$$

or $$\left(\begin{array}{c|c} \boldsymbol{\alpha} & \boldsymbol{\beta} \\ \hline \boldsymbol{\gamma} & \boldsymbol{\delta} \end{array}\right) \begin{pmatrix} \mathbf{p} \\ \hline \mathbf{q} \end{pmatrix} = \begin{pmatrix} \mathbf{0} \\ \hline \mathbf{0} \end{pmatrix},$$

where, in particular, $\boldsymbol{\alpha}$ is the *non-singular* matrix

$$\boldsymbol{\alpha} \equiv \begin{pmatrix} 1 & 1 \\ 1 & 2 \end{pmatrix}.$$

By 'block multiplication', the equation becomes

$$\begin{pmatrix} \boldsymbol{\alpha}\mathbf{p}+\boldsymbol{\beta}\mathbf{q} \\ \hline \boldsymbol{\gamma}\mathbf{p}+\boldsymbol{\delta}\mathbf{q} \end{pmatrix} = \begin{pmatrix} \mathbf{0} \\ \hline \mathbf{0} \end{pmatrix},$$

so that, equating the two 'horizontal halves',

$$\begin{aligned} \boldsymbol{\alpha}\mathbf{p}+\boldsymbol{\beta}\mathbf{q} &= \mathbf{0},\\ \boldsymbol{\gamma}\mathbf{p}+\boldsymbol{\delta}\mathbf{q} &= \mathbf{0}. \end{aligned}$$

Since $\boldsymbol{\alpha}$ is non-singular,

$$\begin{aligned} &\boldsymbol{\alpha}\mathbf{p}+\boldsymbol{\beta}\mathbf{q} = \mathbf{0}\\ \Rightarrow\ &\boldsymbol{\alpha}^{-1}\boldsymbol{\alpha}\mathbf{p}+\boldsymbol{\alpha}^{-1}\boldsymbol{\beta}\mathbf{q} = \mathbf{0}\\ \Rightarrow\ &\mathbf{p} = -\boldsymbol{\alpha}^{-1}\boldsymbol{\beta}\mathbf{q}; \end{aligned}$$

and so, from the second equation,

$$-\boldsymbol{\gamma}\boldsymbol{\alpha}^{-1}\boldsymbol{\beta}\mathbf{q}+\boldsymbol{\delta}\mathbf{q} = \mathbf{0},$$

or

$$(\boldsymbol{\delta}-\boldsymbol{\gamma}\boldsymbol{\alpha}^{-1}\boldsymbol{\beta})\mathbf{q} = \mathbf{0}.$$

Now

$$\boldsymbol{\alpha} \equiv \begin{pmatrix} 1 & 1 \\ 1 & 2 \end{pmatrix},$$

so that

$$\boldsymbol{\alpha}^{-1} \equiv \begin{pmatrix} 2 & -1 \\ -1 & 1 \end{pmatrix}.$$

Hence

$$\boldsymbol{\alpha}^{-1}\boldsymbol{\beta} \equiv \begin{pmatrix} 2 & -1 \\ -1 & 1 \end{pmatrix} \begin{pmatrix} -2 & 5 & -3 \\ 1 & -2 & 4 \end{pmatrix}$$

$$\equiv \begin{pmatrix} -5 & 12 & -10 \\ 3 & -7 & 7 \end{pmatrix},$$

$$\boldsymbol{\gamma}\boldsymbol{\alpha}^{-1}\boldsymbol{\beta} \equiv (3, 4) \begin{pmatrix} -5 & 12 & -10 \\ 3 & -7 & 7 \end{pmatrix}$$

$$\equiv (-3, 8, -2).$$

But

$$\boldsymbol{\delta} \equiv (-3, 8, -2)$$

and so

$$\boldsymbol{\delta}-\boldsymbol{\gamma}\boldsymbol{\alpha}^{-1}\boldsymbol{\beta} = \mathbf{0}.$$

This result, surprising when first met, holds perfectly generally in the circumstances of the problem. The result of it is that *the equation for* $\mathbf{q}$, *namely*

$$(\boldsymbol{\delta}-\boldsymbol{\gamma}\boldsymbol{\alpha}^{-1}\boldsymbol{\beta})\mathbf{q} = \mathbf{0},$$

is satisfied when $\mathbf{q}$ *is any arbitrary vector whatsoever of type* 3×1.

The value of $\mathbf{p}$ is then obtained from this *arbitrary* vector $\mathbf{q}$ by means of the formula

$$\mathbf{p} = -\boldsymbol{\alpha}^{-1}\boldsymbol{\beta}\mathbf{q}.$$

Hence the solution of the given system of equations is obtained *explicitly* in the partitioned form

$$\mathbf{x} = \left(\frac{-\boldsymbol{\alpha}^{-1}\boldsymbol{\beta}\mathbf{q}}{\mathbf{q}}\right)$$

for arbitrary $\mathbf{q}$.

If $\mathbf{q}$ is taken in the form

$$\mathbf{q} \equiv \{\lambda, \mu, \nu\} \qquad (\lambda, \mu, \nu \text{ arbitrary}),$$

then

$$\boldsymbol{\beta}\mathbf{q} \equiv \begin{pmatrix} -2 & 5 & -3 \\ 1 & -2 & 4 \end{pmatrix} \begin{pmatrix} \lambda \\ \mu \\ \nu \end{pmatrix}$$

$$\equiv \begin{pmatrix} -2\lambda+5\mu-3\nu \\ \lambda-2\mu+4\nu \end{pmatrix},$$

$$\boldsymbol{\alpha}^{-1}\boldsymbol{\beta}\mathbf{q} \equiv \begin{pmatrix} 2 & -1 \\ -1 & 1 \end{pmatrix} \begin{pmatrix} -2\lambda+5\mu-3\nu \\ \lambda-2\mu+4\nu \end{pmatrix}$$

$$\boldsymbol{\alpha}^{-1}\boldsymbol{\beta}\mathbf{q} \equiv \begin{pmatrix} -5\lambda+12\mu-10\nu \\ 3\lambda-\ 7\mu+\ 7\nu \end{pmatrix}$$

so that
$$\mathbf{x} \equiv \begin{pmatrix} 5\lambda-12\mu+10\nu \\ -3\lambda+\ 7\mu-\ 7\nu \\ \lambda \\ \mu \\ \nu \end{pmatrix}.$$

COROLLARY. In the language of the preceding chapter (p. 204), the *null-space* (*kernel*) for the mapping $\mathbf{y} = \mathbf{ax}$ has basis $\{5,-3,1,0,0\}, \{-12,7,0,1,0\}, \{10,-7,0,0,1\}$. It is of dimension 3.

Examples

Solve *similarly* the following sets of equations, each given to be of rank 2:

1.
$$\begin{aligned} 2x+\ y+z+\ u &= 0, \\ x+\ y-z+3u &= 0, \\ 4x+3y-z+7u &= 0. \end{aligned}$$

2.
$$\begin{aligned} 3x+2y+\ z+\ u-v &= 0, \\ 2x+\ y-\ z-\ u+v &= 0, \\ x+\ y+2z+2u-2v &= 0. \end{aligned}$$

3.
$$\begin{aligned} 5x+2y+3z &= 0, \\ 2x+\ y-\ z &= 0, \\ 3x+\ y+4z &= 0. \end{aligned}$$

4.
$$\begin{aligned} x+\ y+\ z+\ u &= 0, \\ x-\ z+2u &= 0, \\ x+2y+3z &= 0. \end{aligned}$$

5. A particular equation. In preparation for the important bookwork to follow in §6, consider a particular case where the manipulation is reasonably direct.

THEOREM. To prove that, *if*

$$\mathbf{h} \equiv (\mathbf{a} \mid \mathbf{b})$$

is a partitioned $m \times n$ matrix $(n > m)$ in which **a** *is a non-singular square matrix of order m, then the general solution of the equation*

$$\mathbf{hx} = \mathbf{0}$$

(where **x** *is a column vector of n elements) is of the form*

$$\mathbf{x} \equiv \left(\frac{-\mathbf{a}^{-1}\mathbf{bv}}{\mathbf{v}}\right)$$

in which **v** *is an arbitrary vector of $n-m$ elements.*

Write $\mathbf{x}$ in partitioned form

$$\mathbf{x} \equiv \left(\frac{\mathbf{u}}{\mathbf{v}}\right),$$

where $\mathbf{u}$, $\mathbf{v}$ are column vectors of m, $n-m$ elements respectively. Then

$$(\mathbf{a} \mid \mathbf{b})\left(\frac{\mathbf{u}}{\mathbf{v}}\right) = \mathbf{0},$$

so that

$$\mathbf{au}+\mathbf{bv} = \mathbf{0}.$$

Now $\mathbf{a}$ is non-singular, and so

$$\mathbf{u} + \mathbf{a}^{-1}\mathbf{bv} = \mathbf{0},$$

or

$$\mathbf{u} = -\mathbf{a}^{-1}\mathbf{bv}.$$

Hence

$$\mathbf{x} \equiv \left(\frac{-\mathbf{a}^{-1}\mathbf{bv}}{\mathbf{v}}\right),$$

where no conditions have been imposed on $\mathbf{v}$, which is, therefore, arbitrary.

CHECK. This value of $\mathbf{x}$ does indeed satisfy the given equation, since

$$\begin{aligned}(\mathbf{a} \mid \mathbf{b})\left(\frac{-\mathbf{a}^{-1}\mathbf{bv}}{\mathbf{v}}\right) &= -\mathbf{a}(\mathbf{a}^{-1}\mathbf{bv})+\mathbf{bv} \\ &= -\mathbf{bv}+\mathbf{bv} \\ &= \mathbf{0}.\end{aligned}$$

6. The relation $\boldsymbol{\delta}-\boldsymbol{\gamma}\boldsymbol{\alpha}^{-1}\boldsymbol{\beta} = \mathbf{0}$. (This work is hard, but important.)

Let

$$\mathbf{a} \equiv \left(\begin{array}{c|c}\boldsymbol{\alpha} & \boldsymbol{\beta} \\ \hline \boldsymbol{\gamma} & \boldsymbol{\delta}\end{array}\right)$$

be an $m \times n$ matrix of rank r ($r < m$, $r < n$), written in partitioned form, where $\boldsymbol{\alpha}$ is a *non-singular* square matrix of rank r. It is required *to prove that, necessarily,*

$$\boldsymbol{\delta}-\boldsymbol{\gamma}\boldsymbol{\alpha}^{-1}\boldsymbol{\beta} = \mathbf{0},$$

where $\mathbf{0}$ *is the zero* $(m-r) \times (n-r)$ *matrix.*

Let δ_{ij} be the element in the ith row and jth column† of $\boldsymbol{\delta}$. The elements $\beta_{\mu j}$, for $\mu = 1,2,\ldots,r$ in the jth column of $\boldsymbol{\beta}$ form a vector conveniently

† That is, in the $(r+i)$th row and $(r+j)$th column of $\mathbf{a}$.

called $\boldsymbol{\beta}_j$; the elements $\gamma_{i\lambda}$, for $\lambda = 1, 2, \ldots, r$ in the ith row of $\boldsymbol{\gamma}$ form a vector conveniently called $\boldsymbol{\gamma}_i$. (See the diagrammatic representation, Fig. 15.)

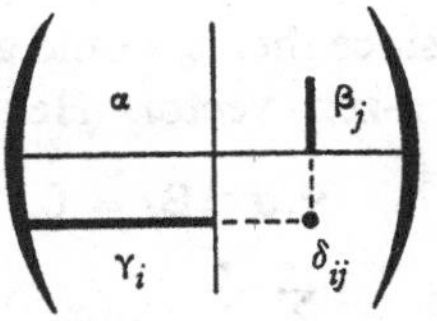

Fig. 15

Consider the partitioned matrix

$$\mathbf{g} \equiv \left(\begin{array}{c|c} \boldsymbol{\alpha} & \boldsymbol{\beta}_j \\ \hline \boldsymbol{\gamma}_i & \delta_{ij} \end{array}\right)$$

of order $r+1$. Since $\mathbf{a}$ is of rank r, it follows that

$$\det \mathbf{g} = 0,$$

and this is precisely the condition for the equation

$$\mathbf{gx} = \mathbf{0}$$

to have a *non-zero* solution, say

$$\mathbf{x} \equiv \left(\frac{\boldsymbol{\xi}}{\lambda}\right),$$

where $\boldsymbol{\xi}$ is an $r \times 1$ column vector and λ a scalar. Hence

$$\left(\begin{array}{c|c} \boldsymbol{\alpha} & \boldsymbol{\beta}_j \\ \hline \boldsymbol{\gamma}_i & \delta_{ij} \end{array}\right) \left(\frac{\boldsymbol{\xi}}{\lambda}\right) = \left(\frac{\mathbf{0}}{0}\right),$$

where $\mathbf{0}$ is the zero vector of r elements.

By 'block multiplication',

$$\left(\frac{\boldsymbol{\alpha\xi}+\boldsymbol{\beta}_j\lambda}{\boldsymbol{\gamma}_i\boldsymbol{\xi}+\delta_{ij}\lambda}\right) = \left(\frac{\mathbf{0}}{0}\right),$$

so that

$$\boldsymbol{\alpha\xi}+\boldsymbol{\beta}_j\lambda = \mathbf{0},$$
$$\boldsymbol{\gamma}_i\boldsymbol{\xi}+\delta_{ij}\lambda = 0.$$

From the first equation, since $\boldsymbol{\alpha}$ is non-singular,

$$\boldsymbol{\xi} = -\boldsymbol{\alpha}^{-1}\boldsymbol{\beta}_j\lambda,$$

and then, from the second equation,

$$-\boldsymbol{\gamma}_i \boldsymbol{\alpha}^{-1} \boldsymbol{\beta}_j \lambda + \delta_{ij} \lambda = 0,$$

or

$$(\delta_{ij} - \boldsymbol{\gamma}_i \boldsymbol{\alpha}^{-1} \boldsymbol{\beta}_j) \lambda = 0.$$

Further, λ cannot be zero, since then $\boldsymbol{\xi}$ would also be zero, contradicting the condition that $\mathbf{x}$ is a non-zero vector. Hence

$$\delta_{ij} - \boldsymbol{\gamma}_i \boldsymbol{\alpha}^{-1} \boldsymbol{\beta}_j = 0.$$

But

$$\boldsymbol{\gamma}_i \boldsymbol{\alpha}^{-1} \boldsymbol{\beta}_j = \sum_{\lambda=1}^{r} \sum_{\mu=1}^{r} \gamma_{i\lambda} (\boldsymbol{\alpha}^{-1})_{\lambda\mu} \beta_{\mu j},$$

which is the element in the ith row and jth column of the matrix $\boldsymbol{\gamma}\boldsymbol{\alpha}^{-1}\boldsymbol{\beta}$; and so *each element* of the matrix $\boldsymbol{\delta} - \boldsymbol{\gamma}\boldsymbol{\alpha}^{-1}\boldsymbol{\beta}$ is zero, so that

$$\boldsymbol{\delta} - \boldsymbol{\gamma}\boldsymbol{\alpha}^{-1}\boldsymbol{\beta} = 0.$$

7. The general solution of the homogeneous equation.

Let

$$\mathbf{a}\mathbf{x} = 0$$

be a given equation in which $\mathbf{a}$ is an $m \times n$ matrix of rank r. By rearrangement, if necessary, this can be partitioned to the form

$$\left(\begin{array}{c|c} \boldsymbol{\alpha} & \boldsymbol{\beta} \\ \hline \boldsymbol{\gamma} & \boldsymbol{\delta} \end{array}\right) \left(\begin{array}{c} \mathbf{u} \\ \hline \mathbf{v} \end{array}\right) = \left(\begin{array}{c} 0 \\ \hline 0 \end{array}\right),$$

where $\boldsymbol{\alpha}$ is a *non-singular* matrix of order r. It is implicit that $r < m$, $r < n$, but the modifications in the work which follows are slight when the inequalities are replaced by equalities.

By 'block multiplication',

$$\boldsymbol{\alpha}\mathbf{u} + \boldsymbol{\beta}\mathbf{v} = 0,$$
$$\boldsymbol{\gamma}\mathbf{u} + \boldsymbol{\delta}\mathbf{v} = 0.$$

Since $\boldsymbol{\alpha}$ is non-singular, the first equation gives

$$\mathbf{u} = -\boldsymbol{\alpha}^{-1}\boldsymbol{\beta}\mathbf{v}$$

and the second equation then gives

$$-\boldsymbol{\gamma}\boldsymbol{\alpha}^{-1}\boldsymbol{\beta}\mathbf{v} + \boldsymbol{\delta}\mathbf{v} = 0$$

or

$$(\boldsymbol{\delta} - \boldsymbol{\gamma}\boldsymbol{\alpha}^{-1}\boldsymbol{\beta})\mathbf{v} = 0.$$

It has just been proved, however, (§6) that

$$\boldsymbol{\delta} - \boldsymbol{\gamma}\boldsymbol{\alpha}^{-1}\boldsymbol{\beta} = 0,$$

and so this equation is satisfied for *arbitrary* $\mathbf{v}$. The general solution of the given equation is thus

$$\mathbf{x} = \left(\frac{-\boldsymbol{\alpha}^{-1}\boldsymbol{\beta}\mathbf{v}}{\mathbf{v}}\right),$$

where $\mathbf{v}$ is an arbitrary column vector of $n-r$ elements.

COROLLARY. *The solution of the set of equations*

$$a_{11}x_1+\ldots+a_{1n}x_n = 0,$$
$$\cdots\cdots\cdots\cdots\cdots\cdots\cdots\cdots$$
$$a_{m1}x_1+\ldots+a_{mn}x_n = 0,$$

where $\mathbf{a}$ *is a matrix of rank* r, *depends on* $n-r$ *parameters.*

Text Example

1. Verify by direct computation that the solution

$$\mathbf{x} = \left(\frac{-\boldsymbol{\alpha}^{-1}\boldsymbol{\beta}\mathbf{v}}{\mathbf{v}}\right)$$

does satisfy the given equation for arbitrary $\mathbf{v}$.

8. The 'column-rank' of a matrix. It is now possible to give a more precise form to some of the work of the preceding chapter.

Let

$$\mathbf{x}, \mathbf{y}, \mathbf{z}, \ldots, \mathbf{w}$$

be p column vectors of n elements each, and form the matrix

$$\mathbf{b} \equiv (\mathbf{x}, \mathbf{y}, \mathbf{z}, \ldots, \mathbf{w})$$

of type $n\times p$ having them as columns. Then (p. 198) any linear relation between the vectors is equivalent to a matrix equation

$$\mathbf{b}\boldsymbol{\xi} = 0.$$

The relation between the p vectors is thus found by solving this matrix equation.

If the matrix $\mathbf{b}$ has rank r, then it has just been proved that $\boldsymbol{\xi}$ depends linearly on $p-r$ parameters (p being the number of columns of $\mathbf{b}$). Hence the p vectors $\mathbf{x}, \mathbf{y}, \ldots, \mathbf{w}$ are subject to $p-r$ linear relations, and so *the number of linearly independent vectors is* $p-(p-r)$, *or* r.

COROLLARY. *In a matrix of rank* r, *the number of linearly independent columns is* r.

REMARK. The treatment has, for definiteness, been confined almost entirely to columns, but exactly analogous results hold for rows. In particular, *in a matrix of rank r, the number of linearly independent rows is also r.*

9. The rank and nullity of a mapping. Given a mapping

$$\mathbf{y} = \mathbf{ax},$$

its domain, range, null-space (kernel), rank and nullity were all defined on p. 203. Here, again, greater precision is now possible.

It is assumed that $\mathbf{a}$ is an $m \times n$ matrix of rank r; then $\mathbf{x}$ is a column vector of n elements and $\mathbf{y}$ is a column vector of m elements.

The null-space U consists of those vectors $\mathbf{x}$ for which $\mathbf{ax} = \mathbf{0}$, and its dimension is called the *nullity s*. Since these vectors depend linearly on $n-r$ parameters, the dimension is also $n-r$, and so *the rank r and nullity s of a matrix* **a** *of type m × n are connected by the relation*

$$r+s = n.$$

The rank of the mapping is the dimension of the range: that is, the number of linearly independent vectors $\mathbf{ax}$. This we have proved (p. 215) to be precisely r, thereby justifying the double use of the word rank for matrix and for associated mapping.

The relation $r+s = n$ can be expressed in the form:

'dimension of range + dimension of null-space = n'.

10. The non-homogeneous equation ax = b. It may be helpful to start with a particular example solved exactly in accordance with the general theory.

***Illustration* 2.** *To solve the equations*

$$\begin{aligned} x+\ y+\ z &= 3, \\ 2x+\ y+4z &= 7, \\ 3x+2y+5z &= k. \end{aligned}$$

In matrix form the equations are

$$\begin{pmatrix} 1 & 1 & 1 \\ 2 & 1 & 4 \\ 3 & 2 & 5 \end{pmatrix} \begin{pmatrix} x \\ y \\ z \end{pmatrix} = \begin{pmatrix} 3 \\ 7 \\ k \end{pmatrix},$$

or

$$\mathbf{ax} = \mathbf{b}.$$

The matrix $\mathbf{a}$ is of rank 2, but the matrix

$$\begin{pmatrix} 1 & 1 \\ 2 & 1 \end{pmatrix}$$

is non-singular, so, following the pattern of preceding work, the matrix equation is written in the form

$$\left(\begin{array}{cc|c} 1 & 1 & 1 \\ 2 & 1 & 4 \\ \hline 3 & 2 & 5 \end{array}\right) \begin{pmatrix} x \\ y \\ \hline z \end{pmatrix} = \begin{pmatrix} 3 \\ 7 \\ \hline k \end{pmatrix},$$

or

$$\left(\begin{array}{c|c} \boldsymbol{\alpha} & \boldsymbol{\beta} \\ \hline \boldsymbol{\gamma} & \delta \end{array}\right) \begin{pmatrix} \mathbf{u} \\ \hline z \end{pmatrix} = \begin{pmatrix} \mathbf{c} \\ \hline k \end{pmatrix}.$$

By 'block multiplication',

$$\boldsymbol{\alpha}\mathbf{u} + \boldsymbol{\beta} z = \mathbf{c},$$
$$\boldsymbol{\gamma}\mathbf{u} + \delta z = k.$$

Since $\boldsymbol{\alpha}$ is non-singular, the first equation is

$$\mathbf{u} = \boldsymbol{\alpha}^{-1}\mathbf{c} - \boldsymbol{\alpha}^{-1}\boldsymbol{\beta} z,$$

and then

$$\boldsymbol{\gamma}(\boldsymbol{\alpha}^{-1}\mathbf{c} - \boldsymbol{\alpha}^{-1}\boldsymbol{\beta} z) + \delta z = k.$$

Hence

$$(\delta - \boldsymbol{\gamma}\boldsymbol{\alpha}^{-1}\boldsymbol{\beta})z = k - \boldsymbol{\gamma}\boldsymbol{\alpha}^{-1}\mathbf{c}.$$

Now, by the fundamental theorem (p. 212), or by direct computation for the particular case,

$$\delta - \boldsymbol{\gamma}\boldsymbol{\alpha}^{-1}\boldsymbol{\beta} = 0,$$

and so, *for general values of k, the equation*

$$(\delta - \boldsymbol{\gamma}\boldsymbol{\alpha}^{-1}\boldsymbol{\beta})z = k - \boldsymbol{\gamma}\boldsymbol{\alpha}^{-1}\mathbf{c}$$

cannot be satisfied for any value of z; that is, *the given equations have no solution for general values of k.*

On the other hand, a different possibility arises when k has the particular value

$$k = \boldsymbol{\gamma}\boldsymbol{\alpha}^{-1}\mathbf{c}$$

(a scalar, despite its form), for then both sides of the equation

$$(\delta - \boldsymbol{\gamma}\boldsymbol{\alpha}^{-1}\boldsymbol{\beta})z = k - \boldsymbol{\gamma}\boldsymbol{\alpha}^{-1}\mathbf{c}$$

vanish automatically, and the equation is satisfied for arbitrary values of z: say $z = \lambda$.

When $z = \lambda$, the vector $\mathbf{u}$ is given by the relation

$$\mathbf{u} = \boldsymbol{\alpha}^{-1}\mathbf{c} - \boldsymbol{\alpha}^{-1}\boldsymbol{\beta}\lambda,$$

and so

$$\mathbf{x} \equiv \begin{pmatrix} \mathbf{u} \\ \hline z \end{pmatrix} = \begin{pmatrix} \boldsymbol{\alpha}^{-1}\mathbf{c} - \boldsymbol{\alpha}^{-1}\boldsymbol{\beta}\lambda \\ \lambda \end{pmatrix},$$

a solution depending on one arbitrary parameter.

The matrices may now be evaluated and the solution completed.

NOTE. This is, of course, no way to work such a simple problem. The first two equations give

$$x + 3z = 4,$$
$$y - 2z = -1,$$

so that
$$x = 4-3z,$$
$$y = -1+2z.$$

Substitution in the third equation gives
$$(12-9z)+(-2+4z)+5z = k,$$
or
$$(-9+4+5)z = k-12+2.$$

Since $-9+4+5=0$, there is no solution at all for general values of k. But when $k=10$ the equation is satisfied by $z=\lambda$, with λ arbitrary. The complete solution is then
$$x = 4-3\lambda, \quad y = -1+2\lambda, \quad z = \lambda.$$

We now approach the general problem:

To solve the equation
$$\mathbf{ax} = \mathbf{b},$$

where $\mathbf{b} \neq \mathbf{0}$ and where $\mathbf{a}$ is an $m \times n$ matrix of rank r (with, for convenience, a *non-singular* submatrix $\boldsymbol{\alpha}$ of r rows and columns in the 'top left-hand' corner).

In partitioned form, the equation is
$$\left(\begin{array}{c|c} \boldsymbol{\alpha} & \boldsymbol{\beta} \\ \hline \boldsymbol{\gamma} & \boldsymbol{\delta} \end{array}\right) \left(\begin{array}{c} \mathbf{u} \\ \hline \mathbf{v} \end{array}\right) = \left(\begin{array}{c} \mathbf{c} \\ \hline \mathbf{d} \end{array}\right).$$

By 'block multiplication',
$$\boldsymbol{\alpha}\mathbf{u}+\boldsymbol{\beta}\mathbf{v} = \mathbf{c},$$
$$\boldsymbol{\gamma}\mathbf{u}+\boldsymbol{\delta}\mathbf{v} = \mathbf{d}.$$

From the first equation, since $\boldsymbol{\alpha}$ is non-singular,
$$\mathbf{u} = \boldsymbol{\alpha}^{-1}\mathbf{c}-\boldsymbol{\alpha}^{-1}\boldsymbol{\beta}\mathbf{v},$$

so that, by the second equation,
$$\boldsymbol{\gamma}\boldsymbol{\alpha}^{-1}\mathbf{c}-\boldsymbol{\gamma}\boldsymbol{\alpha}^{-1}\boldsymbol{\beta}\mathbf{v}+\boldsymbol{\delta}\mathbf{v} = \mathbf{d},$$
or
$$(\boldsymbol{\delta}-\boldsymbol{\gamma}\boldsymbol{\alpha}^{-1}\boldsymbol{\beta})\mathbf{v} = \mathbf{d}-\boldsymbol{\gamma}\boldsymbol{\alpha}^{-1}\mathbf{c}.$$

Now, by the fundamental theorem (p. 212),
$$\boldsymbol{\delta}-\boldsymbol{\gamma}\boldsymbol{\alpha}^{-1}\boldsymbol{\beta} = \mathbf{0},$$
so that, *if*
$$\mathbf{d} \neq \boldsymbol{\gamma}\boldsymbol{\alpha}^{-1}\mathbf{c},$$

there is no solution.

On the other hand, when
$$\mathbf{d} = \boldsymbol{\gamma}\boldsymbol{\alpha}^{-1}\mathbf{c},$$

the vector **v** can take any arbitrary value. Then

$$\mathbf{u} = \boldsymbol{\alpha}^{-1}\mathbf{c} - \boldsymbol{\alpha}^{-1}\boldsymbol{\beta}\mathbf{v},$$

and so, *the given equation, when*

$$\mathbf{d} = \boldsymbol{\gamma}\boldsymbol{\alpha}^{-1}\mathbf{c},$$

is satisfied by

$$\mathbf{x} = \left(\frac{\boldsymbol{\alpha}^{-1}\mathbf{c} - \boldsymbol{\alpha}^{-1}\boldsymbol{\beta}\mathbf{v}}{\mathbf{v}}\right)$$

for arbitrary $(n-r)\times 1$ *column vector* **v**.

11. The rank test for solubility. It was proved in § 10 that the equation

$$\mathbf{a}\mathbf{x} = \mathbf{b},$$

or

$$\left(\begin{array}{c|c} \boldsymbol{\alpha} & \boldsymbol{\beta} \\ \hline \boldsymbol{\gamma} & \boldsymbol{\delta} \end{array}\right) \left(\frac{\mathbf{u}}{\mathbf{v}}\right) = \left(\frac{\mathbf{c}}{\mathbf{d}}\right),$$

has solutions if, and only if,

$$\mathbf{d} = \boldsymbol{\gamma}\boldsymbol{\alpha}^{-1}\mathbf{c}.$$

It has also been proved that

$$\boldsymbol{\delta} = \boldsymbol{\gamma}\boldsymbol{\alpha}^{-1}\boldsymbol{\beta}.$$

When the two matrices

$$\left(\begin{array}{c|c} \boldsymbol{\alpha} & \boldsymbol{\beta} \\ \hline \boldsymbol{\gamma} & \boldsymbol{\delta} \end{array}\right), \quad \left(\begin{array}{c|c|c} \boldsymbol{\alpha} & \boldsymbol{\beta} & \mathbf{c} \\ \hline \boldsymbol{\gamma} & \boldsymbol{\delta} & \mathbf{d} \end{array}\right)$$

are compared, it may thus be observed that there are solutions if, and only if, they can be written in the forms

$$\left(\begin{array}{c|c} \boldsymbol{\alpha} & \boldsymbol{\beta} \\ \hline \boldsymbol{\gamma} & \boldsymbol{\gamma}\boldsymbol{\alpha}^{-1}\boldsymbol{\beta} \end{array}\right), \quad \left(\begin{array}{c|c|c} \boldsymbol{\alpha} & \boldsymbol{\beta} & \mathbf{c} \\ \hline \boldsymbol{\gamma} & \boldsymbol{\gamma}\boldsymbol{\alpha}^{-1}\boldsymbol{\beta} & \boldsymbol{\gamma}\boldsymbol{\alpha}^{-1}\mathbf{c} \end{array}\right),$$

where *all the matrices on the 'bottom' line are* $\boldsymbol{\gamma}\boldsymbol{\alpha}^{-1}$ *times the corresponding matrices on the 'top' line.* Hence (by analogy with the corresponding result proved for columns):

The last $m-r$ *rows of the two matrices under consideration are all linear combinations of the first r rows, formed, for each of the two matrices, by the same rules of combination.* That is, the two matrices have the same number of linearly independent rows, and therefore the same rank.

This establishes the important basic theorem:

A necessary and sufficient condition for the equations written in matrix form

$$\mathbf{a}\mathbf{x} = \mathbf{b}$$

to be soluble is that the two matrices

$$\mathbf{a}, \quad (\mathbf{a} \mid \mathbf{b})$$

have the same rank.

This particular treatment leads readily to a specific form of solution.

Revision Examples

1. The matrices **a**, **p**, **q** are conformable for multiplication to yield the product **paq**. Prove that, if **a** is a non-singular matrix of r rows and columns, then the matrix

$$\left(\begin{array}{c|c} \mathbf{a} & \mathbf{aq} \\ \hline \mathbf{pa} & \mathbf{paq} \end{array}\right)$$

is of rank r.

Prove that, if **c** is an $r \times 1$ column vector, then the equation

$$\left(\begin{array}{c|c} \mathbf{a} & \mathbf{aq} \\ \hline \mathbf{pa} & \mathbf{paq} \end{array}\right) \mathbf{x} = \left(\begin{array}{c} \mathbf{c} \\ \hline \mathbf{pc} \end{array}\right)$$

is satisfied by every vector **x** of the form

$$\mathbf{x} \equiv \left(\begin{array}{c} \mathbf{a}^{-1}\mathbf{c} - \mathbf{qv} \\ \hline \mathbf{v} \end{array}\right).$$

Can every vector **x** satisfying the equation be expressed in this form?

2. Using the method of the text, solve the following systems of equations:

(i)
$$\begin{aligned} 2x+3y+z &= 6, \\ x+2y-z &= 2, \\ 4x+7y-z &= 10. \end{aligned}$$

(ii)
$$\begin{aligned} x+y+z &= 6, \\ x+2y+3z &= 14, \\ 3x+4y+5z &= k. \end{aligned}$$

(iii)
$$\begin{aligned} x+y+z+u &= 4, \\ x-y+z-u &= 0, \\ 5x+3y+5z+3u &= 16, \\ 7x+y+7z+u &= 16. \end{aligned}$$

(iv)
$$\begin{aligned} 2x+y+z &= 6, \\ x+y+3z &= 6. \end{aligned}$$

3. Investigate the following sets of equations by the methods of the text, all numbers being real:

(i)
$$\begin{aligned} x+y-2z &= 0,\\ ax+by+cz &= 0,\\ bx+cy+az &= d. \end{aligned}$$

(ii)
$$\begin{aligned} ax+by+cz &= p,\\ bx+cy+az &= q,\\ ax+by+cz &= r. \end{aligned}$$

(iii)
$$\begin{aligned} x+ay+2z-t &= 1,\\ -2x+2y+3z+2t &= 5b,\\ ax+y+z+t &= 2. \end{aligned}$$

(iv)
$$\begin{aligned} x+2y+\lambda z &= 0,\\ 2x+3y-2z &= \lambda,\\ \lambda x+y+z &= 3. \end{aligned}$$

(v)
$$\begin{aligned} 2y+3z &= a,\\ -2x+bz &= -2,\\ -3x-by &= 1,\\ ax-2y+z &= 0. \end{aligned}$$

4. Find the value of k for which the equations

$$\begin{aligned} 3x+4y-2z+7w &= 4,\\ x+3y-2z+5w &= 1,\\ x-2y+2z-3w &= k,\\ 3x-y+2z-w &= 5 \end{aligned}$$

are consistent and, for this value of k, obtain the general solution.

5. The polynomials

$$a_0x^3+a_1x^2+a_2x+a_3, \quad b_0x^3+b_1x^2+b_2x+b_3$$

have a common factor of degree 2. Prove that the rank of the matrix

$$\begin{pmatrix} a_0 & a_1 & a_2 & a_3 & 0 \\ 0 & a_0 & a_1 & a_2 & a_3 \\ b_0 & b_1 & b_2 & b_3 & 0 \\ 0 & b_0 & b_1 & b_2 & b_3 \end{pmatrix}$$

is at most 3.

6. Prove that, if λ satisfies the equation

$$\det(\mathbf{a}-\lambda\mathbf{I}) = 0,$$

where

$$\mathbf{a} \equiv \begin{pmatrix} 1 & 2 & 2 \\ 0 & 2 & 1 \\ 0 & -1 & 0 \end{pmatrix},$$

then the rank of the matrix $\mathbf{a}-\lambda\mathbf{I}$ is unity.

APPENDIX TO CHAPTER 16

Some harder theorems on rank.

1. Algebraic lemma. Let

$$\mathbf{a} \equiv \left(\begin{array}{c|c} \alpha & \beta \\ \hline \gamma & \delta \end{array}\right)$$

be a matrix of type $m \times n$, partitioned as shown so that α is a square matrix of order r, assumed non-singular.

It is required *to prove the fundamental formula*

$$\left(\begin{array}{c|c} \mathbf{I}_r & 0 \\ \hline -\gamma\alpha^{-1} & \mathbf{I}_{m-r} \end{array}\right) \left(\begin{array}{c|c} \alpha & \beta \\ \hline \gamma & \delta \end{array}\right) \left(\begin{array}{c|c} \mathbf{I}_r & -\alpha^{-1}\beta \\ \hline 0 & \mathbf{I}_{n-r} \end{array}\right) = \left(\begin{array}{c|c} \alpha & 0 \\ \hline 0 & \delta-\gamma\alpha^{-1}\beta \end{array}\right).$$

The left-hand side is

$$\left(\begin{array}{c|c} \mathbf{I}_r & 0 \\ \hline -\gamma\alpha^{-1} & \mathbf{I}_{m-r} \end{array}\right) \left(\begin{array}{c|c} \alpha & -\alpha\alpha^{-1}\beta+\beta \\ \hline \gamma & -\gamma\alpha^{-1}\beta+\delta \end{array}\right)$$

$$= \left(\begin{array}{c|c} \mathbf{I}_r & 0 \\ \hline -\gamma\alpha^{-1} & \mathbf{I}_{m-r} \end{array}\right) \left(\begin{array}{c|c} \alpha & 0 \\ \hline \gamma & \delta-\gamma\alpha^{-1}\beta \end{array}\right)$$

$$= \left(\begin{array}{c|c} \alpha & 0 \\ \hline -\gamma\alpha^{-1}\alpha+\gamma & \delta-\gamma\alpha^{-1}\beta \end{array}\right)$$

$$= \begin{pmatrix} \alpha & 0 \\ 0 & \delta-\gamma\alpha^{-1}\beta \end{pmatrix}.$$

2. Standard reduction of a matrix of rank *r*. Suppose now that the matrix $\mathbf{a}$ is of rank r. Then (p. 212)

$$\delta-\gamma\alpha^{-1}\beta = 0.$$

Thus

$$\left(\begin{array}{c|c} \mathbf{I}_r & 0 \\ \hline -\gamma\alpha^{-1} & \mathbf{I}_{m-r} \end{array}\right) \left(\begin{array}{c|c} \alpha & \beta \\ \hline \gamma & \delta \end{array}\right) \left(\begin{array}{c|c} \mathbf{I}_r & -\alpha^{-1}\beta \\ \hline 0 & \mathbf{I}_{n-r} \end{array}\right) = \left(\begin{array}{c|c} \alpha & 0 \\ \hline 0 & 0 \end{array}\right).$$

Further, *each of the matrices*

$$\begin{pmatrix} \mathbf{I}_r & \mathbf{0} \\ -\boldsymbol{\gamma}\boldsymbol{\alpha}^{-1} & \mathbf{I}_{m-r} \end{pmatrix}, \quad \begin{pmatrix} \mathbf{I}_r & -\boldsymbol{\alpha}^{-1}\boldsymbol{\beta} \\ \mathbf{0} & \mathbf{I}_{n-r} \end{pmatrix}$$

is non-singular, so that *the matrix* **a**, *of type* $m \times n$ *and rank* r, *has been reduced to the form*

$$\left(\begin{array}{c|c} \boldsymbol{\alpha} & \mathbf{0} \\ \hline \mathbf{0} & \mathbf{0} \end{array}\right)$$

on pre-multiplying by a non-singular matrix of order m and post-multiplying by a non-singular matrix of order n, where the matrix $\boldsymbol{\alpha}$ *in the final form is a non-singular matrix of order r.*

This reduction may be carried one stage further. Post-multiply by the non-singular matrix of order n,

$$\left(\begin{array}{c|c} \boldsymbol{\alpha}^{-1} & \mathbf{0} \\ \hline \mathbf{0} & \mathbf{I}_{n-r} \end{array}\right).$$

Then

$$\left(\begin{array}{c|c} \boldsymbol{\alpha} & \mathbf{0} \\ \hline \mathbf{0} & \mathbf{0} \end{array}\right) \left(\begin{array}{c|c} \boldsymbol{\alpha}^{-1} & \mathbf{0} \\ \hline \mathbf{0} & \mathbf{I}_{n-r} \end{array}\right) = \left(\begin{array}{c|c} \mathbf{I}_r & \mathbf{0} \\ \hline \mathbf{0} & \mathbf{0} \end{array}\right).$$

The net result, since the product of the two non-singular post-multiplying matrices of order n is a matrix, also non-singular, of order n, is that *the matrix* **a** *of rank r may be pre-multiplied by a non-singular matrix* **p** *of order m and post-multiplied by a non-singular matrix* **q** *of order n so that*

$$\mathbf{paq} = \left(\begin{array}{c|c} \mathbf{I}_r & \mathbf{0}_{r,n-r} \\ \hline \mathbf{0}_{m-r,r} & \mathbf{0}_{m-r,n-r} \end{array}\right);$$

briefly, *a matrix of rank r can be reduced to the form*

$$\left(\begin{array}{c|c} \mathbf{I}_r & \mathbf{0} \\ \hline \mathbf{0} & \mathbf{0} \end{array}\right).$$

3. The rank-constancy theorem. Suppose that an $m \times n$ matrix **a** of rank r is premultiplied by a non-singular square matrix **u** of order m.

THEOREM. *The rank of* **ua** *is also r.*

Since **a** is of rank r, it has r linearly independent columns, and so there are $n-r$ linear relations connecting the columns of **a**. Each such relation can be expressed in the matrix form (p. 198)

$$\mathbf{a}\boldsymbol{\xi} = \mathbf{0},$$

where $\boldsymbol{\xi}$ is an $n \times 1$ vector. But

$$\begin{aligned} \mathbf{a}\boldsymbol{\xi} &= \mathbf{0} \\ \Rightarrow \mathbf{ua}\boldsymbol{\xi} &= \mathbf{0}, \end{aligned}$$

and so each linear relation among the columns of **a** is reflected in one among the columns of **ua** with the same multiplying vector $\boldsymbol{\xi}$. Hence the columns of **ua** have at least $n-r$ linear relations, and there may conceivably be more. Hence

$$\operatorname{rank}(\mathbf{ua}) \leqslant \operatorname{rank}(\mathbf{a}).$$

But **u** is non-singular, so $\mathbf{u}^{-1}$ exists and is non-singular. Identical argument thus shows that

$$\operatorname{rank}(\mathbf{u}^{-1}\mathbf{ua}) \leqslant \operatorname{rank}(\mathbf{ua}),$$

or

$$\operatorname{rank}(\mathbf{a}) \leqslant \operatorname{rank}(\mathbf{ua}).$$

Hence

$$\operatorname{rank}(\mathbf{ua}) = \operatorname{rank}(\mathbf{a}).$$

In the same way, with minor variations, it can be proved that, *if* **a** *is postmultiplied by a non-singular* $n \times n$ *matrix* **v**, *then the rank of* **av** *is also* r.

COROLLARIES. (i) *The rank of* **uav** *is* r.

(ii) *The rank of a product* **ab** *is not greater than the ranks of the individual matrices* **a** *and* **b**.

The proof of (ii) is similar to that of the main theorem.

4. The converse of the standard-reduction theorem. It was proved in §2 that a matrix of rank r can be reduced to the standard form

$$\mathbf{u}\left(\begin{array}{c|c} \mathbf{I}_r & \mathbf{0} \\ \hline \mathbf{0} & \mathbf{0} \end{array}\right)\mathbf{v},$$

where the square matrices **u**, **v** are non-singular. The work of §3 establishes the converse property that *every matrix of this form has, for non-singular* **u**, **v**, *the rank* r.

5. Sylvester's law of nullity. Let **a**, **b** be two given matrices, of types $m \times n$, $n \times p$ and of ranks r, s.

THEOREM. *The rank of the product matrix* **ab** *satisfies the relation*

$$\rho \equiv \operatorname{rank}(\mathbf{ab}) \geqslant r+s-n.$$

Since the rank of **b** is s, there exist non-singular matrices **u**, **v** such that

$$\mathbf{b} = \mathbf{u}\left(\begin{array}{c|c} \mathbf{I}_s & \mathbf{0} \\ \hline \mathbf{0} & \mathbf{0} \end{array}\right)\mathbf{v}.$$

Also the rank of $\mathbf{a}$ is r, and so the rank of

$$\mathbf{au}$$

is also r.

Express the matrix $\mathbf{au}$ in partitioned form

$$\mathbf{au} \equiv (\boldsymbol{\alpha}, \boldsymbol{\beta})$$

where $\boldsymbol{\alpha}$, $\boldsymbol{\beta}$ are of types $m \times s$ and $m \times (n-s)$ respectively. Now $\mathbf{au}$ has r linearly independent columns, so that $\boldsymbol{\alpha}$ has at least $r-(n-s)$ linearly independent columns. (If this number is negative or zero, then, since $\rho \geqslant 0$, the result

$$\rho \geqslant r+s-n$$

follows automatically. Assume therefore that it is positive.) Since $\boldsymbol{\alpha}$ has at least $r+s-n$ linearly independent columns, the rank of the matrix

$$(\boldsymbol{\alpha}, \mathbf{0})$$

is at least $r+s-n$. Further,

$$\begin{aligned}(\boldsymbol{\alpha}, \mathbf{0}) &= (\boldsymbol{\alpha}, \boldsymbol{\beta})\left(\begin{array}{c|c}\mathbf{I}_s & \mathbf{0} \\ \hline \mathbf{0} & \mathbf{0}\end{array}\right) \\ &= \mathbf{au}\left(\begin{array}{c|c}\mathbf{I}_s & \mathbf{0} \\ \hline \mathbf{0} & \mathbf{0}\end{array}\right) \\ &= \mathbf{a}(\mathbf{bv}^{-1}) \\ &= \mathbf{abv}^{-1}\end{aligned}$$

Thus

$$\begin{aligned}\rho &\equiv \operatorname{rank}(\mathbf{ab}) \\ &= \operatorname{rank}(\mathbf{abv}^{-1}) \qquad (\mathbf{v}^{-1} \text{ is non-singular}) \\ &= \operatorname{rank}(\boldsymbol{\alpha}, \mathbf{0}) \\ &\geqslant r+s-n.\end{aligned}$$

REMARK. The selection of $m \times s$ and $m \times (n-s)$ as the types for $\boldsymbol{\alpha}, \boldsymbol{\beta}$ ensures the existence of the product

$$(\boldsymbol{\alpha}, \boldsymbol{\beta})\left(\begin{array}{c|c}\mathbf{I}_s & \mathbf{0} \\ \hline \mathbf{0} & \mathbf{0}\end{array}\right)$$

used in the proof.

***Illustration* 1.** *The rank of the adjoint.* Let a be a given $n \times n$ square matrix. It is required *to find the rank of the adjoint* $\mathbf{A}$.

The fundamental formula is (p. 160)

$$\mathbf{aA} = \alpha \mathbf{I}_n.$$

where $\alpha = \det \mathbf{a}$.

Thus

$$\det \mathbf{a} \det \mathbf{A} = \det(\mathbf{aA}) = \det(\alpha \mathbf{I}_n),$$

so that

$$\alpha \det \mathbf{A} = \alpha^n.$$

(i) If $\alpha \neq 0$, then

$$\det \mathbf{A} = \alpha^{n-1} \neq 0.$$

Hence the rank of $\mathbf{A}$ is n.

(ii) If the rank of $\mathbf{a}$ is $n-1$, then at least one cofactor (of order $n-1$) in $\mathbf{a}$ is not zero, and so

$$\text{rank}(\mathbf{A}) \geqslant 1.$$

By Sylvester's law of nullity,

$$\text{rank}(\mathbf{aA}) \geqslant \text{rank}(\mathbf{a}) + \text{rank}(\mathbf{A}) - n,$$

or

$$\text{rank}(\mathbf{0}) \geqslant \text{rank}(\mathbf{a}) + \text{rank}(\mathbf{A}) - n,$$

or

$$0 \geqslant (n-1) + \text{rank}(\mathbf{A}) - n = \text{rank}(\mathbf{A}) - 1,$$

so that

$$\text{rank}(\mathbf{A}) \leqslant 1.$$

Hence when

$$\text{rank}(\mathbf{a}) = n-1,$$

$$\text{rank}(\mathbf{A}) = 1.$$

(iii) If rank $(\mathbf{a}) < n-1$, all the cofactors of $\mathbf{a}$ vanish, so that $\mathbf{A}$ is the zero matrix. Hence

$$\text{rank}(\mathbf{A}) = 0.$$

Revision Examples

1. Two matrices $\mathbf{a}$, $\mathbf{b}$, of types $m \times n$ and $n \times m$ respectively, are such that $m < n$ and

$$\mathbf{ab} = \mathbf{I},$$

where $\mathbf{I}$ is the unit matrix of order m. Prove that $\mathbf{a}$ is of rank m.

2. Reduce to the form

$$\left(\begin{array}{c|c} \mathbf{I}_2 & \mathbf{0} \\ \hline \mathbf{0} & \mathbf{0} \end{array}\right)$$

the matrices

$$\text{(i)} \begin{pmatrix} 2 & 3 & 1 & 4 \\ 1 & 2 & 2 & 3 \\ 0 & -1 & -3 & -2 \end{pmatrix}, \qquad \text{(ii)} \begin{pmatrix} 5 & 2 & 1 \\ 2 & 1 & 3 \\ 9 & 4 & 7 \\ -1 & 0 & 5 \end{pmatrix},$$

$$\text{(iii)} \begin{pmatrix} 3 & 1 & 1 & 1 & 1 \\ 2 & 1 & 4 & 3 & 2 \\ 1 & 1 & 7 & 5 & 3 \end{pmatrix}.$$

3. Verify the relation $\rho \geqslant r+s-n$ for the pairs of matrices:

(i) $$\begin{pmatrix} 1 & 2 & 3 & 4 \\ 4 & 3 & 2 & 1 \\ 3 & 1 & -1 & -3 \end{pmatrix} \begin{pmatrix} 0 & 3 \\ 2 & 1 \\ 4 & 4 \\ 8 & 9 \end{pmatrix},$$ (ii) $$\begin{pmatrix} 1 & 2 & 3 \\ 1 & 1 & 1 \\ 2 & 1 & 0 \\ 4 & 4 & 4 \end{pmatrix} \begin{pmatrix} 1 & 1 \\ 2 & 2 \\ 3 & 3 \end{pmatrix}.$$

4. Estimate by inspection the rank of the products

(i) $$\begin{pmatrix} 1 & 3 & 5 \\ 4 & 3 & 2 \\ 5 & 6 & 0 \end{pmatrix} \begin{pmatrix} 1 & 1 \\ 1 & 2 \\ 2 & 3 \end{pmatrix} \begin{pmatrix} 3 & 2 \\ 1 & 1 \end{pmatrix},$$

(ii) $$\begin{pmatrix} 1 & 0 \\ 2 & 1 \end{pmatrix} \begin{pmatrix} 1 & 2 & 3 & 4 \\ 2 & 4 & 6 & 8 \end{pmatrix} \begin{pmatrix} 1 & 0 & 0 & 0 \\ 0 & 1 & 0 & 0 \\ 3 & 2 & -2 & -3 \\ 2 & 3 & -1 & -2 \end{pmatrix},$$

and then multiply the matrices together and find independently the ranks of the products.

5. In a certain motion of a rigid body, it is found that the components of velocity of that point of the body, whose coordinates are (x,y,z) referred to fixed rectangular axes, are given by the relations

$$u = \omega_3 y - \omega_2(z-c), \quad v = \omega_1 z - \omega_3(x-a), \quad w = \omega_2 x - \omega_1(y-b).$$

Prove that there is no point of the body at rest unless

$$a\omega_2\omega_3 + b\omega_3\omega_1 + c\omega_1\omega_2 = 0.$$

Prove also that, if this condition is satisfied, there is a line of points at rest, and that the plane through this line and the origin is given by the equation

$$c\omega_2 x + a\omega_3 y + b\omega_1 z = 0.$$

6. The column vectors **x**, **y**, **z**, **t**, of four elements each, satisfy two linear relations. Prove that the matrix

$$\begin{pmatrix} p_{01} & q_{01} & r_{01} & \cdots \\ p_{02} & q_{02} & r_{02} & \cdots \\ p_{03} & q_{03} & r_{03} & \cdots \\ p_{12} & q_{12} & r_{12} & \cdots \\ p_{13} & q_{13} & r_{13} & \cdots \\ p_{23} & q_{23} & r_{23} & \cdots \end{pmatrix}$$

is of rank 1, where

$$p_{ij} = x_i y_j - x_j y_i,\ q_{ij} = x_i z_j - x_j z_i,\ r_{ij} = x_i t_j - x_j t_1,\ s_{ij} = y_i z_j - y_j z_i, \ldots$$

7. Find the rank of the matrix

$$\mathbf{a} \equiv \begin{pmatrix} 1 & 2 & 3 & 4 \\ 1 & 3 & 5 & 7 \\ 1 & 4 & 7 & 10 \\ 1 & 5 & 9 & 13 \end{pmatrix}.$$

Find a 4×4 matrix $\mathbf{b}$ such that

$$\mathbf{ba} = \mathbf{0},$$

where $\mathbf{0}$ is the zero 4×4 matrix.

8. A partitioned matrix $\mathbf{u}$ is given by the relation

$$\mathbf{u} \equiv \left(\begin{array}{c|c} \mathbf{a} & \mathbf{b} \\ \hline \mathbf{c} & \mathbf{d} \end{array}\right),$$

where $\mathbf{a}$, $\mathbf{b}$, $\mathbf{c}$, $\mathbf{d}$ are square matrices of order n and $\mathbf{a}$ is non-singular. By multiplying $\mathbf{u}$ on the right by a suitable matrix of the form

$$\left(\begin{array}{c|c} \mathbf{I} & \mathbf{x} \\ \hline \mathbf{0} & \mathbf{I} \end{array}\right),$$

or otherwise, prove that

$$\det \mathbf{u} = \det(\mathbf{ad} - \mathbf{aca}^{-1}\mathbf{b}).$$

9. If $\mathbf{a}$, $\mathbf{b}$ are two $m \times n$ matrices of ranks r, s, prove that the rank of $\mathbf{a}+\mathbf{b}$ cannot exceed the least of m, n and $r+s$; show by example that it can be equal to it.

Prove also that, if $r > s$, the rank of $\mathbf{a}+\mathbf{b}$ cannot be less than $r-s$.

10. Prove that the system of equations

$$\begin{aligned} x+\ y-\ 3z+\ u-\ 2w &= 0, \\ 2x-\ y-\ 3z-2u+\ 5w &= 0, \\ x-3y+\ z+4u+10w &= 0, \\ 3x+4y-10z-7u-\ 9w &= 0 \end{aligned}$$

is consistent when any values are assigned to z and w, and x, y, u are to be found.

Prove that the equations are inconsistent if arbitrary values are assigned to u and w, and x, y, z are to be found.

11. Find the rank of the matrix

$$\mathbf{a} \equiv \begin{pmatrix} 2 & -3 & -1 & 1 \\ 3 & 4 & -4 & -3 \\ 0 & 17 & -5 & -9 \end{pmatrix}.$$

Obtain a formula for the complete set of solutions of the equation

$$\mathbf{ax} = \mathbf{0}.$$

Which, if any, of these solutions

(i) satisfy the equations

$$x_1+x_2+x_3+x_4+1 = 0$$

and

$$x_1-x_2-x_3-x_4-3 = 0,$$

where

$$\mathbf{x} \equiv \{x_1, x_2, x_3, x_4\},$$

(ii) are linearly dependent on the pair $\{0, 1, 2, 3\}$ and $\{3, 2, 1, 0\}$?

12. Find non-singular matrices $\mathbf{p}$, $\mathbf{q}$ such that

$$\mathbf{paq} \equiv \left(\begin{array}{c|c} \mathbf{I}_r & \mathbf{0} \\ \hline \mathbf{0} & \mathbf{0} \end{array}\right)$$

when $\mathbf{a}$ is given to be the matrix

(1) $\begin{pmatrix} 2 & 1 & 4 \\ 3 & 2 & 1 \\ 7 & 4 & 9 \end{pmatrix}$, (ii) $\begin{pmatrix} 4 & 3 & -2 \\ 1 & 1 & 3 \\ 7 & 6 & 7 \end{pmatrix}$, (iii) $\begin{pmatrix} 3 & 5 & -2 \\ 2 & 3 & 4 \\ 1 & 1 & 10 \end{pmatrix}$.

SECTION III

INTRODUCTION TO MORE ADVANCED WORK

The aim of this section is to give an introductory account of topics which are interesting in themselves and which the student should know about, but which would, for detailed treatment, demand more experience than it seems right to assume here. The beginner may like to know about them; the serious student of the subject should consult more advanced textbooks.

The theme of much of the section is the reduction of quadratic forms to 'sums of squares'. This is important in many branches of mathematics, notably in geometry. For example, a conic may be given by the equation

$$41x^2 - 24xy + 34y^2 = 100.$$

This version can hardly be called informative; but the work to be given on p. 238 shows that the equation can be put in the alternative form

$$25u^2 + 50v^2 = 100,$$

or

$$\frac{u^2}{4} + \frac{v^2}{2} = 1,$$

in which the conic is recognized as an ellipse whose axes are 2 and $\sqrt{2}$ in length.

17

QUADRATIC FORMS: EIGENVECTORS

1. The x′ay notation. Let **a** be a given square matrix of order n (usually taken here to be 2 or 3) and let **x**, **y**, ... be column vectors of n elements. Thus ($n = 3$)

$$\mathbf{a} \equiv \begin{pmatrix} a_{11} & a_{12} & a_{13} \\ a_{21} & a_{22} & a_{23} \\ a_{31} & a_{32} & a_{33} \end{pmatrix}, \quad \mathbf{x} \equiv \begin{pmatrix} x_1 \\ x_2 \\ x_3 \end{pmatrix}.$$

Then **ax** is the *column vector*.

$$\mathbf{ax} \equiv \begin{pmatrix} a_{11}x_1 + a_{12}x_2 + a_{13}x_3 \\ a_{21}x_1 + a_{22}x_2 + a_{23}x_3 \\ a_{31}x_1 + a_{32}x_2 + a_{33}x_3 \end{pmatrix}.$$

The transpose **y′** of a column vector **y** is the row-vector

$$\mathbf{y}' \equiv (y_1, y_2, y_3),$$

and so **y′ ax** is the *scalar*

$$\begin{aligned} &y_1(a_{11}x_1 + a_{12}x_2 + a_{13}x_3) \\ &+y_2(a_{21}x_1 + a_{22}x_2 + a_{23}x_3) \\ &+y_3(a_{31}x_1 + a_{32}x_2 + a_{33}x_3). \end{aligned}$$

This scalar function is linear in x_1, x_2, x_3 and also linear in y_1, y_2, y_3. It is called a *bilinear form* in the two sets of variables x_i and y_i.

Since transposition does not affect a scalar, it follows that

$$\begin{aligned} \mathbf{y}'\mathbf{ax} &= (\mathbf{y}'\mathbf{ax})' \\ &= \mathbf{x}'\mathbf{a}'\mathbf{y}, \end{aligned}$$

as can also be verified by direct computation.

When **x** and **y** are the same vector, say **x**, then

$$\begin{aligned} \mathbf{x}'\mathbf{ax} \equiv\ &x_1(a_{11}x_1 + a_{12}x_2 + a_{13}x_3) \\ &+x_2(a_{21}x_1 + a_{22}x_2 + a_{23}x_3) \\ &+x_3(a_{31}x_1 + a_{32}x_2 + a_{33}x_3) \\ \equiv\ &a_{11}x_1^2 + a_{22}x_2^2 + a_{33}x_3^2 \\ &+(a_{23}+a_{32})x_2x_3 + (a_{31}+a_{13})x_3x_1 + (a_{12}+a_{21})x_1x_2. \end{aligned}$$

This is a *homogeneous quadratic form* in the variables x_1, x_2, x_3.

If

$$\mathbf{a} \equiv \begin{pmatrix} a_{11} & a_{12} \\ a_{21} & a_{22} \end{pmatrix}, \quad \mathbf{x} \equiv \begin{pmatrix} x_1 \\ x_2 \end{pmatrix},$$

the corresponding quadratic form is

$$\mathbf{x}'\mathbf{a}\mathbf{x} \equiv a_{11}x_1^2 + a_{22}x_2^2 + (a_{12}+a_{21})x_1x_2.$$

Warning Example. If

$$\mathbf{a} \equiv \begin{pmatrix} 2 & 0 \\ 0 & 3 \end{pmatrix}, \quad \mathbf{b} \equiv \begin{pmatrix} 2 & 5 \\ -5 & 3 \end{pmatrix},$$

then

$$\mathbf{x}'\mathbf{a}\mathbf{x} \equiv 2x_1^2 + (0+0)x_1x_2 + 3x_2^2,$$
$$\mathbf{x}'\mathbf{b}\mathbf{x} \equiv 2x_1^2 + (5-5)x_1x_2 + 3x_2^2,$$

and each of these is the quadratic form

$$2x_1^2 + 3x_2^2.$$

Hence *it is possible for the same quadratic form to arise from different matrices.*

To mitigate the effects of the Warning Example, note that the coefficient of a product $x_i x_j$ has the composite form

$$a_{ij} + a_{ji}.$$

For a given quadratic form, the coefficients $a_{11}, a_{22}, a_{33}, \ldots$ are completely determined, but only the *sum* of elements $a_{ij} + a_{ji}$.

To avoid ambiguity, it is agreed, by convention, that *the matrix* **a** *of a quadratic form* **x'ax** *is to be taken as symmetric, so that*

$$\mathbf{a} = \mathbf{a}'.$$

On this basis, the matrix of a quadratic form

$$3x_1^2 + 5x_1x_2 + 7x_2^2$$

is unambiguously

$$\begin{pmatrix} 3 & \frac{5}{2} \\ \frac{5}{2} & 7 \end{pmatrix}.$$

DEFINITION. The matrix **a** (**a**$'$ = **a**) is called the *matrix of the quadratic form* **x'ax**.

Examples

1. Write down the matrices of the quadratic forms:

(i) $x_1^2 - 2x_2^2 + 4x_1x_2$,

(ii) $x_1^2 - 2x_2^2 + 3x_3^2 + 4x_2x_3 - 6x_3x_1 + 4x_1x_2$,

(iii) $2x_1^2 - 4x_1x_2$,

(iv) $x_2x_3 + 3x_3x_1 + 5x_1x_2$.

2. Write down the quadratic forms whose matrices are:

$$\begin{pmatrix} 0 & 2 & 3 \\ 2 & 4 & -1 \\ 3 & -1 & 2 \end{pmatrix}, \quad \begin{pmatrix} 3 & 0 \\ 0 & 4 \end{pmatrix}, \quad \begin{pmatrix} 0 & -2 \\ 3 & 0 \end{pmatrix}, \quad \begin{pmatrix} 5 & -3 & -2 \\ -3 & 3 & -1 \\ -2 & -1 & 1 \end{pmatrix}.$$

3. Find $\mathbf{x}'\mathbf{a}\mathbf{x}$ when

(i) $$\mathbf{a} \equiv \begin{pmatrix} a & h \\ h & b \end{pmatrix}, \quad \mathbf{x} \equiv \begin{pmatrix} x \\ y \end{pmatrix},$$

(ii) $$\mathbf{a} \equiv \begin{pmatrix} a & h & g \\ h & b & f \\ g & f & c \end{pmatrix}, \quad \mathbf{x} \equiv \begin{pmatrix} x \\ y \\ z \end{pmatrix}.$$

4. Prove that, if $\mathbf{a}$ is skew-symmetric, so that $\mathbf{a}' = -\mathbf{a}$, then, identically,

$$\mathbf{x}'\mathbf{a}\mathbf{x} = 0.$$

5. Write down the matrices of the quadratic forms

$$2x_1x_2 + 2x_2x_3 + 2x_3x_4,$$
$$2x_1x_2 + 2x_2x_3 + 2x_3x_4 + 2x_4x_5$$

and prove that the rank of each matrix is 4.

Quadratic forms in two variables

2. Linear mappings (transformations). Write

$$u \equiv a_{11}x_1^2 + 2a_{12}x_1x_2 + a_{22}x_2^2$$
$$\equiv \mathbf{x}'\mathbf{a}\mathbf{x},$$

where

$$\mathbf{a} \equiv \begin{pmatrix} a_{11} & a_{12} \\ a_{21} & a_{22} \end{pmatrix},$$

with $\mathbf{a}' = \mathbf{a}$ so that $a_{12} = a_{21}$, and where

$$\mathbf{x} \equiv \{x_1, x_2\}.$$

The immediate problem is *to study the effect on u of the linear mapping*

$$\mathbf{x} = \mathbf{p}\mathbf{y},$$

where $$\mathbf{p} \equiv \begin{pmatrix} p_{11} & p_{12} \\ p_{21} & p_{22} \end{pmatrix}, \quad \mathbf{y} \equiv \begin{pmatrix} y_1 \\ y_2 \end{pmatrix}.$$

It is *not* assumed that $\mathbf{p}$ is symmetric.

Since
$$\mathbf{x}' = \mathbf{y}'\mathbf{p}',$$
the transform of u is given by the relation
$$\begin{aligned} u &= (\mathbf{y}'\mathbf{p}')\mathbf{a}(\mathbf{p}\mathbf{y}) \\ &= \mathbf{y}'(\mathbf{p}'\mathbf{a}\mathbf{p})\mathbf{y} \\ &= \mathbf{y}'\mathbf{b}\mathbf{y}, \end{aligned}$$
say, where
$$\mathbf{b} = \mathbf{p}'\mathbf{a}\mathbf{p}.$$

Note that
$$\mathbf{b}' = \mathbf{p}'\mathbf{a}'(\mathbf{p}')' = \mathbf{p}'\mathbf{a}\mathbf{p} = \mathbf{b},$$
so that $\mathbf{b}$ *is also symmetric.*

The quadratic form, in terms of $\mathbf{y}$, is
$$b_{11}y_1^2 + 2b_{12}y_1y_2 + b_{22}y_2^2.$$

What is proposed to be done is to choose $\mathbf{p}$ in such a way that $\mathbf{b}$ and its associated quadratic form are as simple as possible. This will require a study of *orthogonal mappings* and *orthogonal matrices* as the next task.

Examples

1. Find the quadratic form in $\{y_1, y_2\}$ into which the form
$$2x_1^2 + 6x_1x_2 + 5x_2^2$$
is mapped by $\mathbf{x} = \mathbf{p}\mathbf{y}$ when $\mathbf{p}$ is

(i) $\begin{pmatrix} 2 & 0 \\ 0 & 3 \end{pmatrix}$, (ii) $\begin{pmatrix} 0 & 2 \\ 3 & 0 \end{pmatrix}$,

(iii) $\begin{pmatrix} 1 & 1 \\ -2 & 3 \end{pmatrix}$, (iv) $\begin{pmatrix} 2 & -3 \\ 1 & 4 \end{pmatrix}$,

(v) $\begin{pmatrix} -2 & -1 \\ 5 & 3 \end{pmatrix}$, (vi) $\begin{pmatrix} -1 & 4 \\ 3 & -2 \end{pmatrix}$.

3. Orthogonal transformations and orthogonal matrices. The elements x_1, x_2 and y_1, y_2 of the vectors $\mathbf{x}$ and $\mathbf{y}$ may be regarded as the Cartesian coordinates of two points $X(x_1, x_2)$ and $Y(y_1, y_2)$. The mapping
$$\mathbf{x} = \mathbf{p}\mathbf{y}$$
then maps a configuration defined by points such as X into one defined by points such as Y.

Particular importance attaches to the behaviour of *distance* under such mappings. Suppose that $X(x_1, x_2)$, $U(u_1, u_2)$ are two given points. The square of their distance apart is

$$(x_1 - u_1)^2 + (x_2 - u_2)^2,$$

or, in matrix notation,

$$(\mathbf{x} - \mathbf{u})'(\mathbf{x} - \mathbf{u}).$$

Under the mapping, these points become $Y(y_1, y_2)$, $V(v_1, v_2)$, where

$$\mathbf{x} = \mathbf{p}\mathbf{y}, \quad \mathbf{u} = \mathbf{p}\mathbf{v},$$

and the square of the distance between Y and V is

$$(\mathbf{y} - \mathbf{v})'(\mathbf{y} - \mathbf{v}).$$

If, as is desired, the distances are unchanged by the mapping, then

$$(\mathbf{x} - \mathbf{u})'(\mathbf{x} - \mathbf{u}) \equiv (\mathbf{y} - \mathbf{v})'(\mathbf{y} - \mathbf{v}),$$

so that

$$\{\mathbf{p}(\mathbf{y} - \mathbf{v})\}'\{\mathbf{p}(\mathbf{y} - \mathbf{v})\} \equiv (\mathbf{y} - \mathbf{v})'(\mathbf{y} - \mathbf{v}).$$

Write $\mathbf{y} - \mathbf{v} = \mathbf{z}$. Then

$$\mathbf{z}'\mathbf{p}'\mathbf{p}\mathbf{z} \equiv \mathbf{z}'\mathbf{z}$$

for all values of $\mathbf{z}$.

These are two *quadratic forms* in $\mathbf{z}$, and, further, the matrix $\mathbf{p}'\mathbf{p}$ is *symmetrical*, since

$$(\mathbf{p}'\mathbf{p})' = \mathbf{p}'(\mathbf{p}')' = \mathbf{p}'\mathbf{p}.$$

Hence *the distances are unaltered by the mapping if, and only if, the matrices of the two quadratic forms are equal; that is, if, and only if,*

$$\mathbf{p}'\mathbf{p} = \mathbf{I}.$$

DEFINITION. A square matrix $\mathbf{p}$ of order n is called *orthogonal* when

$$\mathbf{p}'\mathbf{p} = \mathbf{I}_n.$$

COROLLARY. If $\mathbf{p}$ *is an orthogonal matrix, then* $\det\mathbf{p} = \pm 1$:

For

$$\begin{aligned} &\mathbf{p}'\mathbf{p} = \mathbf{I}_n \\ \Rightarrow\ &\det(\mathbf{p}'\mathbf{p}) = \det\mathbf{I}_n = 1 \\ \Rightarrow\ &\det\mathbf{p}'\det\mathbf{p} = 1 \\ \Rightarrow\ &\{\det\mathbf{p}\}^2 = 1 \\ \Rightarrow\ &\det\mathbf{p} = \pm 1. \end{aligned}$$

Note that *an orthogonal matrix cannot be singular.*

Examples

1. Prove that the following matrices are orthogonal:

(i) $\begin{pmatrix} \frac{4}{5} & \frac{3}{5} \\ -\frac{3}{5} & \frac{4}{5} \end{pmatrix}$, (ii) $\begin{pmatrix} -\frac{5}{13} & \frac{12}{13} \\ -\frac{12}{13} & -\frac{5}{13} \end{pmatrix}$,

(iii) $\begin{pmatrix} \frac{1}{\sqrt{2}} & -\frac{1}{\sqrt{2}} \\ \frac{1}{\sqrt{2}} & \frac{1}{\sqrt{2}} \end{pmatrix}$, (iv) $\begin{pmatrix} \cos\theta & -\sin\theta \\ \sin\theta & \cos\theta \end{pmatrix}$.

2. Find the value of ρ if the matrix

$$\begin{pmatrix} 8\rho & -15\rho \\ 15\rho & 8\rho \end{pmatrix}$$

is orthogonal.

3. Find values of a, x, y which make the matrix

$$\begin{pmatrix} a & 2a \\ x & y \end{pmatrix}$$

orthogonal.

4. Prove that the product of two (conformable) orthogonal matrices is an orthogonal matrix.

4. Orthogonal reduction to a 'sum of squares': particular example. Let

$$u \equiv 41x_1^2 - 24x_1x_2 + 34x_2^2$$
$$\equiv \mathbf{x}'\mathbf{a}\mathbf{x},$$

where

$$\mathbf{a} \equiv \begin{pmatrix} 41 & -12 \\ -12 & 34 \end{pmatrix}.$$

The problem is *to find an orthogonal transformation*

$$\mathbf{x} = \mathbf{p}\mathbf{y},$$

where

$$\mathbf{p}'\mathbf{p} = \mathbf{I},$$

reducing u to the form

$$u \equiv b_{11}y_1^2 + b_{22}y_2^2$$
$$\equiv \mathbf{y}'\mathbf{b}\mathbf{y},$$

where

$$\mathbf{b} \equiv \begin{pmatrix} b_{11} & 0 \\ 0 & b_{22} \end{pmatrix}.$$

As before (p. 236), **b** is given by the relation

$$\mathbf{b} = \mathbf{p}'\mathbf{a}\mathbf{p}.$$

Since $\mathbf{p}'\mathbf{p} = \mathbf{I}$, it follows that

$$\mathbf{p}' = \mathbf{p}^{-1},$$

so that

$$\mathbf{b} = \mathbf{p}^{-1}\mathbf{a}\mathbf{p},$$

or

$$\mathbf{p}\mathbf{b} = \mathbf{a}\mathbf{p}.$$

This requires

$$\begin{pmatrix} p_{11} & p_{12} \\ p_{21} & p_{22} \end{pmatrix} \begin{pmatrix} b_{11} & 0 \\ 0 & b_{22} \end{pmatrix} = \begin{pmatrix} 41 & -12 \\ -12 & 34 \end{pmatrix} \begin{pmatrix} p_{11} & p_{12} \\ p_{21} & p_{22} \end{pmatrix},$$

or, equating corresponding elements in the expanded matrices,

$$\begin{aligned} p_{11}b_{11} &= 41p_{11} - 12p_{21}, \\ p_{12}b_{22} &= 41p_{12} - 12p_{22}, \\ p_{21}b_{11} &= -12p_{11} + 34p_{21}, \\ p_{22}b_{22} &= -12p_{12} + 34p_{22}. \end{aligned}$$

Hence, rearranging,

$$\left.\begin{aligned} (41-b_{11})p_{11} - 12p_{21} &= 0, \\ -12p_{11} + (34-b_{11})p_{21} &= 0 \end{aligned}\right\}$$

and

$$\left.\begin{aligned} (41-b_{22})p_{12} - 12p_{22} &= 0, \\ -12p_{12} + (34-b_{22}) &= 0. \end{aligned}\right\}$$

Since $p_{11}, p_{12}, p_{21}, p_{22}$ cannot all be zero, elimination gives

$$\begin{vmatrix} 41-b_{11} & -12 \\ -12 & 34-b_{11} \end{vmatrix} = 0, \quad \begin{vmatrix} 41-b_{22} & -12 \\ -12 & 34-b_{22} \end{vmatrix} = 0.$$

Hence b_{11} *and* b_{22} *are the two roots of the quadratic equation*

$$\begin{vmatrix} 41-\lambda & -12 \\ -12 & 34-\lambda \end{vmatrix} = 0.$$

This equation is

$$\lambda^2 - 75\lambda + 1250 = 0,$$

so that

$$\lambda = 25 \quad \text{or} \quad 50.$$

When $\lambda = 25$, so that $b_{11} = 25$, the first pair of equations give

$$\frac{p_{11}}{3} = \frac{p_{21}}{4}$$

$$= \rho,$$

say; and when $\lambda = 50$, so that $b_{22} = 50$, the second pair of equations give

$$\frac{p_{12}}{4} = \frac{p_{22}}{-3}$$

$$= \sigma,$$

say. The matrix $\mathbf{p}$ is therefore of the form

$$\mathbf{p} \equiv \begin{pmatrix} 3\rho, & 4\sigma \\ 4\rho, & -3\sigma \end{pmatrix}.$$

The relation $\mathbf{p}'\mathbf{p} = \mathbf{I}$ thus gives

$$\begin{pmatrix} 3\rho & 4\rho \\ 4\sigma & -3\sigma \end{pmatrix} \begin{pmatrix} 3\rho & 4\sigma \\ 4\rho & -3\sigma \end{pmatrix} = \begin{pmatrix} 1 & 0 \\ 0 & 1 \end{pmatrix},$$

or, equating corresponding elements,

$$25\rho^2 = 1, \quad 25\sigma^2 = 1.$$

Either sign may be taken for ρ, σ for present purposes; say $\rho = \frac{1}{5}$, $\sigma = \frac{1}{5}$. Then

$$\mathbf{p} \equiv \begin{pmatrix} \frac{3}{5} & \frac{4}{5} \\ \frac{4}{5} & -\frac{3}{5} \end{pmatrix}.$$

Finally, $\mathbf{b}$ is obtained from the formula

$$\begin{aligned} \mathbf{b} &\equiv \mathbf{p}'\mathbf{a}\mathbf{p} \\ &\equiv \begin{pmatrix} \frac{3}{5} & \frac{4}{5} \\ \frac{4}{5} & -\frac{3}{5} \end{pmatrix} \begin{pmatrix} 41 & -12 \\ -12 & 34 \end{pmatrix} \begin{pmatrix} \frac{3}{5} & \frac{4}{5} \\ \frac{4}{5} & -\frac{3}{5} \end{pmatrix} \\ &\equiv \begin{pmatrix} \frac{3}{5} & \frac{4}{5} \\ \frac{4}{5} & -\frac{3}{5} \end{pmatrix} \begin{pmatrix} 15 & 40 \\ 20 & -30 \end{pmatrix} \\ &\equiv \begin{pmatrix} 25 & 0 \\ 0 & 50 \end{pmatrix}, \end{aligned}$$

so that the quadratic form becomes

$$25y_1^2 + 50y_2^2.$$

Note that the coefficients 25, 50 are precisely the two solutions of the quadratic equation for λ. This result is true in general; see p. 248.

Examples

1. Find mappings to reduce the following quadratics to the form $b_{11}y_1^2 + b_{22}y_2^2$:

(i) $x_1^2 + 4x_1x_2 + x_2^2$, (ii) $3x_1^2 - 4x_1x_2 + 6x_2^2$,
(iii) $5x_1^2 + 6x_1x_2 - 3x_2^2$, (iv) $2x_1^2 - 4x_1x_2 + 5x_2^2$,
(v) $7x_1^2 - 6x_1x_2 - x_2^2$, (vi) $7x_1^2 + 12x_1x_2 + 2x_2^2$,
(vii) $33x_1^2 + 20x_1x_2 + 54x_2^2$.

2. Devise a method to ensure that, in proposing a quadratic form

$$a_{11}x_1^2+2a_{12}x_1x_2+a_{22}x_2^2$$

with integral coefficients, the values of a_{11}, a_{12}, a_{22} may be selected so that the roots of the equation

$$\begin{vmatrix} a_{11}-\lambda & a_{12} \\ a_{12} & a_{22}-\lambda \end{vmatrix} = 0$$

are integers.

Quadratic forms in more than two variables

5. Orthogonal reduction to a sum of squares: particular example in three variables. Let

$$\begin{aligned} u &\equiv 5x_1^2+3x_2^2+3x_3^2+2x_2x_3+2x_3x_1+2x_1x_2 \\ &\equiv \mathbf{x}'\mathbf{a}\mathbf{x}, \end{aligned}$$

where

$$\mathbf{a} \equiv \begin{pmatrix} 5 & 1 & 1 \\ 1 & 3 & 1 \\ 1 & 1 & 3 \end{pmatrix}, \quad \mathbf{x} \equiv \begin{pmatrix} x_1 \\ x_2 \\ x_3 \end{pmatrix}.$$

The problem is *to find an orthogonal transformation*

$$\mathbf{x} = \mathbf{p}\mathbf{y},$$

where

$$\mathbf{p}'\mathbf{p} = \mathbf{I},$$

reducing u to the form

$$\begin{aligned} u &\equiv b_{11}y_1^2+b_{22}y_2^2+b_{33}y_3^2 \\ &\equiv \mathbf{y}'\mathbf{b}\mathbf{y}, \end{aligned}$$

where

$$\mathbf{b} \equiv \begin{pmatrix} b_{11} & 0 & 0 \\ 0 & b_{22} & 0 \\ 0 & 0 & b_{33} \end{pmatrix}.$$

As before (p. 239)

$$\begin{aligned} \mathbf{b} &= \mathbf{p}'\mathbf{a}\mathbf{p} \\ \Rightarrow \mathbf{p}\mathbf{b} &= \mathbf{a}\mathbf{p}, \end{aligned}$$

or

$$\begin{pmatrix} p_{11} & p_{12} & p_{13} \\ p_{21} & p_{22} & p_{23} \\ p_{31} & p_{32} & p_{33} \end{pmatrix} \begin{pmatrix} b_{11} & 0 & 0 \\ 0 & b_{22} & 0 \\ 0 & 0 & b_{33} \end{pmatrix} = \begin{pmatrix} 5 & 1 & 1 \\ 1 & 3 & 1 \\ 1 & 1 & 3 \end{pmatrix} \begin{pmatrix} p_{11} & p_{12} & p_{13} \\ p_{21} & p_{22} & p_{23} \\ p_{31} & p_{32} & p_{33} \end{pmatrix}.$$

This time there are 9 relations, of which a typical set of 3 (the first column of the products) is

$$p_{11}b_{11} = 5p_{11}+p_{21}+p_{31},$$
$$p_{21}b_{11} = p_{11}+3p_{21}+p_{31},$$
$$p_{31}b_{11} = p_{11}+p_{21}+3p_{31},$$

or

$$(5-b_{11})p_{11}+p_{21}+p_{31} = 0,$$
$$p_{11}+(3-b_{11})p_{21}+p_{31} = 0,$$
$$p_{11}+p_{21}+(3-b_{11})p_{31} = 0.$$

Eliminate $p_{11}:p_{21}:p_{31}$ determinantally†:

$$\begin{vmatrix} 5-b_{11} & 1 & 1 \\ 1 & 3-b_{11} & 1 \\ 1 & 1 & 3-b_{11} \end{vmatrix} = 0.$$

On expansion the determinant is

$$36-36b_{11}+11b_{11}^2-b_{11}^3,$$

so the equation for b_{11}, on factorization, is

$$(2-b_{11})(3-b_{11})(6-b_{11}) = 0.$$

Take b_{11} to have any one of these values; say

$$b_{11} = 2.$$

[The values 3, 6 give b_{22} and b_{33} by analogous argument. See immediately below.]

When $b_{11}=2$, the equations for $p_{11}:p_{21}:p_{31}$ are

$$3p_{11}+p_{21}+p_{31} = 0,$$
$$p_{11}+p_{21}+p_{31} = 0,$$

so that

$$p_{11} = 0, \quad p_{21} = \rho, \quad p_{31} = -\rho$$

for arbitrary ρ.

In the same way, with $b_{22}=3$, equations obtained for $p_{12}:p_{22}:p_{32}$ are

$$2p_{12}+p_{22}+p_{32} = 0,$$
$$p_{12}+p_{32} = 0,$$
$$p_{12}+p_{22} = 0,$$

† They cannot all be zero, since then $\mathbf{p}$ would have a column of zeros, giving $\det\mathbf{p} = 0$, in contradiction of the condition $\det\mathbf{p} = \pm 1$.

so that

$$p_{12} = -\sigma, \quad p_{22} = \sigma, \quad p_{32} = \sigma$$

for arbitrary σ.

Finally, with $b_{33} = 6$, equations for $p_{13}:p_{23}:p_{33}$ are

$$\begin{aligned} -p_{13}+p_{23}+p_{33} &= 0,\\ p_{13}-3p_{23}+p_{33} &= 0,\\ p_{13}+p_{23}-3p_{33} &= 0, \end{aligned}$$

so that

$$p_{13} = 2\tau, \quad p_{23} = \tau, \quad p_{33} = \tau$$

for arbitrary τ.

The matrix $\mathbf{p}$ is thus obtained in the form

$$\mathbf{p} \equiv \begin{pmatrix} 0 & -\sigma & 2\tau \\ \rho & \sigma & \tau \\ -\rho & \sigma & \tau \end{pmatrix}.$$

But $\mathbf{p}'\mathbf{p} = \mathbf{I}$, so that

$$\begin{pmatrix} 0 & \rho & -\rho \\ -\sigma & \sigma & \sigma \\ 2\tau & \tau & \tau \end{pmatrix} \begin{pmatrix} 0 & -\sigma & 2\tau \\ \rho & \sigma & \tau \\ -\rho & \sigma & \tau \end{pmatrix} = \begin{pmatrix} 1 & 0 & 0 \\ 0 & 1 & 0 \\ 0 & 0 & 1 \end{pmatrix},$$

or

$$\begin{pmatrix} 2\rho^2 & 0 & 0 \\ 0 & 3\sigma^2 & 0 \\ 0 & 0 & 6\tau^2 \end{pmatrix} = \begin{pmatrix} 1 & 0 & 0 \\ 0 & 1 & 0 \\ 0 & 0 & 1 \end{pmatrix}.$$

Hence (taking, say, positive signs for the square roots)

$$\mathbf{p} \equiv \begin{pmatrix} 0 & -\dfrac{1}{\sqrt{3}} & \dfrac{2}{\sqrt{6}} \\ \dfrac{1}{\sqrt{2}} & \dfrac{1}{\sqrt{3}} & \dfrac{1}{\sqrt{6}} \\ -\dfrac{1}{\sqrt{2}} & \dfrac{1}{\sqrt{3}} & \dfrac{1}{\sqrt{6}} \end{pmatrix},$$

so the transformation is

$$\mathbf{x} = \mathbf{p}\mathbf{y},$$

or

$$\begin{aligned} x_1 &= -\frac{1}{\sqrt{3}}y_2+\frac{2}{\sqrt{6}}y_3,\\ x_2 &= \frac{1}{\sqrt{2}}y_1+\frac{1}{\sqrt{3}}y_2+\frac{1}{\sqrt{6}}y_3,\\ x_3 &= -\frac{1}{\sqrt{2}}y_1+\frac{1}{\sqrt{3}}y_2+\frac{1}{\sqrt{6}}y_3. \end{aligned}$$

The matrix **b** is **p′ap**, so that

$$\mathbf{b} \equiv \begin{pmatrix} 0 & \frac{1}{\sqrt{2}} & -\frac{1}{\sqrt{2}} \\ -\frac{1}{\sqrt{3}} & \frac{1}{\sqrt{3}} & \frac{1}{\sqrt{3}} \\ \frac{2}{\sqrt{6}} & \frac{1}{\sqrt{6}} & \frac{1}{\sqrt{6}} \end{pmatrix} \begin{pmatrix} 5 & 1 & 1 \\ 1 & 3 & 1 \\ 1 & 1 & 3 \end{pmatrix} \begin{pmatrix} 0 & -\frac{1}{\sqrt{3}} & \frac{2}{\sqrt{6}} \\ \frac{1}{\sqrt{2}} & \frac{1}{\sqrt{3}} & \frac{1}{\sqrt{6}} \\ -\frac{1}{\sqrt{2}} & \frac{1}{\sqrt{3}} & \frac{1}{\sqrt{6}} \end{pmatrix}$$

$$= \begin{pmatrix} 0 & \frac{2}{\sqrt{2}} & -\frac{2}{\sqrt{2}} \\ -\frac{3}{\sqrt{3}} & +\frac{3}{\sqrt{3}} & \frac{3}{\sqrt{3}} \\ \frac{12}{\sqrt{6}} & \frac{6}{\sqrt{6}} & \frac{6}{\sqrt{6}} \end{pmatrix} \begin{pmatrix} 0 & -\frac{1}{\sqrt{3}} & \frac{2}{\sqrt{6}} \\ \frac{1}{\sqrt{2}} & \frac{1}{\sqrt{3}} & \frac{1}{\sqrt{6}} \\ -\frac{1}{\sqrt{2}} & \frac{1}{\sqrt{3}} & \frac{1}{\sqrt{6}} \end{pmatrix}$$

$$= \begin{pmatrix} 2 & 0 & 0 \\ 0 & 3 & 0 \\ 0 & 0 & 6 \end{pmatrix}, \text{ a matrix of DIAGONAL form;}$$

and the transformed quadratic is thus

$$2y_1^2+3y_2^2+6y_3^2.$$

Once again, the coefficients are the roots of the equation

$$\begin{vmatrix} 5-\lambda & 1 & 1 \\ 1 & 3-\lambda & 1 \\ 1 & 1 & 3-\lambda \end{vmatrix} = 0.$$

Example

1. Find a non-singular matrix **p** such that **p′ ap** is diagonal, where

(i) $\mathbf{a} \equiv \begin{pmatrix} 0 & 1 & 0 \\ 1 & 0 & 2 \\ 0 & 2 & 0 \end{pmatrix}$, (ii) $\mathbf{a} \equiv \begin{pmatrix} 7 & -1 & -1 \\ -1 & 5 & 1 \\ -1 & 1 & 5 \end{pmatrix}$,

(iii) $\mathbf{a} \equiv \begin{pmatrix} 1 & 1 & 1 \\ 1 & -1 & 1 \\ 1 & 1 & -1 \end{pmatrix}$, (iv) $\mathbf{a} \equiv \begin{pmatrix} 3 & -1 & 0 \\ -1 & 3 & 0 \\ 0 & 0 & 5 \end{pmatrix}$.

What are the corresponding quadratic forms before and after the related mappings?

6. More general orthogonal reduction. With particular examples for 2, 3 variables as guides, consider the quadratic form

$$u \equiv \mathbf{x}'\mathbf{a}\mathbf{x},$$

where $\mathbf{a}$ is a *symmetric* $n \times n$ matrix. It is required *to find an orthogonal mapping*

$$\mathbf{x} = \mathbf{p}\mathbf{y},$$

where $$\mathbf{p}'\mathbf{p} = \mathbf{I},$$

reducing u *to the form*

$$u \equiv \mathbf{y}'\mathbf{b}\mathbf{y},$$

where $b_{ij} = 0$ *when* $i \neq j$.

Since $\mathbf{x}' = \mathbf{y}'\mathbf{p}'$, it follows that

$$u \equiv \mathbf{x}'\mathbf{a}\mathbf{x} \equiv \mathbf{y}'\mathbf{p}'\mathbf{a}\mathbf{p}\mathbf{y} \equiv \mathbf{y}'\mathbf{b}\mathbf{y},$$

and so $$\mathbf{b} = \mathbf{p}'\mathbf{a}\mathbf{p},$$

from which, in virtue of the relation

$$\mathbf{p}'\mathbf{p} = \mathbf{I},$$

or $$\mathbf{p}' = \mathbf{p}^{-1},$$

we obtain the relation

$$\mathbf{p}\mathbf{b} = \mathbf{a}\mathbf{p}.$$

The left-hand product is

$$\begin{pmatrix} p_{11} & p_{12} & \cdots & p_{1n} \\ p_{21} & p_{22} & \cdots & p_{2n} \\ \cdots & \cdots & \cdots & \cdots \\ p_{n1} & p_{n2} & & p_{nn} \end{pmatrix} \begin{pmatrix} b_{11} & 0 & \cdots & 0 \\ 0 & b_{22} & \cdots & 0 \\ \cdots & \cdots & \cdots & \cdots \\ 0 & 0 & \cdots & b_{nn} \end{pmatrix}$$

$$= \begin{pmatrix} p_{11}b_{11}, & p_{12}b_{22}, & \ldots, & p_{1n}b_{nn} \\ p_{21}b_{11}, & p_{22}b_{22}, & \ldots, & p_{2n}b_{nn} \\ \cdots & \cdots & \cdots & \cdots \\ p_{n1}b_{11}, & p_{n2}b_{22}, & \ldots, & p_{nn}b_{nn} \end{pmatrix},$$

so that, if the column vectors of $\mathbf{p}$ are

$$\xi_1, \xi_2, \ldots, \xi_n,$$

then, in 'partition' notation,

$$\mathbf{p}\mathbf{b} \equiv (b_{11}\xi_1, b_{22}\xi_2, \ldots, b_{nn}\xi_n).$$

Further, by an elementary example of 'block' multiplication, the right-hand side is

$$\begin{aligned} \mathbf{a}\mathbf{p} &= \mathbf{a}(\xi_1, \xi_2, \ldots, \xi_n) \\ &= (\mathbf{a}\xi_1, \mathbf{a}\xi_2, \ldots, \mathbf{a}\xi_n), \end{aligned}$$

and so the relation

$$\mathbf{ap} = \mathbf{pb}$$

gives

$$(\mathbf{a}\boldsymbol{\xi}_1, \mathbf{a}\boldsymbol{\xi}_2, \ldots, \mathbf{a}\boldsymbol{\xi}_n) = (b_{11}\boldsymbol{\xi}_1, b_{22}\boldsymbol{\xi}_2, \ldots, b_{nn}\boldsymbol{\xi}_n).$$

Hence, for all values of i,

$$\mathbf{a}\boldsymbol{\xi}_i = b_{ii}\boldsymbol{\xi}_i,$$

or

$$(\mathbf{a} - b_{ii}\mathbf{I})\boldsymbol{\xi}_i = 0.$$

Now no $\boldsymbol{\xi}_i$ can be a zero vector, otherwise $\det \mathbf{p}$ would be zero, whereas, necessarily, $\det \mathbf{p} = \pm 1$. The scalar b_{ii} therefore satisfies the relation

$$\det(\mathbf{a} - b_{ii}\mathbf{I}) = 0;$$

in other words, $b_{11}, b_{22}, \ldots, b_{nn}$ are roots of the equation

$$\det(\mathbf{a} - \lambda\mathbf{I}) = 0.$$

DEFINITION. A scalar λ satisfying the *characteristic* equation

$$\det(\mathbf{a} - \lambda\mathbf{I}) = 0$$

is called an *eigenvalue* (*characteristic* value, *latent* value) of the matrix $\mathbf{a}$; and a vector $\boldsymbol{\xi}$ such that

$$\mathbf{a}\boldsymbol{\xi} = \lambda\boldsymbol{\xi}$$

is called an *eigenvector corresponding to* λ.

Each eigenvector is undetermined to the extent of a scalar multiplier, since also

$$\mathbf{a}(k\boldsymbol{\xi}) = \lambda(k\boldsymbol{\xi}).$$

When necessary, the multiplier is selected so that $\boldsymbol{\xi}$ is *normalized*; that is, selected to have that value which makes the scalar quantity $\boldsymbol{\xi}'\boldsymbol{\xi}$ to be of *unit* value, so that

$$\boldsymbol{\xi}'\boldsymbol{\xi} = 1.$$

Consider now the characteristic equation

$$\det(\mathbf{a} - \lambda\mathbf{I}) \equiv \begin{vmatrix} a_{11}-\lambda & a_{12} & \ldots & a_{1n} \\ a_{21} & a_{22}-\lambda & \ldots & a_{2n} \\ \ldots & \ldots & \ldots & \ldots \\ a_{n1} & a_{n2} & \ldots & a_{nn}-\lambda \end{vmatrix} = 0.$$

On expansion, it is of degree n in λ, with n roots $\lambda_1, \lambda_2, \ldots, \lambda_n$. *We confine attention to the case when these n eigenvalues are all distinct.*

Let $\boldsymbol{\xi}_1, \boldsymbol{\xi}_2, \ldots, \boldsymbol{\xi}_n$ be the eigenvectors corresponding to $\lambda_1, \lambda_2, \ldots, \lambda_n$ respectively, normalized so that

$$\boldsymbol{\xi}_i'\boldsymbol{\xi}_i = 1 \qquad (i = 1, 2, \ldots, n).$$

Then, by definition,

$$\mathbf{a}\boldsymbol{\xi}_i = \lambda_i \boldsymbol{\xi}_i,$$

so that

$$\boldsymbol{\xi}_j' \mathbf{a}\boldsymbol{\xi}_i = \lambda_i \boldsymbol{\xi}_j' \boldsymbol{\xi}_i;$$

also

$$\mathbf{a}\boldsymbol{\xi}_j = \lambda_j \boldsymbol{\xi}_j,$$

so that

$$\boldsymbol{\xi}_i' \mathbf{a}\boldsymbol{\xi}_j = \lambda_j \boldsymbol{\xi}_i' \boldsymbol{\xi}_j.$$

But scalars are unchanged by transposition, and the matrix **a** is symmetric; hence

$$\boldsymbol{\xi}_j' \mathbf{a}\boldsymbol{\xi}_i = \boldsymbol{\xi}_i' \mathbf{a}\boldsymbol{\xi}_j$$

and

$$\boldsymbol{\xi}_j' \boldsymbol{\xi}_i = \boldsymbol{\xi}_i' \boldsymbol{\xi}_j.$$

Hence there are two relations

$$\boldsymbol{\xi}_i' \mathbf{a}\boldsymbol{\xi}_j = \lambda_i \boldsymbol{\xi}_i' \boldsymbol{\xi}_j,$$
$$\boldsymbol{\xi}_i' \mathbf{a}\boldsymbol{\xi}_j = \lambda_j \boldsymbol{\xi}_i' \boldsymbol{\xi}_j,$$

where $\lambda_i \neq \lambda_j$ if $i \neq j$. It follows that

$$\boldsymbol{\xi}_i' \mathbf{a}\boldsymbol{\xi}_j = 0$$
$$\boldsymbol{\xi}_i' \boldsymbol{\xi}_j = 0$$

whenever $i \neq j$.

It should also be noted (the case $i = j$) that

$$\boldsymbol{\xi}_i' \mathbf{a}\boldsymbol{\xi}_i = \lambda_i \boldsymbol{\xi}_i' \boldsymbol{\xi}_i = \lambda_i,$$

since $\boldsymbol{\xi}_i' \boldsymbol{\xi}_i = 1$ for normalized eigenvectors.

Finally, it has already been established that the columns of the mapping matrix **p** are the eigenvectors of **a**; so, in 'partition' notation,

$$\mathbf{p} \equiv (\boldsymbol{\xi}_1, \boldsymbol{\xi}_2, \ldots, \boldsymbol{\xi}_n).$$

Also **p**′ is the matrix whose *rows* are the transposes of the columns of **p**, so that

$$\mathbf{p}' \equiv \{\boldsymbol{\xi}_1', \boldsymbol{\xi}_2', \ldots, \boldsymbol{\xi}_n'\}.$$

Hence, by 'block' multiplication,

$$\mathbf{p}'\mathbf{p} \equiv \begin{pmatrix} \boldsymbol{\xi}_1' \\ \boldsymbol{\xi}_2' \\ \vdots \\ \boldsymbol{\xi}_n' \end{pmatrix} (\boldsymbol{\xi}_1, \boldsymbol{\xi}_2, \ldots, \boldsymbol{\xi}_n)$$

$$\equiv \begin{pmatrix} \boldsymbol{\xi}_1'\boldsymbol{\xi}_1 & \boldsymbol{\xi}_1'\boldsymbol{\xi}_2 & \ldots & \boldsymbol{\xi}_1'\boldsymbol{\xi}_n \\ \boldsymbol{\xi}_2'\boldsymbol{\xi}_1 & \boldsymbol{\xi}_2'\boldsymbol{\xi}_2 & \ldots & \boldsymbol{\xi}_2'\boldsymbol{\xi}_n \\ \cdots & \cdots & \cdots & \cdots \\ \boldsymbol{\xi}_n'\boldsymbol{\xi}_1 & \boldsymbol{\xi}_n'\boldsymbol{\xi}_2 & \ldots & \boldsymbol{\xi}_n'\boldsymbol{\xi}_n \end{pmatrix} = \begin{pmatrix} 1 & 0 & \ldots & 0 \\ 0 & 1 & \ldots & 0 \\ \cdots & \cdots & \cdots & \cdots \\ 0 & 0 & \ldots & 1 \end{pmatrix}$$

$$= \mathbf{I},$$

so that $\mathbf{p}$ is orthogonal; and

$$\mathbf{p}'\mathbf{a}\mathbf{p} \equiv \mathbf{p}'(\mathbf{a}\mathbf{p})$$

$$\equiv \begin{pmatrix} \boldsymbol{\xi}_1' \\ \boldsymbol{\xi}_2' \\ \vdots \\ \boldsymbol{\xi}_n' \end{pmatrix} (\mathbf{a}\boldsymbol{\xi}_1, \mathbf{a}\boldsymbol{\xi}_2, \ldots, \mathbf{a}\boldsymbol{\xi}_n)$$

$$\equiv \begin{pmatrix} \boldsymbol{\xi}_1'\mathbf{a}\boldsymbol{\xi}_1 & \boldsymbol{\xi}_1'\mathbf{a}\boldsymbol{\xi}_2 & \ldots & \boldsymbol{\xi}_1'\mathbf{a}\boldsymbol{\xi}_n \\ \boldsymbol{\xi}_2'\mathbf{a}\boldsymbol{\xi}_1 & \boldsymbol{\xi}_2'\mathbf{a}\boldsymbol{\xi}_2 & \ldots & \boldsymbol{\xi}_2'\mathbf{a}\boldsymbol{\xi}_n \\ \cdots & \cdots & \cdots & \cdots \\ \boldsymbol{\xi}_n'\mathbf{a}\boldsymbol{\xi}_1 & \boldsymbol{\xi}_n'\mathbf{a}\boldsymbol{\xi}_2 & \ldots & \boldsymbol{\xi}_n'\mathbf{a}\boldsymbol{\xi}_n \end{pmatrix} \equiv \begin{pmatrix} \lambda_1 & 0 & \ldots & 0 \\ 0 & \lambda_2 & \ldots & 0 \\ \cdots & \cdots & \cdots & \cdots \\ 0 & 0 & \ldots & \lambda_n \end{pmatrix}.$$

Hence

$$\mathbf{b} \equiv \mathbf{p}'\mathbf{a}\mathbf{p}$$

is a diagonal matrix whose diagonal elements are the n distinct eigenvalues of $\mathbf{a}$, and the quadratic form

$$\mathbf{x}'\mathbf{a}\mathbf{x}$$

is reduced by the orthogonal mapping

$$\mathbf{x} = \mathbf{p}\mathbf{y}$$

to the sum of squares

$$\mathbf{y}'\mathbf{b}\mathbf{y}$$

$$\equiv \lambda_1 y_1^2 + \lambda_2 y_2^2 + \ldots + \lambda_n y_n^2.$$

Revision Examples

1. Show that the matrix

$$\mathbf{p} \equiv \begin{pmatrix} \cos\theta & \sin\theta \\ -\sin\theta & \cos\theta \end{pmatrix}$$

is orthogonal, and interpret the mapping

$$\mathbf{x} = \mathbf{p}\mathbf{y}$$

geometrically.

2. Show that, if $\mathbf{p}$ is an orthogonal 2×2 matrix, then the lines

$$p_{11}x_1 + p_{12}x_2 = 0,$$
$$p_{21}x_1 + p_{22}x_2 = 0$$

are perpendicular.

3. Find an orthogonal matrix **p** such that the matrix $\mathbf{p}^{-1}\mathbf{ap}$ is diagonal, where

$$\mathbf{a} \equiv \begin{pmatrix} 4 & -12 \\ -12 & 11 \end{pmatrix}$$

and find the lengths of the axes of the conic

$$4x^2-24xy+11y^2 = 25.$$

4. Find a real orthogonal mapping which maps the conic

$$2x^2+4xy+5y^2 = 1$$

to the form $$\lambda\xi^2+\mu\eta^2 = 1,$$

where λ, μ are to be determined.

5. Find a real orthogonal mapping which reduces the quadratic form

$$5x^2+2y^2+2z^2+5t^2+2yz+2zx+2xy-2xt-2yt-2zt$$

to a sum of squares.

6. Given that one eigenvalue of the matrix

$$\mathbf{a} \equiv \begin{pmatrix} 11 & -6 & 2 \\ -6 & 10 & -4 \\ 2 & -4 & 6 \end{pmatrix}$$

is 18, find the others.

Find also a matrix **p** for which $\mathbf{p}^{-1}\mathbf{ap}$ is diagonal.

7. Find an orthogonal mapping to reduce the quadratic form

$$3x_1^2+5x_2^2+3x_3^2-2x_2x_3+2x_3x_1-2x_1x_2$$

to a sum of squares.

8. Find a matrix **p** such that the matrix $\mathbf{p}^{-1}\mathbf{ap}$ is diagonal, where

$$\mathbf{a} \equiv \begin{pmatrix} 3 & 2 \\ 2 & 6 \end{pmatrix},$$

and deduce an expression for $\mathbf{a}^n$, where n is a positive integer.

NOTE. The general result is worth comment. Suppose that **a** is a given matrix and that the orthogonal mapping $\mathbf{x} = \mathbf{py}$ gives, after the manner of the text, the *diagonal* matrix $\mathbf{p}^{-1}\mathbf{ap}$. Then, denoting this matrix by $\boldsymbol{\delta}$, where

$$\boldsymbol{\delta} \equiv \begin{pmatrix} \lambda & 0 \\ 0 & \mu \end{pmatrix},$$

we have, with n factors,

$$(\mathbf{p}^{-1}\mathbf{ap})(\mathbf{p}^{-1}\mathbf{ap})(\mathbf{p}^{-1}\mathbf{ap})\ldots(\mathbf{p}^{-1}\mathbf{ap}) = \boldsymbol{\delta}^n = \begin{pmatrix} \lambda^n & 0 \\ 0 & \mu^n \end{pmatrix},$$

so that
$$\mathbf{p}^{-1}\mathbf{a}^n\mathbf{p} = \begin{pmatrix} \lambda^n & 0 \\ 0 & \mu^n \end{pmatrix},$$

and so
$$\mathbf{a}^n = \mathbf{p}\begin{pmatrix} \lambda^n & 0 \\ 0 & \mu^n \end{pmatrix}\mathbf{p}^{-1},$$

which is easily calculated.

9. If
$$\mathbf{a} \equiv \begin{pmatrix} -1 & 2 \\ 4 & 1 \end{pmatrix},$$
find a matrix $\mathbf{p}$ of the form
$$\mathbf{p} \equiv \begin{pmatrix} 1 & 1 \\ \lambda & \mu \end{pmatrix}$$
such that $\mathbf{p}^{-1}\mathbf{ap}$ is a diagonal matrix.

Hence or otherwise express $\mathbf{a}^n$ in the form of a 2×2 matrix, where n is a positive integer.

10. Define the eigenvalues of an $n\times n$ matrix $\mathbf{a}$ and prove that the matrix $\mathbf{u}^{-1}\mathbf{au}$ has the same eigenvalues.

Find all the 2×2 matrices $\mathbf{x}$ such that
$$\mathbf{x}^2+4\mathbf{x}+3\mathbf{I} = \mathbf{0}.$$

11. Find the eigenvalues and eigenvectors of the matrix
$$\mathbf{a} \equiv \begin{pmatrix} 0 & 1 & 0 \\ 0 & 0 & 1 \\ 1 & 0 & 0 \end{pmatrix}.$$

12. If
$$\mathbf{a} \equiv \begin{pmatrix} 0 & 1 & 0 \\ 1 & 0 & 0 \\ 0 & 0 & 2 \end{pmatrix},$$
find a set of numbers α, β, γ and an orthogonal matrix $\mathbf{b}$ such that
$$\mathbf{ba} = \begin{pmatrix} \alpha & 0 & 0 \\ 0 & \beta & 0 \\ 0 & 0 & \gamma \end{pmatrix}\mathbf{b}.$$

13. An examiner reading Question 7 observes that the corresponding matrix
$$\begin{pmatrix} 3 & -1 & 1 \\ -1 & 5 & -1 \\ 1 & -1 & 3 \end{pmatrix}$$
has integral eigenvalues. Show that he can produce an indefinite number

of matrices with integral eigenvalues by increasing or decreasing the diagonal elements 3, 5, 3 by the same amount each.

Simultaneous reduction of two quadratic forms

The work of this section is definitely harder than the standard envisaged for the book as a whole, but it seems useful to include it here for completeness, restricting the treatment to the case of three variables and particular numerical values. The method is general.

Illustration 1. *To find a mapping*

$$\mathbf{x} = \mathbf{q}\mathbf{y}$$

(where $\mathbf{q}$ need *not* be orthogonal) *to reduce the two quadratic forms*

$$u \equiv 3x_1^2+5x_2^2+5x_3^2+2x_2x_3+6x_3x_1-2x_1x_2,$$
$$v \equiv 5x_1^2+12x_2^2+8x_2x_3+4x_3x_1$$

simultaneously to sums of squares.

The forms are

$$\mathbf{x}'\mathbf{a}\mathbf{x} = 0, \quad \mathbf{x}'\mathbf{b}\mathbf{x} = 0,$$

where $$\mathbf{a} \equiv \begin{pmatrix} 3 & -1 & 3 \\ -1 & 5 & 1 \\ 3 & 1 & 5 \end{pmatrix}, \quad \mathbf{b} \equiv \begin{pmatrix} 5 & 0 & 2 \\ 0 & 12 & 4 \\ 2 & 4 & 0 \end{pmatrix}.$$

Consider the equation

$$\det(\lambda\mathbf{a}-\mathbf{b}) \equiv \begin{vmatrix} 3\lambda-5 & -\lambda & 3\lambda-2 \\ -\lambda & 5\lambda-12 & \lambda-4 \\ 3\lambda-2 & \lambda-4 & 5\lambda \end{vmatrix} = 0.$$

On reduction, this is

$$\lambda^3-5\lambda^2+2\lambda+8 = 0,$$

or $$(\lambda-4)(\lambda-2)(\lambda+1) = 0.$$

Form the vector $\boldsymbol{\xi}$ arising from the equations

$$(3\lambda-5)\xi_1-\lambda\xi_2+(3\lambda-2)\xi_3 = 0,$$
$$-\lambda\xi_1+(5\lambda-12)\xi_2+(\lambda-4)\xi_3 = 0,$$
$$(3\lambda-2)\xi_1+(\lambda-4)\xi_2+5\lambda\xi_3 = 0$$

for $\lambda = 4, 2, -1$ in turn. Since the determinant vanishes, only two of these equations will be independent.

(i) When $\lambda = 4$,

$$7\xi_1-4\xi_2+10\xi_3 = 0,$$
$$-4\xi_1+8\xi_2 = 0,$$
$$10\xi_1 \qquad +20\xi_3 = 0.$$

Hence
$$\frac{\xi_1}{2} = \frac{\xi_2}{1} = \frac{\xi_3}{-1}$$
$$= \rho,$$
say. Thus a vector is
$$\mathbf{q}_1 \equiv \{2\rho, \rho, -\rho\}.$$

(ii) When $\lambda = 2$,
$$\begin{aligned} \xi_1 - 2\xi_2 + 4\xi_3 &= 0, \\ -2\xi_1 - 2\xi_2 - 2\xi_3 &= 0, \\ 4\xi_1 - 2\xi_2 + 10\xi_3 &= 0. \end{aligned}$$

Hence
$$\frac{\xi_1}{2} = \frac{\xi_2}{-1} = \frac{\xi_3}{-1}$$
$$= \sigma,$$
say. Thus a vector is
$$\mathbf{q}_2 \equiv \{2\sigma, -\sigma, -\sigma\}.$$

(iii) When $\lambda = -1$,
$$\begin{aligned} -8\xi_1 + \xi_2 - 5\xi_3 &= 0, \\ \xi_1 - 17\xi_2 - 5\xi_3 &= 0, \\ -5\xi_1 - 5\xi_2 - 5\xi_3 &= 0. \end{aligned}$$

Hence
$$\frac{\xi_1}{2} = \frac{\xi_2}{1} = \frac{\xi_3}{-3}$$
$$= \tau,$$
say. Thus a vector is
$$\mathbf{q}_3 \equiv \{2\tau, \tau, -3\tau\}.$$

Bearing in mind the analogous work for the orthogonal transformation of a single variable, form the matrix $\mathbf{q}$ whose columns are $\mathbf{q}_1, \mathbf{q}_2, \mathbf{q}_3$:
$$\mathbf{q} \equiv (\mathbf{q}_1, \mathbf{q}_2, \mathbf{q}_3) \equiv \begin{pmatrix} 2\rho & 2\sigma & 2\tau \\ \rho & -\sigma & \tau \\ -\rho & -\sigma & -3\tau \end{pmatrix}.$$

The mapping
$$\mathbf{x} = \mathbf{q}\mathbf{y}$$
gives
$$\mathbf{x}'\mathbf{a}\mathbf{x} \equiv \mathbf{y}'\mathbf{q}'\mathbf{a}\mathbf{q}\mathbf{y}, \quad \mathbf{x}'\mathbf{b}\mathbf{x} \equiv \mathbf{y}'\mathbf{q}'\mathbf{b}\mathbf{q}\mathbf{y}$$
so it is necessary to form the matrices $\mathbf{q}'\mathbf{a}\mathbf{q}$, $\mathbf{q}'\mathbf{b}\mathbf{q}$. By direct computation,
$$\mathbf{q}'\mathbf{a}\mathbf{q} \equiv \begin{pmatrix} 2\rho & \rho & -\rho \\ 2\sigma & -\sigma & -\sigma \\ 2\tau & \tau & -3\tau \end{pmatrix} \begin{pmatrix} 3 & -1 & 3 \\ -1 & 5 & 1 \\ 3 & 1 & 5 \end{pmatrix} \begin{pmatrix} 2\rho & 2\sigma & 2\tau \\ \rho & -\sigma & \tau \\ -\rho & -\sigma & -3\tau \end{pmatrix}$$
$$\equiv \begin{pmatrix} 2\rho & \rho & -\rho \\ 2\sigma & -\sigma & -\sigma \\ 2\tau & \tau & -3\tau \end{pmatrix} \begin{pmatrix} 2\rho & 4\sigma & -4\tau \\ 2\rho & -8\sigma & 0 \\ 2\rho & 0 & -8\tau \end{pmatrix}$$
$$\equiv \begin{pmatrix} 4\rho^2 & 0 & 0 \\ 0 & 16\sigma^2 & 0 \\ 0 & 0 & 16\tau^2 \end{pmatrix}$$

and
$$\mathbf{q}'\mathbf{b}\mathbf{q} \equiv \begin{pmatrix} 2\rho & \rho & -\rho \\ 2\sigma & -\sigma & -\sigma \\ 2\tau & \tau & -3\tau \end{pmatrix} \begin{pmatrix} 5 & 0 & 2 \\ 0 & 12 & 4 \\ 2 & 4 & 0 \end{pmatrix} \begin{pmatrix} 2\rho & 2\sigma & 2\tau \\ \rho & -\sigma & \tau \\ -\rho & -\sigma & -3\tau \end{pmatrix}$$
$$\equiv \begin{pmatrix} 2\rho & \rho & -\rho \\ 2\sigma & -\sigma & -\sigma \\ 2\tau & \tau & -3\tau \end{pmatrix} \begin{pmatrix} 8\rho & 8\sigma & 4\tau \\ 8\rho & -16\sigma & 0 \\ 8\rho & 0 & 8\tau \end{pmatrix}$$
$$\equiv \begin{pmatrix} 16\rho^2 & 0 & 0 \\ 0 & 32\sigma^2 & 0 \\ 0 & 0 & -16\tau^2 \end{pmatrix}.$$

The transformed quadratics are thus
$$u \equiv 4\rho^2 y_1^2 + 16\sigma^2 y_2^2 + 16\tau^2 y_3^2,$$
$$v \equiv 16\rho^2 y_1^2 + 32\sigma^2 y_2^2 - 16\tau^2 y_3^2.$$

The constants ρ, σ, τ are still at disposal. Now all the coefficients in u are positive (u is then called a *positive definite quadratic form*) and they can all be made $+1$ by taking
$$\rho = \tfrac{1}{2}, \quad \sigma = \tfrac{1}{4}, \quad \tau = \tfrac{1}{4}.$$

Then the mapping
$$\mathbf{x} = \mathbf{q}\mathbf{y},$$
where
$$\mathbf{q} \equiv \begin{pmatrix} 1 & \tfrac{1}{2} & \tfrac{1}{2} \\ \tfrac{1}{2} & -\tfrac{1}{4} & \tfrac{1}{4} \\ -\tfrac{1}{2} & -\tfrac{1}{4} & -\tfrac{3}{4} \end{pmatrix}$$
reduces the quadratics to the forms
$$u \equiv y_1^2 + y_2^2 + y_3^2,$$
$$v \equiv 4y_1^2 + 2y_2^2 - y_3^2.$$

Note that the coefficients of v are 4, 2, -1 which are the three values previously obtained for λ.

Examples

Reduce the following pairs of quadratic forms (at least one in each case being positive definite) to sums of squares, the positive definite one being $y_1^2 + y_2^2 + y_3^2$:

1. $2x_1^2 + 5x_2^2 + 4x_3^2 + 14x_2x_3 + 8x_3x_1 + 6x_1x_2$,
   $2x_1^2 + 14x_2^2 + 3x_3^2 - 4x_2x_3 + 10x_1x_2$.
2. $3x_1^2 + 6x_2^2 + 14x_3^2 - 6x_2x_3$,
   $3x_1^2 + 6x_2^2 + 9x_3^2 + 4x_2x_3 + 10x_3x_1$.
3. $5x_1^2 + 10x_2^2 + 6x_3^2 + 8x_2x_3 + 10x_3x_1 + 6x_1x_2$,
   $3x_1^2 + 7x_2^2 + x_3^2 + 2x_2x_3 + 6x_3x_1 + 6x_1x_2$.
4. $2x_1^2 + x_2^2 + 2x_3^2 + 2x_2x_3 - 2x_3x_1$,
   $x_1^2 + 2x_2^2 + 2x_3^2 + 4x_2x_3$.

5. $2x_1^2+3x_2^2+9x_3^2-10x_2x_3$,
$3x_1^2-5x_2^2+4x_3^2+4x_2x_3+4x_3x_1-2x_1x_2$.

6. $x_1^2+5x_2^2+3x_3^2-6x_2x_3-2x_3x_1-2x_1x_2$,
$2x_1^2+6x_2^2+5x_3^2-10x_2x_3+2x_3x_1-4x_1x_2$.

7. $2x_1^2+2x_2^2+3x_3^2-4x_2x_3-4x_3x_1+2x_1x_2$,
$-x_1^2+2x_3x_1-2x_1x_2$.

8. $5x_1^2+2x_2^2+3x_3^2+6x_3x_1+2x_1x_2$,
$x_2^2+14x_2x_3+4x_3x_1+8x_1x_2$.

When the values of λ in the equation $\det(\lambda\mathbf{a}-\mathbf{b})=0$ are not all distinct, the general treatment is much more complicated, but the particular case of three variables happens to lend itself to special treatment.

Illustration 2. Let

$$u \equiv 3x_1^2+2x_2^2+6x_3^2+8x_3x_1,$$
$$v \equiv x_1^2+2x_2^2-2x_3^2.$$

Then $$\lambda u-v \equiv (3\lambda-1)x_1^2+(2\lambda-2)x_2^2+(6\lambda+2)x_3^2+8\lambda x_3x_1.$$

The determinant $\det(\lambda\mathbf{a}-\mathbf{b})$ vanishes when

$$\begin{vmatrix} 3\lambda-1 & 0 & 4\lambda \\ 0 & 2\lambda-2 & 0 \\ 4\lambda & 0 & 6\lambda+2 \end{vmatrix} = 0,$$

or, on reduction,

$$\lambda^3-\lambda^2-\lambda+1 = 0,$$

or $$(\lambda-1)^2(\lambda+1) = 0.$$

Thus $$\lambda = 1,\ 1,\ -1.$$

The simplification arises from the fact that, *when $\lambda=\alpha$ is any repeated root, the quadratic form $\alpha u-v$ is a multiple of the square of a form linear in x_1, x_2, x_3.* Here,

$$u-v \equiv 2x_1^2+8x_3^2+8x_3x_1$$
$$\equiv 2(x_1+2x_3)^2.$$

Further, with $\lambda=-1$,

$$u+v \equiv 4x_1^2+4x_2^2+4x_3^2+8x_3x_1,$$

and this *must* be the sum of two squares of forms linear in x_1, x_2, x_3. In this simple case,

$$u+v = 4(x_1+x_3)^2+4x_2^2.$$

Solve these two equations to give u and v:

$$u = (x_1+2x_3)^2+2(x_1+x_3)^2+2x_2^2,$$
$$v = -(x_1+2x_3)^2+2(x_1+x_3)^2+2x_2^2.$$

Note that, written in the form

$$u \equiv (x_1+2x_3)^2+(\sqrt{2}x_1+\sqrt{2}x_3)^2+(\sqrt{2}x_2)^2,$$
$$v \equiv -(x_1+2x_3)^2+(\sqrt{2}x_1+\sqrt{2}x_3)^2+(\sqrt{2}x_2)^2,$$

the forms are

$$u \equiv y_1^2+y_2^2+y_3^2,$$
$$v \equiv -y_1^2+y_2^2+y_3^2$$

where the coefficients -1, $+1$, $+1$ in v are the roots of the determinantal equation in λ.

Example

1. Apply the method just illustrated to the pairs of quadratic forms

(i) $u \equiv x_1^2+x_2^2+x_3^2,$
$v \equiv 2x_2x_3+2x_3x_1+2x_1x_2;$

(ii) $u \equiv x_1^2+2x_2^2+x_3^2+2x_1x_2,$
$v \equiv 2x_2^2+4x_2x_3+2x_3x_1+2x_1x_2;$

(iii) $2x_1^2+2x_2^2+2x_3^2+2x_2x_3+2x_3x_1+2x_1x_2,$
$2x_2^2+2x_2x_3-2x_3x_1+2x_1x_2;$

(iv) $x_1^2+2x_2^2+3x_3^2+4x_2x_3+2x_3x_1+2x_1x_2,$
$x_1^2+2x_2^2+5x_3^2+4x_2x_3+2x_3x_1+2x_1x_2.$

18

EIGENVALUES AND EIGENVECTORS: SPECIAL MATRICES

Some elementary properties of eigenvalues and eigenvectors deserve to be recorded briefly here. For convenience, most of the work is restricted to matrices of type 3×3, but the results are usually general.

1. Eigenvalues and eigenvectors. Let $\mathbf{a}$ be a given square matrix of type 3×3 (more generally, $n \times n$). The eigenvectors $\mathbf{x}$ and eigenvalues λ have already been defined (p. 246) by the property

$$\mathbf{ax} = \lambda\mathbf{x},$$

where $\mathbf{x}$ is a *non-zero* column vector. The values of λ are the 3 (more generally, n) roots of the equation

$$\det(\mathbf{a} - \lambda\mathbf{I}) = 0.$$

It will be assumed throughout this chapter that the eigenvalues $\lambda_1, \lambda_2, \lambda_3$ *are all different.*

2. The independence of the eigenvectors. To prove *that the three eigenvectors* $\boldsymbol{\xi}_1, \boldsymbol{\xi}_2, \boldsymbol{\xi}_3$ *are linearly independent*:

If not, there must be a relation

$$\alpha_1 \boldsymbol{\xi}_1 + \alpha_2 \boldsymbol{\xi}_2 + \alpha_3 \boldsymbol{\xi}_3 \equiv \mathbf{0},$$

where $\alpha_1, \alpha_2, \alpha_3$ are not all zero.

Multiply on the left by $\mathbf{a}$:

$$\alpha_1 \mathbf{a}\boldsymbol{\xi}_1 + \alpha_2 \mathbf{a}\boldsymbol{\xi}_2 + \alpha_3 \mathbf{a}\boldsymbol{\xi}_3 = \mathbf{0},$$

and so, since $\mathbf{a}\boldsymbol{\xi}_i = \lambda_i \boldsymbol{\xi}_i$,

$$\alpha_1 \lambda_1 \boldsymbol{\xi}_1 + \alpha_2 \lambda_2 \boldsymbol{\xi}_2 + \alpha_3 \lambda_3 \boldsymbol{\xi}_3 = \mathbf{0}.$$

Hence

$$\lambda_1(-\alpha_2 \boldsymbol{\xi}_2 - \alpha_3 \boldsymbol{\xi}_3) + \alpha_2 \lambda_2 \boldsymbol{\xi}_2 + \alpha_3 \lambda_3 \boldsymbol{\xi}_3 = \mathbf{0},$$

or
$$(\lambda_2-\lambda_1)\alpha_2\xi_2+(\lambda_3-\lambda_1)\alpha_3\xi_3 = 0;$$
say
$$\beta_2\xi_2+\beta_3\xi_3 = 0.$$

By similar argument,
$$\beta_2\lambda_2\xi_2+\beta_3\lambda_3\xi_3 = 0,$$
so that
$$(\lambda_3-\lambda_2)\beta_3\xi_3 = 0.$$
But
$$\lambda_3 \neq \lambda_2,$$
so that
$$\beta_3 = 0,$$
or
$$(\lambda_3-\lambda_1)\alpha_3 = 0.$$
But
$$\lambda_3 \neq \lambda_1,$$
so that
$$\alpha_3 = 0.$$

Similar proof gives $\alpha_1 = 0$, $\alpha_2 = 0$, contradicting the hypothesis that $\alpha_1, \alpha_2, \alpha_3$ are not all zero.

Hence ξ_1, ξ_2, ξ_3 are linearly independent.

COROLLARIES. (i) *The eigenvectors are necessarily distinct*, an eigenvector corresponding to a value λ_i not being any multiple of one corresponding to $\lambda_j (i \neq j)$.

(ii) When $\lambda_1, \lambda_2, \lambda_3$ are distinct, *each of the matrices* $\mathbf{a}-\lambda_1\mathbf{I}$, $\mathbf{a}-\lambda_2\mathbf{I}$, $\mathbf{a}-\lambda_3\mathbf{I}$ *is of rank* 2.

Examples

Find eigenvalues and eigenvectors for the matrices:

1. $\begin{pmatrix} 6 & -2 & 0 \\ -2 & 4 & -2 \\ 0 & -2 & 2 \end{pmatrix}$; **2.** $\begin{pmatrix} 2 & -2 & 1 \\ 2 & -8 & -2 \\ 1 & 2 & 2 \end{pmatrix}$;

3. $\begin{pmatrix} 1 & 0 & 4 \\ 0 & 5 & 4 \\ 4 & 4 & 3 \end{pmatrix}$; **4.** $\begin{pmatrix} 17 & -2 & -2 \\ -2 & 14 & -4 \\ -2 & -4 & 14 \end{pmatrix}$

given that, in Question 4, one eigenvalue is 18.

5. Prove that the eigenvalues of the matrix
$$\begin{pmatrix} -a_1 & 1 & 0 & 0 \\ -a_2 & 0 & 1 & 0 \\ -a_3 & 0 & 0 & 1 \\ -a_4 & 0 & 0 & 0 \end{pmatrix}$$
are the roots of the equation
$$x^4+a_1x^3+a_2x^2+a_3x+a_4 = 0.$$

6. Find the eigenvalues and eigenvectors of the matrix

$$\begin{pmatrix} 0 & a & a \\ -a & 0 & a \\ -a & -a & 0 \end{pmatrix}.$$

3. Matrix polynomials. Let $\mathbf{a}$ be a given square matrix of order, say, 3 and

$$f(u) \equiv pu^2 + qu + r$$

a given polynomial of order, say, 2. Then the matrix

$$\mathbf{b} \equiv p\mathbf{a}^2 + q\mathbf{a} + r\mathbf{I}$$

is a matrix of order 3, conveniently denoted by the notation

$$\mathbf{b} \equiv f(\mathbf{a}).$$

THEOREM. To prove that, *if* λ *is an eigenvalue of* $\mathbf{a}$ *with eigenvector* $\boldsymbol{\xi}$, *then* $f(\lambda)$ *is an eigenvalue of* $\mathbf{b}$, *with eigenvector also* $\boldsymbol{\xi}$:

Since λ is an eigenvalue of $\mathbf{a}$ with eigenvector $\boldsymbol{\xi}$,

$$\mathbf{a}\boldsymbol{\xi} = \lambda\boldsymbol{\xi}$$
$$\Rightarrow \mathbf{a}^2\boldsymbol{\xi} = \lambda(\mathbf{a}\boldsymbol{\xi}) = \lambda^2\boldsymbol{\xi}.$$

(More generally, $\mathbf{a}^n\boldsymbol{\xi} = \lambda^n\boldsymbol{\xi}$.) Thus,

$$\begin{aligned} \mathbf{b}\boldsymbol{\xi} &\equiv (p\mathbf{a}^2 + q\mathbf{a} + r\mathbf{I})\boldsymbol{\xi} \equiv p\mathbf{a}^2\boldsymbol{\xi} + q\mathbf{a}\boldsymbol{\xi} + r\boldsymbol{\xi} \\ &\equiv p\lambda^2\boldsymbol{\xi} + q\lambda\boldsymbol{\xi} + r\boldsymbol{\xi} \\ &\equiv (p\lambda^2 + q\lambda + r)\boldsymbol{\xi} \\ &\equiv f(\lambda)\boldsymbol{\xi}, \end{aligned}$$

so that $\boldsymbol{\xi}$ is an eigenvector of $\mathbf{b}$, with eigenvalue $f(\lambda)$.

THEOREM (the *Cayley–Hamilton Theorem*): To prove that *a square matrix satisfies its characteristic equation*:

The characteristic equation is (p. 246)

$$\det(\mathbf{a} - \lambda\mathbf{I}) = 0;$$

say

$$g(\lambda) = 0.$$

If λ_1 is any eigenvalue, with eigenvector $\boldsymbol{\xi}_1$, then, by the preceding theorem, suitably modified,

$$\begin{aligned} g(\mathbf{a})\boldsymbol{\xi}_1 &= g(\lambda_1)\boldsymbol{\xi}_1 \\ &= \mathbf{0}, \end{aligned}$$

since $g(\lambda_1)$ vanishes when the eigenvalue λ_1 is a root of the characteristic equation.

If, then, $\boldsymbol{\xi}_1, \boldsymbol{\xi}_2, \boldsymbol{\xi}_3$ are the eigenvectors corresponding to the distinct values $\lambda_1, \lambda_2, \lambda_3$, it follows (§2) that they are linearly independent, so that the determinant whose columns are $\boldsymbol{\xi}_1, \boldsymbol{\xi}_2, \boldsymbol{\xi}_3$ does not vanish: that is, the square 3×3 matrix

$$\mathbf{p} \equiv (\boldsymbol{\xi}_1, \boldsymbol{\xi}_2, \boldsymbol{\xi}_3)$$

is non-singular. But

$$\begin{aligned} g(\mathbf{a})\,\mathbf{p} &\equiv (g(\mathbf{a})\,\boldsymbol{\xi}_1, g(\mathbf{a})\,\boldsymbol{\xi}_2, g(\mathbf{a})\,\boldsymbol{\xi}_3) \\ &\equiv (0, 0, 0), \end{aligned}$$

so that

$$g(\mathbf{a})\,\mathbf{p} = \mathbf{0},$$

where $\mathbf{0}$ is the zero 3×3 matrix. Multiply on the right by $\mathbf{p}^{-1}$, which exists since $\mathbf{p}$ is non-singular. Hence

$$g(\mathbf{a}) = \mathbf{0},$$

as required.

NOTE. This proof is general for any $n \times n$ matrix $\mathbf{a}$ whose eigenvalues are all different.

***Illustration* 1.** Consider the matrix

$$\mathbf{a} \equiv \begin{pmatrix} 1 & 1 \\ 6 & 2 \end{pmatrix},$$

whose characteristic equation is

$$\begin{vmatrix} 1-\lambda & 1 \\ 6 & 2-\lambda \end{vmatrix} = 0$$

or

$$\lambda^2 - 3\lambda - 4 = 0,$$

so that

$$\lambda + 1 = 0 \quad \text{or} \quad \lambda - 4 = 0.$$

Now

$$\begin{aligned} \mathbf{a} + \mathbf{I} &= \begin{pmatrix} 1 & 1 \\ 6 & 2 \end{pmatrix} + \begin{pmatrix} 1 & 0 \\ 0 & 1 \end{pmatrix} = \begin{pmatrix} 2 & 1 \\ 6 & 3 \end{pmatrix} \\ &\neq \mathbf{0}; \end{aligned}$$

$$\begin{aligned} \mathbf{a} - 4\mathbf{I} &= \begin{pmatrix} 1 & 1 \\ 6 & 2 \end{pmatrix} - \begin{pmatrix} 4 & 0 \\ 0 & 4 \end{pmatrix} = \begin{pmatrix} -3 & 1 \\ 6 & -2 \end{pmatrix} \\ &\neq \mathbf{0}. \end{aligned}$$

But

$$\begin{aligned} (\mathbf{a} + \mathbf{I})(\mathbf{a} - 4\mathbf{I}) &= \begin{pmatrix} 2 & 1 \\ 6 & 3 \end{pmatrix} \begin{pmatrix} -3 & 1 \\ 6 & -2 \end{pmatrix} \\ &= \begin{pmatrix} 0 & 0 \\ 0 & 0 \end{pmatrix} \\ &= \mathbf{0}. \end{aligned}$$

This emphasizes the fact that

$$\begin{aligned} &(\mathbf{a} - \lambda_i \mathbf{I})\,\boldsymbol{\xi}_i = \mathbf{0} \\ &\not\Rightarrow \mathbf{a} - \lambda_i \mathbf{I} = \mathbf{0} \end{aligned}$$

when ξ_i is merely a *vector*. It was necessary in the proof to build up the 3×3 *matrix* (ξ_1, ξ_2, ξ_3) in order to obtain the multiplier $\mathbf{p}^{-1}$.

Examples

1. Find the characteristic equation of the matrix

$$\mathbf{u} \equiv \begin{pmatrix} 0 & -c & b \\ c & 0 & -a \\ -b & a & 0 \end{pmatrix},$$

and deduce that

$$\mathbf{u}^3 + (a^2+b^2+c^2)\,\mathbf{u} = 0.$$

2. Find the eigenvalues of the matrix

$$\mathbf{a} \equiv \begin{pmatrix} -1 & -5 & 7 \\ -3 & -5 & 9 \\ -3 & -9 & 13 \end{pmatrix}.$$

Prove that

$$\mathbf{a}^{-1} = \tfrac{1}{8}(\mathbf{a}^2 - 7\mathbf{a} + 14\mathbf{I}).$$

4. Some particular square matrices. Some of the particular square matrices defined below have been met already, but repetition here is convenient.

(i) A matrix $\mathbf{a}$ is *symmetric* if

$$\mathbf{a}' = \mathbf{a}$$

and skew-symmetric if

$$\mathbf{a}' = -\mathbf{a}.$$

(ii) A matrix is *orthogonal* if

$$\mathbf{a}'\mathbf{a} = \mathbf{I}.$$

It follows (p. 237) that

$$\det \mathbf{a} = \pm 1$$

so that (since $\det \mathbf{a} \neq 0$) $\mathbf{a}^{-1}$ exists. Thus

$$\mathbf{a}' = \mathbf{a}^{-1}.$$

It also follows that

$$\mathbf{a}\mathbf{a}' = \mathbf{I}.$$

(iii) A matrix whose elements are *complex* numbers is called *Hermitian* if

$$\mathbf{a}' = \bar{\mathbf{a}},$$

where $\bar{\mathbf{a}}$ is the matrix formed from $\mathbf{a}$ by replacing each element a_{ij} by its conjugate complex $\bar{a}_{ij}$.

It follows that, with similar notation

$$\bar{\mathbf{a}}' = \mathbf{a}.$$

(iv) A matrix whose elements are *complex* numbers is *unitary* if

$$\bar{\mathbf{a}}'\mathbf{a} = \mathbf{I}.$$

It follows then that

$$\bar{\mathbf{a}}' = \mathbf{a}^{-1},$$
$$\mathbf{a}' = \bar{\mathbf{a}}^{-1}.$$

The Hermitian and unitary matrices reduce to the symmetric and orthogonal matrices respectively for the particular cases in which the elements are real.

Examples

1. Prove that, if $\mathbf{a}$ is a matrix with complex coefficients, then $\bar{\mathbf{a}}'\mathbf{a}$ and $\mathbf{a}\bar{\mathbf{a}}'$ are both Hermitian.

Prove also that, if $\mathbf{a}$ is square, then

$$\bar{\mathbf{a}}' + \mathbf{a}$$

is also Hermitian.

2. Prove that the matrix

$$\begin{pmatrix} \sqrt{(1-|z|^2)} & z \\ \bar{z} & -\sqrt{(1-|z|^2)} \end{pmatrix}$$

is both Hermitian and unitary, where $|z|$ is the modulus of z.

5. Some typical eigenvector properties.

(i) Let $\mathbf{a}$ be a *symmetric matrix with real coefficients*. Then *the eigenvalues are real and the eigenvectors may be chosen to be real*:

[The phrase 'may be chosen' is inserted merely because of the fact that, if, say, $\{1,2,3\}$ is an eigenvector, so also is $\{1+i, 2+2i, 3+3i\}$.]

Let λ be any eigenvalue and $\bar{\lambda}$ its conjugate complex. The corresponding vectors are $\boldsymbol{\xi}$, $\bar{\boldsymbol{\xi}}$ (the latter being the conjugate complex of the former since the calculations for it merely involve replacing $-i$ for i in the calculations of $\boldsymbol{\xi}$ from λ). Thus

$$\mathbf{a}\boldsymbol{\xi} = \lambda\boldsymbol{\xi},$$
$$\mathbf{a}\bar{\boldsymbol{\xi}} = \bar{\lambda}\bar{\boldsymbol{\xi}}.$$

From the first equation,

$$\bar{\boldsymbol{\xi}}'\mathbf{a}\boldsymbol{\xi} = \lambda\bar{\boldsymbol{\xi}}'\boldsymbol{\xi}$$

and, from the second,

$$\boldsymbol{\xi}'\mathbf{a}\bar{\boldsymbol{\xi}} = \bar{\lambda}\boldsymbol{\xi}'\bar{\boldsymbol{\xi}}.$$

Transpose the latter equation:

$$\boldsymbol{\xi}'\mathbf{a}\boldsymbol{\xi} = \bar{\lambda}\boldsymbol{\xi}'\boldsymbol{\xi} \qquad (\mathbf{a}' = \mathbf{a}).$$

Comparing the two equations

$$\boldsymbol{\xi}'\mathbf{a}\boldsymbol{\xi} = \lambda\boldsymbol{\xi}'\boldsymbol{\xi},$$
$$\boldsymbol{\xi}'\mathbf{a}\boldsymbol{\xi} = \bar{\lambda}\boldsymbol{\xi}'\boldsymbol{\xi},$$

it follows that *either*

$$\boldsymbol{\xi}'\mathbf{a}\boldsymbol{\xi} = 0 \quad \text{and} \quad \boldsymbol{\xi}'\boldsymbol{\xi} = 0$$

or $$\lambda = \bar{\lambda}.$$

But

$$\boldsymbol{\xi} \equiv \{\xi_1, \xi_2, \ldots, \xi_n\},$$
$$\boldsymbol{\xi}' \equiv (\bar{\xi}_1, \bar{\xi}_2, \ldots, \bar{\xi}_n),$$

so that $$\boldsymbol{\xi}'\boldsymbol{\xi} \equiv \xi_1\bar{\xi}_1+\xi_2\bar{\xi}_2+\ldots+\xi_n\bar{\xi}_n.$$

Moreover, if $\xi_k = \alpha_k + i\beta_k$, then $\bar{\xi}_k = \alpha_k - i\beta_k$, so that

$$\bar{\xi}_k\xi_k = \alpha_k^2+\beta_k^2 = |\xi_k|^2.$$

Hence $$\boldsymbol{\xi}'\boldsymbol{\xi} = |\xi_1|^2+|\xi_2|^2+\ldots+|\xi_n|^2,$$

and this vanishes if, and only if,

$$|\xi_1| = |\xi_2| = \ldots = |\xi_n| = 0,$$

so that, also,

$$\xi_1 = \xi_2 = \ldots = \xi_n = 0,$$

in which case $$\boldsymbol{\xi} = \mathbf{0}.$$

But the definition of an eigenvector $\boldsymbol{\xi}$ excludes the possibility $\boldsymbol{\xi} = \mathbf{0}$. Hence this case is excluded, so that

$$\lambda = \bar{\lambda},$$

and the eigenvalue λ is therefore real.

(ii) Let $\mathbf{a}$ be a *Hermitian* matrix. Then *the eigenvalues are all real*:

The argument is very similar to that just given, and may be stated more concisely. Let λ be any eigenvalue, with eigenvector $\boldsymbol{\xi}$, so that

$$\mathbf{a}\boldsymbol{\xi} = \lambda\boldsymbol{\xi}.$$

It follows that

$$\bar{\mathbf{a}}\boldsymbol{\xi} = \bar{\lambda}\boldsymbol{\xi}.$$

Hence $$\boldsymbol{\xi}'\mathbf{a}\boldsymbol{\xi} = \lambda\boldsymbol{\xi}'\boldsymbol{\xi}, \quad \boldsymbol{\xi}'\bar{\mathbf{a}}\boldsymbol{\xi} = \bar{\lambda}\boldsymbol{\xi}'\boldsymbol{\xi},$$

and, transposing these *scalar* relations,

$$\boldsymbol{\xi}'\mathbf{a}'\boldsymbol{\xi} = \lambda\boldsymbol{\xi}'\boldsymbol{\xi}, \quad \boldsymbol{\xi}'\bar{\mathbf{a}}'\boldsymbol{\xi} = \bar{\lambda}\boldsymbol{\xi}'\boldsymbol{\xi}.$$

But, for a Hermitian matrix,

$$\mathbf{a}' = \bar{\mathbf{a}}, \quad \bar{\mathbf{a}}' = \mathbf{a},$$

so that we have the two equations

$$\bar{\boldsymbol{\xi}}' \mathbf{a} \boldsymbol{\xi} = \lambda \bar{\boldsymbol{\xi}}' \boldsymbol{\xi},$$
$$\bar{\boldsymbol{\xi}}' \mathbf{a} \boldsymbol{\xi} = \bar{\lambda} \bar{\boldsymbol{\xi}}' \boldsymbol{\xi}.$$

Hence *either*

$$\bar{\boldsymbol{\xi}}' \mathbf{a} \boldsymbol{\xi} = 0 \quad \text{and} \quad \bar{\boldsymbol{\xi}}' \boldsymbol{\xi} = 0,$$

or

$$\lambda = \bar{\lambda};$$

and, as in (i), $\bar{\boldsymbol{\xi}}' \boldsymbol{\xi} = 0 \Rightarrow \boldsymbol{\xi} = \mathbf{0}$, which is impossible. It therefore follows that

$$\lambda = \bar{\lambda},$$

so that λ is real.

COROLLARY. *The eigenvalues of a skew-symmetric matrix* **a** *with real coefficients are all pure imaginary.*

(Consider the *Hermitian* matrix $i\mathbf{a}$.)

(iii) Let **a** be a *unitary matrix*. Then *all the eigenvalues have unit modulus.*

Since **a** is unitary, then

$$\bar{\mathbf{a}}' \mathbf{a} = \mathbf{I}.$$

If **x** is an eigenvector corresponding to the eigenvalue λ, then

$$\mathbf{a}\mathbf{x} = \lambda \mathbf{x} \qquad (\mathbf{x} \neq \mathbf{0}),$$

so that

$$\bar{\mathbf{a}}' \mathbf{a}\mathbf{x} = \lambda \bar{\mathbf{a}}' \mathbf{x},$$

or, since $\bar{\mathbf{a}}' \mathbf{a} = \mathbf{I}$,

$$\mathbf{x} = \lambda \bar{\mathbf{a}}' \mathbf{x}.$$

Taking conjugate complex values,

$$\bar{\mathbf{x}} = \bar{\lambda} \mathbf{a}' \bar{\mathbf{x}},$$

and so

$$\bar{\mathbf{x}}' = \bar{\lambda} \bar{\mathbf{x}}' \mathbf{a}.$$

Hence

$$\bar{\mathbf{x}}' \mathbf{a}\mathbf{x} = (\bar{\lambda} \bar{\mathbf{x}}' \mathbf{a})(\lambda \mathbf{x}) = \lambda \bar{\lambda} \bar{\mathbf{x}}' \mathbf{a}\mathbf{x}.$$

Further,

$$\bar{\mathbf{x}}' \mathbf{a}\mathbf{x} = \bar{\mathbf{x}}'(\lambda \mathbf{x}) = \lambda \bar{\mathbf{x}}' \mathbf{x},$$

and the right-hand side is not zero since (i) $\bar{\mathbf{x}}' \mathbf{x} \neq 0$, otherwise each element of **x** would be zero, so that **x** would be the zero vector **0**,

(ii) $\lambda \neq 0$, otherwise zero would be a root of the characteristic equation $\det(\mathbf{a}-\lambda\mathbf{I})=0$, giving $\det\mathbf{a}=0$, whereas

$$\det(\bar{\mathbf{a}}')\det\mathbf{a} = \det(\bar{\mathbf{a}}'\mathbf{a}) = \det\mathbf{I} = 1.$$

The position thus is that

$$\lambda\bar{\lambda}\bar{\mathbf{x}}'\mathbf{ax} = \bar{\mathbf{x}}'\mathbf{ax},$$

$$\bar{\mathbf{x}}'\mathbf{ax} \neq 0.$$

Hence $$\lambda\bar{\lambda} = 1,$$

so that $$|\lambda| = 1.$$

6. Orthogonal matrices and rotation. Confining attention to matrices of order 3, let $\mathbf{a}$ be a given orthogonal matrix, so that

$$\mathbf{a}'\mathbf{a} = \mathbf{I}.$$

Since $\det\mathbf{a}' = \det\mathbf{a}$, it follows that

$$\{\det\mathbf{a}\}^2 = 1,$$

so that $$\det\mathbf{a} = \pm 1.$$

Each sign is possible; elementary examples are

$$\begin{vmatrix} 1 & 0 & 0 \\ 0 & 1 & 0 \\ 0 & 0 & 1 \end{vmatrix} = +1, \quad \begin{vmatrix} -1 & 0 & 0 \\ 0 & 1 & 0 \\ 0 & 0 & 1 \end{vmatrix} = -1.$$

The work of this paragraph is concerned with orthogonal matrices of determinant $+1$.

The characteristic equation is

$$\det(\mathbf{a}-\lambda\mathbf{I}) = 0$$

which, expanding the determinantal form

$$\begin{vmatrix} a_{11}-\lambda & a_{12} & a_{13} \\ a_{21} & a_{22}-\lambda & a_{23} \\ a_{31} & a_{32} & a_{33}-\lambda \end{vmatrix} = 0,$$

assumes the polynomial form

$$(\det\mathbf{a})+\alpha\lambda+\beta\lambda^2-\lambda^3 = 0,$$

or, in more usual monic form,

$$\lambda^3-\beta\lambda^2-\alpha\lambda-(\det\mathbf{a}) = 0.$$

The values of the numerical coefficients α, β are irrelevant for immediate purposes.

Note that $\lambda = 0$ cannot be an eigenvector, since $\det \mathbf{a} \neq 0$.

The characteristic equation

$$\det(\mathbf{a} - \lambda \mathbf{I}) = 0$$

may be written

$$\det(\mathbf{a} - \lambda \mathbf{a}\mathbf{a}') = 0$$

or

$$\det\{\mathbf{a}(\mathbf{I} - \lambda \mathbf{a}')\} = 0,$$

or, since $\lambda \neq 0$ and $\det \mathbf{a} \neq 0$,

$$\det(\mathbf{a}' - \lambda^{-1}\mathbf{I}) = 0,$$

or, transposing,

$$\det(\mathbf{a} - \lambda^{-1}\mathbf{I}) = 0.$$

The characteristic equation is thus also given by

$$\lambda^{-3} - \beta\lambda^{-2} - \alpha\lambda^{-1} - (\det \mathbf{a}) = 0,$$

or

$$(\det \mathbf{a})\,\lambda^3 + \alpha\lambda^2 + \beta\lambda - 1 = 0.$$

For the case $\det \mathbf{a} = +1$, the two equations

$$\lambda^3 - \beta\lambda^2 - \alpha\lambda - 1 = 0,$$

$$\lambda^3 + \alpha\lambda^2 + \beta\lambda - 1 = 0$$

are therefore identical, and so

$$\beta = -\alpha.$$

Hence the characteristic equation is of the form

$$\lambda^3 + \alpha\lambda^2 - \alpha\lambda - 1 = 0,$$

or

$$(\lambda - 1)\{\lambda^2 + (\alpha + 1)\,\lambda + 1\} = 0.$$

Thus one eigenvalue is

$$\lambda = +1$$

and the two others, λ_1 and λ_2, are such that

$$\lambda_1 \lambda_2 = 1,$$

so that each is the reciprocal of the other.

At this point, the result (p. 263) of the preceding paragraph becomes important; that, for a unitary matrix and therefore for its real counterpart an orthogonal matrix, the modulus of an eigenvalue is necessarily unity. Hence

$$|\lambda_1| = |\lambda_2| = 1.$$

The two eigenvalues λ_1, λ_2 are thus of the form

$$\lambda_1 = e^{i\theta}, \quad \lambda_2 = e^{-i\theta},$$

being real when $\theta = 0$ or π (in the range $0, 2\pi$) and complex conjugates otherwise. The characteristic equation is then

$$(\lambda-1)(\lambda^2-2\lambda\cos\theta+1) = 0,$$

or

$$\lambda^3-(1+2\cos\theta)\lambda^2+(1+2\cos\theta)\lambda-1 = 0.$$

We are concerned particularly with the eigenvector corresponding to the eigenvalue $+1$. Denoting it by the symbol $\boldsymbol{\xi}$, the defining relation is

$$\mathbf{a}\boldsymbol{\xi} = \boldsymbol{\xi},$$

and, for convenience, $\boldsymbol{\xi}$ will be *normalized* by the relation

$$\boldsymbol{\xi}'\boldsymbol{\xi} = 1,$$

so that, regarded as a vector of three-dimensional space, $\boldsymbol{\xi}$ is a *unit* vector.

Text Examples

1. Prove that

$$\mathbf{a}'\boldsymbol{\xi} = \boldsymbol{\xi}, \quad \boldsymbol{\xi}'\mathbf{a} = \boldsymbol{\xi}', \quad \boldsymbol{\xi}'\mathbf{a}' = \boldsymbol{\xi}'.$$

2. Prove that, if n is any integer (positive, negative or zero),

$$\mathbf{a}^n\boldsymbol{\xi} = \boldsymbol{\xi}.$$

3. Prove that

$$(\mathbf{I}-\boldsymbol{\xi}\boldsymbol{\xi}')' = \mathbf{I}-\boldsymbol{\xi}\boldsymbol{\xi}'$$

and that

$$(\mathbf{I}-\boldsymbol{\xi}\boldsymbol{\xi}')^2 = \mathbf{I}-\boldsymbol{\xi}\boldsymbol{\xi}'.$$

The next problem is to consider the *orthogonal mapping*

$$\mathbf{y} = \mathbf{ax}$$

of the vector $\mathbf{x}$ to the vector $\mathbf{ax}$. The relation

$$\boldsymbol{\xi} = \mathbf{a}\boldsymbol{\xi}$$

shows that *all vectors that are scalar multiples of* $\boldsymbol{\xi}$ *are unchanged by the mapping*.

The work which follows† is directed towards establishing the fact that *the mapping, for* $\det\mathbf{a} = +1$, *is described geometrically by the rotation*

† The result is usually proved by a transformation argument, but there seems to be interest in obtaining the rotation from its elementary properties.

(*in ordinary Euclidean three-dimensional space*) *of the point* **x** *about the line* l *through the origin in direction* $\boldsymbol{\xi}$ (*p*. 174), *to the position* **ax**. Briefly, 'the mapping is a rotation about $\boldsymbol{\xi}$'.

Let **p** be an arbitrary point of space. The foot of the perpendicular from **p** to the line l is (p. 176) the point

$$\boldsymbol{\xi}\boldsymbol{\xi}'\mathbf{p}.$$

In the same way, the foot of the perpendicular from the map **ap** to the line l is

$$\boldsymbol{\xi}\boldsymbol{\xi}'(\mathbf{ap})$$

or

$$\boldsymbol{\xi}(\boldsymbol{\xi}'\mathbf{a})\mathbf{p}.$$

But

$$\begin{aligned} \mathbf{a}\boldsymbol{\xi} &= \boldsymbol{\xi} \\ \Rightarrow \mathbf{a}'\mathbf{a}\boldsymbol{\xi} &= \mathbf{a}'\boldsymbol{\xi} \\ \Rightarrow \quad \boldsymbol{\xi} &= \mathbf{a}'\boldsymbol{\xi} \\ \Rightarrow \quad \boldsymbol{\xi}' &= \boldsymbol{\xi}'\mathbf{a}, \end{aligned}$$

so that

$$\boldsymbol{\xi}(\boldsymbol{\xi}'\mathbf{a})\mathbf{p} = \boldsymbol{\xi}\boldsymbol{\xi}'\mathbf{p}.$$

Hence *the feet of the perpendiculars from* **p** *and* **ap** *to the line* l *coincide.* Call this point **q**.

Further (p. 154) the square of the distance from **p** to the line is

$$\begin{aligned} &(\mathbf{p}-\mathbf{q})'(\mathbf{p}-\mathbf{q}) \\ &= (\mathbf{p}-\boldsymbol{\xi}\boldsymbol{\xi}'\mathbf{p})'(\mathbf{p}-\boldsymbol{\xi}\boldsymbol{\xi}'\mathbf{p}) \\ &= \mathbf{p}'(\mathbf{I}-\boldsymbol{\xi}\boldsymbol{\xi}')'(\mathbf{I}-\boldsymbol{\xi}\boldsymbol{\xi}')\mathbf{p} \\ &= \mathbf{p}'(\mathbf{I}-\boldsymbol{\xi}\boldsymbol{\xi}')(\mathbf{I}-\boldsymbol{\xi}\boldsymbol{\xi}')\mathbf{p} \\ &= \mathbf{p}'\{\mathbf{I}^2-2\boldsymbol{\xi}\boldsymbol{\xi}'+\boldsymbol{\xi}\boldsymbol{\xi}'\boldsymbol{\xi}\boldsymbol{\xi}'\}\mathbf{p} \\ &= \mathbf{p}'(\mathbf{I}-\boldsymbol{\xi}\boldsymbol{\xi}')\mathbf{p}, \end{aligned}$$

since $\boldsymbol{\xi}'\boldsymbol{\xi} = 1$; and the square of the distance from **ap** to the line is, similarly,

$$\begin{aligned} &(\mathbf{ap}-\boldsymbol{\xi}\boldsymbol{\xi}'\mathbf{ap})'(\mathbf{ap}-\boldsymbol{\xi}\boldsymbol{\xi}'\mathbf{ap}) \\ &= \mathbf{p}'\mathbf{a}'(\mathbf{I}-\boldsymbol{\xi}\boldsymbol{\xi}')'(\mathbf{I}-\boldsymbol{\xi}\boldsymbol{\xi}')\mathbf{ap} \\ &= \mathbf{p}'\mathbf{a}'(\mathbf{I}-\boldsymbol{\xi}\boldsymbol{\xi}')\mathbf{ap} \qquad \text{(as before)} \\ &= \mathbf{p}'(\mathbf{a}'\mathbf{a}-\mathbf{a}'\boldsymbol{\xi}\boldsymbol{\xi}'\mathbf{a})\mathbf{p} \\ &= \mathbf{p}'(\mathbf{I}-\boldsymbol{\xi}\boldsymbol{\xi}')\mathbf{p} \end{aligned}$$

since $\mathbf{a}'\boldsymbol{\xi} = \boldsymbol{\xi}$ and $\boldsymbol{\xi}'\mathbf{a} = \boldsymbol{\xi}'$.

Hence **p** *and* **ap** *are equidistant from* l.

The mapping will therefore be a rotation about $\boldsymbol{\xi}$ if we can establish one further fact: that the angle between the perpendicular to l from $\mathbf{p}$ and the perpendicular to l from $\mathbf{ap}$ is independent of the point $\mathbf{p}$ selected.

Now the square of the distance between $\mathbf{p}$ and $\mathbf{ap}$ is equal to

$$\begin{aligned}&(\mathbf{p}-\mathbf{ap})'(\mathbf{p}-\mathbf{ap})\\ &= \mathbf{p}'(\mathbf{I}-\mathbf{a}')(\mathbf{I}-\mathbf{a})\mathbf{p}\\ &= \mathbf{p}'(\mathbf{I}^2-\mathbf{a}-\mathbf{a}'+\mathbf{a}'\mathbf{a})\mathbf{p}\\ &= \mathbf{p}'(2\mathbf{I}-\mathbf{a}-\mathbf{a}')\mathbf{p}.\end{aligned}$$

The triangle with vertices $\mathbf{p}$, $\mathbf{ap}$, $\mathbf{q}$ is therefore an isosceles triangle whose shape is independent of $\mathbf{p}$ *provided* that the two quadratic forms

$$\mathbf{p}'(\mathbf{I}-\boldsymbol{\xi}\boldsymbol{\xi}')\mathbf{p},$$
$$\mathbf{p}'(2\mathbf{I}-\mathbf{a}-\mathbf{a}')\mathbf{p}$$

have a ratio independent of $\mathbf{p}$; and this result must now be proved.

We begin, somewhat unexpectedly, perhaps, by establishing the identity

$$2(1-\cos\theta)\,\boldsymbol{\xi}\boldsymbol{\xi}' = \mathbf{a}^2-2\mathbf{a}\cos\theta+\mathbf{I},$$

where θ is the angle defined earlier for the characteristic equation. In fact, the earlier expression

$$(\lambda-1)(\lambda^2-2\lambda\cos\theta+1) = 0$$

for the characteristic equation, coupled with the Cayley–Hamilton equation (p. 258), shows that

$$(\mathbf{a}-\mathbf{I})(\mathbf{a}^2-2\mathbf{a}\cos\theta+\mathbf{I}) = \mathbf{0},$$

so that, if $\mathbf{b}_1, \mathbf{b}_2, \mathbf{b}_3$ are the column vectors of $\mathbf{a}^2-2\mathbf{a}\cos\theta+\mathbf{I}$, it follows that

$$(\mathbf{a}-\mathbf{I})(\mathbf{b}_1, \mathbf{b}_2, \mathbf{b}_3) = \mathbf{0},$$

so that

$$(\mathbf{a}-\mathbf{I})\,\mathbf{b}_i = \mathbf{0} \qquad (i = 1,2,3).$$

But the only non-zero vectors satisfying the relation

$$(\mathbf{a}-\mathbf{I})\,\mathbf{x} = \mathbf{0}$$

are the scalar multiples of $\boldsymbol{\xi}$, so that

$$\mathbf{b}_i = b_i\,\boldsymbol{\xi} \qquad (i = 1,2,3).$$

Hence

$$\begin{aligned}\mathbf{a}^2-2\mathbf{a}\cos\theta+\mathbf{I} &\equiv (b_1\,\boldsymbol{\xi},\ b_2\,\boldsymbol{\xi},\ b_3\,\boldsymbol{\xi})\\ &\equiv \boldsymbol{\xi}\mathbf{b}',\end{aligned}$$

where **b** is the column vector $\{b_1, b_2, b_3\}$, so that

$$\boldsymbol{\xi}'(\mathbf{a}^2 - 2\mathbf{a}\cos\theta + \mathbf{I}) = \boldsymbol{\xi}'\boldsymbol{\xi}\mathbf{b}'.$$

But

$$\boldsymbol{\xi}'\mathbf{a}^2 = \boldsymbol{\xi}'\mathbf{a} = \boldsymbol{\xi}'$$

and

$$\boldsymbol{\xi}'\boldsymbol{\xi} = 1,$$

and so

$$\boldsymbol{\xi}'(1 - 2\cos\theta + 1) = \mathbf{b}';$$

thus

$$\mathbf{b}' \equiv 2(1 - \cos\theta)\boldsymbol{\xi}'.$$

Hence

$$\mathbf{a}^2 - 2\mathbf{a}\cos\theta + \mathbf{I} = 2(1 - \cos\theta)\boldsymbol{\xi}\boldsymbol{\xi}'.$$

The final result now follows; for

$$\mathbf{a}'(\mathbf{a}^2 - 2\mathbf{a}\cos\theta + \mathbf{I}) = 2(1 - \cos\theta)\mathbf{a}'\boldsymbol{\xi}\boldsymbol{\xi}',$$

so that

$$(\mathbf{a} - 2\mathbf{I}\cos\theta + \mathbf{a}') = 2(1 - \cos\theta)\boldsymbol{\xi}\boldsymbol{\xi}',$$

or

$$\mathbf{I} - \tfrac{1}{2}(\mathbf{a} + \mathbf{a}') = (1 - \cos\theta)(\mathbf{I} - \boldsymbol{\xi}\boldsymbol{\xi}').$$

The isosceles triangle with vertices **p**, **ap**, **q** thus has the ratio

$$\frac{\text{square of sides}}{\text{square of base}} = \frac{\mathbf{p}'(\mathbf{I} - \boldsymbol{\xi}\boldsymbol{\xi}')\mathbf{p}}{\mathbf{p}'(2\mathbf{I} - \mathbf{a} - \mathbf{a}')\mathbf{p}}$$

$$= \frac{1}{2(1 - \cos\theta)}$$

$$= \frac{1}{4\sin^2\frac{1}{2}\theta},$$

so that

$$\frac{\frac{1}{2}\text{ base}}{\text{side}} = \sin\tfrac{1}{2}\theta.$$

The angle between the sides is thus θ, and so, since this is independent of **p**, the mapping

$$\mathbf{y} = \mathbf{ax} \qquad (\det \mathbf{a} = 1)$$

represents a rotation through an angle θ about the line through the origin in the direction $\boldsymbol{\xi}$.

[The continuity of the mapping shows that the rotation is in the same *sense* for near points, and therefore for all points.]

Text Examples

Write

$$\mathbf{m} = \mathbf{I}-\tfrac{1}{2}(\mathbf{a}+\mathbf{a}'), \quad \mathbf{n} = \mathbf{I}-\boldsymbol{\xi}\boldsymbol{\xi}'$$

and, in the notation of the text, prove the following results:

1. $$\mathbf{m}\boldsymbol{\xi} = 0, \quad \mathbf{n}\boldsymbol{\xi} = 0.$$

2. $$\mathbf{mn} = \mathbf{nm} = \mathbf{m},$$
$$\mathbf{n}^2 = \mathbf{n}.$$
$$\mathbf{m}^2 = (1-\cos\theta)\,\mathbf{m}.$$

3. Prove that **m** and **n** are both singular matrices.

4. Criticize the argument

$$\begin{aligned}\mathbf{m}^2 &= (1-\cos\theta)\,\mathbf{m}\\ &= (1-\cos\theta)\,\mathbf{mn},\end{aligned}$$

so that

$$\mathbf{m}\{\mathbf{m}-(1-\cos\theta)\,\mathbf{n}\} = 0.$$

Hence $$\mathbf{m} = (1-\cos\theta)\,\mathbf{n}.$$

Revision Examples

1. Prove that, if the matrix **a** is skew-symmetric, with λ as an eigenvalue, then $-\lambda$ is also an eigenvalue.

2. Prove that, if **a** is an orthogonal 3×3 matrix, then

$$\det \mathbf{a} = \pm 1.$$

Prove also that, if $\lambda_1, \lambda_2, \lambda_3$ are the eigenvalues, then

$$\lambda_1\lambda_2\lambda_3 = \det \mathbf{a}.$$

3. The matrix **a** has distinct eigenvalues. Prove that, if **b** is a matrix such that $\mathbf{ab} = \mathbf{ba}$, then every eigenvector of **a** is an eigenvector of **b**.

4. The matrix **a**, with complex elements, is Hermitian. Prove that $\det\mathbf{a}$ is real and that the scalar form $\bar{\mathbf{x}}'\mathbf{ax}$ is real.

5. Prove that, if **x** is an eigenvector of **a** and of **b**, then it is an eigenvector of **ab**.

6. The 3×3 matrices **a**, **b** have each the linearly independent column vectors $\mathbf{x}_1, \mathbf{x}_2, \mathbf{x}_3$ as eigenvectors. Prove that the matrix

$$\mathbf{p} \equiv (\mathbf{x}_1, \mathbf{x}_2, \mathbf{x}_3)$$

with columns $\mathbf{x}_1, \mathbf{x}_2, \mathbf{x}_3$ is non-singular; and that, if

$$\mathbf{u} \equiv \begin{pmatrix} \lambda_1 & 0 & 0 \\ 0 & \lambda_2 & 0 \\ 0 & 0 & \lambda_3 \end{pmatrix}, \quad \mathbf{v} \equiv \begin{pmatrix} \mu_1 & 0 & 0 \\ 0 & \mu_2 & 0 \\ 0 & 0 & \mu_3 \end{pmatrix}$$

are diagonal matrices in which $\lambda_1, \lambda_2, \lambda_3$ are eigenvalues of $\mathbf{a}$ and μ_1, μ_2, μ_3 are eigenvalues of $\mathbf{b}$, then

$$\mathbf{ap} = \mathbf{pu}, \quad \mathbf{bp} = \mathbf{pv}.$$

Deduce that $$\mathbf{ab} = \mathbf{ba}.$$

7. Show that the product of two orthogonal matrices is orthogonal. Is the inverse of an orthogonal matrix orthogonal?

8. The 3×3 matrix $\mathbf{a}$ has distinct eigenvalues $\lambda_1, \lambda_2, \lambda_3$ and corresponding eigenvectors $\mathbf{x}_1, \mathbf{x}_2, \mathbf{x}_3$. By expressing an arbitrary column vector $\mathbf{y}$ is the form $\alpha_1\mathbf{x}_1+\alpha_2\mathbf{x}_2+\alpha_3\mathbf{x}_3$, show that the ratio $\mathbf{y}'\mathbf{a}\mathbf{y}/\mathbf{y}'\mathbf{y}$ cannot be greater than the largest of $\lambda_1, \lambda_2, \lambda_3$.

9. Explain what is meant by the eigenvalues of a square matrix and show that the eigenvalues of a real 2×2 symmetric matrix are real.

The equation of a conic S in rectangular coordinates is

$$ax^2+2hxy+by^2 = c,$$

where a, b, c, h are all real and $c > 0$. Prove that, if S contains some real points, the matrix

$$\mathbf{a} \equiv \begin{pmatrix} a & h \\ h & b \end{pmatrix}$$

has an eigenvalue λ_1 which is positive.

If the other eigenvalue is λ_2, describe the conic S in the cases

$$\lambda_2 < 0, \quad \lambda_2 = 0, \quad \lambda_2 > 0, \quad \lambda_2 = -\lambda_1, \quad \lambda_2 = \lambda_1.$$

Show that the equation

$$8x^2-2\sqrt{6}xy+7y^2 = 10$$

represents an ellipse, and find the lengths of its semi-axes.

10. Prove that the matrix

$$\begin{pmatrix} \frac{1}{2}+\frac{1}{4}\sqrt{2} & \frac{1}{2}-\frac{1}{4}\sqrt{2} & \frac{1}{2} \\ \frac{1}{2}-\frac{1}{4}\sqrt{2} & \frac{1}{2}+\frac{1}{4}\sqrt{2} & -\frac{1}{2} \\ -\frac{1}{2} & \frac{1}{2} & \frac{1}{2}\sqrt{2} \end{pmatrix}$$

is orthogonal. Find the eigenvectors and normalize them (so that $\boldsymbol{\xi}'\boldsymbol{\xi} = 1$).

11. Show that the matrix

$$\mathbf{p} \equiv \begin{pmatrix} \sin\theta & \cos\theta & 0 \\ -\cos\theta & \sin\theta & 0 \\ 0 & 0 & 1 \end{pmatrix}$$

is orthogonal.

A square matrix $\mathbf{a}$ is of order 3, and $\mathbf{b}$ is the matrix $\mathbf{pap'}$. Show that, by a suitable choice of θ, one of the off-diagonal terms in $\mathbf{b}$ can be made zero, the sum of the squares of the remaining off-diagonal terms being equal to the sum of the squares of the corresponding terms in $\mathbf{a}$.

12. Find the characteristic equation of the matrix

$$\mathbf{a} \equiv \begin{pmatrix} b & c & a \\ c & a & b \\ a & b & c \end{pmatrix},$$

and prove that the matrices

$$\mathbf{b} \equiv \begin{pmatrix} c & a & b \\ a & b & c \\ b & c & a \end{pmatrix}, \quad \mathbf{c} \equiv \begin{pmatrix} a & b & c \\ b & c & a \\ c & a & b \end{pmatrix}$$

have the same characteristic equation as $\mathbf{a}$.

Prove also that, if

$$\mathbf{bc} = \mathbf{cb},$$

then two of the roots of the characteristic equation are zero.

13. Prove that, if $\mathbf{a}$ is a real skew-symmetric matrix, then the matrix $\mathbf{I}+\mathbf{a}$ is non-singular.

[If you wish, consider only the cases when the order of the matrix is 2, 3, 4.]

Prove also that the matrix

$$\mathbf{b} \equiv (\mathbf{I}-\mathbf{a})(\mathbf{I}+\mathbf{a})^{-1}$$

is orthogonal.

Obtain $\mathbf{b}$ when the order is 2 and

$$\mathbf{a} \equiv \begin{pmatrix} 0 & p \\ -p & 0 \end{pmatrix}.$$

14. Prove that, if $\mathbf{a}$ is skew-symmetric, then

$$(\mathbf{I}-\mathbf{a})^{-1}(\mathbf{I}+\mathbf{a}) = (\mathbf{I}+\mathbf{a})(\mathbf{I}-\mathbf{a})^{-1},$$

each expression defining an orthogonal matrix $\mathbf{u}$.

Prove also that, if $\mathbf{x}$ is a vector such that

$$\mathbf{ax} = \mathbf{0},$$

then

$$\mathbf{ux} = \mathbf{x}.$$

15. If $\mathbf{a}$ is a non-singular square matrix, show that the eigenvectors of $\mathbf{a}^{-1}$ are the same as those of $\mathbf{a}$ and find how the eigenvalues of $\mathbf{a}^{-1}$ are related to those of $\mathbf{a}$.

By considering its inverse, find the eigenvalues and eigenvectors of the matrix

$$\begin{pmatrix} 1 & -16 & 20 \\ -16 & 13 & 4 \\ 20 & 4 & -5 \end{pmatrix}.$$

16. Prove that, if $\lambda_1, \lambda_2, \ldots, \lambda_n$ are the eigenvalues of an $n \times n$ matrix $\mathbf{a}$, then $\det \mathbf{a} = \lambda_1 \lambda_2 \ldots \lambda_n$.

17. Prove that the matrices $\mathbf{a}$ and $\mathbf{b}^{-1}\mathbf{ab}$ have the same eigenvalues.

Deduce that, if $\mathbf{u}$, $\mathbf{v}$ are arbitrary non-singular matrices, then $\mathbf{uv}$ and $\mathbf{vu}$ have the same eigenvalues.

18. Given a polynomial

$$f(x) \equiv ax^2 + bx + c$$

and a matrix $\mathbf{u}$ with eigenvalues $\lambda_1, \lambda_2, \ldots, \lambda_n$ prove that

$$\det(a\mathbf{u}^2 + b\mathbf{u} + c\mathbf{I}) = f(\lambda_1) f(\lambda_2) \ldots f(\lambda_n).$$

19. Prove that, if $\mathbf{a}$ is a square matrix such that

$$\mathbf{a}'\mathbf{a} = \mathbf{0},$$

then

$$\mathbf{a} = \mathbf{0}.$$

20. Prove that, if the matrix $\mathbf{a}$ is symmetric and the matrix $\mathbf{s}$ is skew-symmetric, both of order n, and if $\mathbf{b}$ is defined by the relation

$$\mathbf{b} = (\mathbf{a}+\mathbf{s})^{-1}(\mathbf{a}-\mathbf{s}),$$

where $\mathbf{a}+\mathbf{s}$ is non-singular, then

$$\mathbf{b}'\mathbf{ab} = \mathbf{a}, \quad \mathbf{b}'\mathbf{sb} = \mathbf{s}.$$

21. Prove that, if $\mathbf{a}$, $\mathbf{b}$ are square matrices of order n with real elements and if $\mathbf{h} \equiv \mathbf{a} + i\mathbf{b}$ is Hermitian, then the matrix

$$\mathbf{u} \equiv \begin{pmatrix} \mathbf{a} & \mathbf{b} \\ -\mathbf{b} & \mathbf{a} \end{pmatrix}$$

is symmetric.

Prove also that, if α is an eigenvalue of $\mathbf{h}$, then α is real, and α is also an eigenvalue of $\mathbf{u}$.

Prove that

$$\det \mathbf{u} = \det(\mathbf{a}+i\mathbf{b}) \det(\mathbf{a}-i\mathbf{b}).$$

22. Given that

$$\mathbf{u} \equiv \begin{pmatrix} 0 & b & -c \\ -b & 0 & a \\ c & -a & 0 \end{pmatrix},$$

prove that

$$\mathbf{u}^3 + (a^2+b^2+c^2)\,\mathbf{u} = 0.$$

Deduce, or prove otherwise, that, if a, b, c are real, the matrix

$$\mathbf{u}^2 + (a^2+b^2+c^2)\,\mathbf{u} + \mathbf{I}$$

is orthogonal if, and only if, *either* $a^2+b^2+c^2 = 1$ *or* $\mathbf{u}$ is a zero matrix.

23. Given two column vectors $\mathbf{x}$, $\mathbf{y}$ of n elements each, prove that the matrix

$$\mathbf{p} \equiv \mathbf{x}\mathbf{y}' - \mathbf{y}\mathbf{x}'$$

is skew-symmetric.

Prove also that, if

$$\boldsymbol{\xi} \equiv \lambda\mathbf{x} + \mu\mathbf{y}, \quad \boldsymbol{\eta} \equiv \nu\mathbf{x} + \rho\mathbf{y},$$

then

$$\boldsymbol{\xi}\boldsymbol{\eta}' - \boldsymbol{\eta}\boldsymbol{\xi}' = (\lambda\rho - \mu\nu)\,\mathbf{p}.$$

Two further $n \times 1$ column vectors $\mathbf{u}$, $\mathbf{v}$ are connected with $\mathbf{x}$, $\mathbf{y}$ by means of the relations

$$\mathbf{u}'\mathbf{x} = \mathbf{u}'\mathbf{y} = \mathbf{v}'\mathbf{x} = \mathbf{v}'\mathbf{y} = 0.$$

Prove that, if

$$\boldsymbol{\pi} \equiv \mathbf{u}\mathbf{v}' - \mathbf{v}\mathbf{u}',$$

then

$$\mathbf{p}\boldsymbol{\pi} = 0.$$

24. Prove that the matrix

$$\begin{pmatrix} \frac{1}{3} & -\frac{2}{3} & \frac{2}{3} \\ \frac{2}{3} & -\frac{1}{3} & -\frac{2}{3} \\ \frac{2}{3} & \frac{2}{3} & \frac{1}{3} \end{pmatrix}$$

is orthogonal, and express it in the form

$$(\mathbf{I}-\mathbf{S})(\mathbf{I}+\mathbf{S})^{-1},$$

where $\mathbf{S}$ is real and skew-symmetrical.

25. Prove that the matrix $\mathbf{a}$ given by

$$\begin{pmatrix} l & m & n & 0 \\ 0 & 0 & 0 & -1 \\ n & l & -m & 0 \\ -m & n & -l & 0 \end{pmatrix}$$

is orthogonal, where $l = \frac{2}{7}$, $m = \frac{3}{7}$, $n = \frac{6}{7}$.

Find the eigenvalues of **a** and deduce that the characteristic equation of the matrix $\mathbf{a}^2+\mathbf{I}$ is

$$(\lambda-2)^2(49\lambda^2-\lambda+1) = 0.$$

26. Given an $n \times n$ matrix

$$\mathbf{a} \equiv (\beta^{ij}),$$

where $\beta = \exp(2\pi i/n)$, show, by considering the eigenvalues of $\mathbf{a}^2$, or otherwise, that the eigenvalues of **a** are all one or other of $\pm\sqrt{n}$, $\pm i\sqrt{n}$.

27. If

$$\mathbf{a} \equiv \begin{pmatrix} 0 & -\cos\alpha & -\sin\alpha \\ \cos\alpha & \sin^2\alpha & -\sin\alpha\cos\alpha \\ \sin\alpha & -\sin\alpha\cos\alpha & \cos^2\alpha \end{pmatrix},$$

find (i) the inverse matrix $\mathbf{a}^{-1}$, (ii) the eigenvalues and eigenvectors of **a**, (iii) a matrix **u** such that $\mathbf{u}^{-1}\mathbf{a}\mathbf{u}$ is a diagonal matrix.

28. The square matrix **a** is such that

$$\bar{\mathbf{a}}'\mathbf{a}^3 = \mathbf{I}.$$

Prove that every eigenvalue of **a** has one of the values $+1$ or -1.

29. Two $n \times n$ matrices **a**, **b** are such that $\mathbf{ab} = \mathbf{ba}$. The polynomial $\det(\lambda\mathbf{a}-\mu\mathbf{b})$ is written in the form $f(\lambda, \mu)$. Prove that $f(\mathbf{b}, \mathbf{a}) = \mathbf{0}$.

30. The square matrices **a**, **b** of order 3 are such that **a** is not singular and the three eigenvalues λ_1, λ_2, λ_3 of the matrix $\mathbf{a}^{-1}\mathbf{b}$ are distinct. Prove that the rank of each of the matrices $\lambda_1\mathbf{a}-\mathbf{b}$, $\lambda_2\mathbf{a}-\mathbf{b}$, $\lambda_3\mathbf{a}-\mathbf{b}$ is 2.

31. An $n \times n$ matrix **a** with complex elements is such that

$$\mathbf{a}\bar{\mathbf{a}}' = \bar{\mathbf{a}}'\mathbf{a},$$

and **x** is a column vector. Prove that, if

$$\mathbf{ax} = \mathbf{0},$$

then $$\bar{\mathbf{a}}'\mathbf{x} = \mathbf{0}.$$

Deduce that, if $$\mathbf{ax} = \lambda\mathbf{x},$$

then $$\bar{\mathbf{a}}'\mathbf{x} = \bar{\lambda}\mathbf{x}.$$

Prove, finally, that, if

$$\mathbf{ax} = \lambda\mathbf{x}, \quad \mathbf{ay} = \mu\mathbf{y} \qquad (\lambda \neq \mu),$$

then $$\bar{\mathbf{x}}'\mathbf{y} = 0.$$

32. Prove that, if **a** is an $n \times n$ matrix with distinct eigenvalues, then a non-singular matrix **u** exists such that the matrix $\mathbf{d} \equiv \mathbf{u}^{-1}\mathbf{a}\mathbf{u}$ is diagonal.

The matrix **b** commutes with **a**, so that $\mathbf{ab} = \mathbf{ba}$. Prove that it is of the form

$$\mathbf{b} = \mathbf{u}\mathbf{c}\mathbf{u}^{-1},$$

where **c** is diagonal, and show that **b** can be expressed as a polynomial in **a**.

33. Show that, if λ is an eigenvalue of the square $n \times n$ matrix **a**, and if f is any polynomial, then $f(\lambda)$ is an eigenvalue of $f(\mathbf{a})$.

Calculate $\mathbf{a}^2$ and $\mathbf{a}^4$ for the case

$$a_{rs} = \omega^{rs},$$

where $\omega = e^{2\pi i/n}$.

Deduce that the eigenvalues of **a** then have the form $n^{\frac{1}{2}} i^m$, where $m = 0, 1, 2$ or 3.

34. If the matrix **a** is skew-symmetric, prove that any vectors **x** and **y** related by $\mathbf{ax} = \mathbf{y}$ are perpendicular.

35. Given a square matrix **a**, prove that it is possible to find a symmetric matrix **u** and a skew-symmetric matrix **v** such that $\mathbf{a} = \mathbf{u} + \mathbf{v}$.

36. Prove that if, given any matrix **b** of type 3×3, vectors **x**, **y**, **z** are such that

$$\mathbf{y} = \mathbf{bx}, \quad \mathbf{z} = \mathbf{b}'\mathbf{x},$$

then $\mathbf{y} - \mathbf{z}$ is perpendicular to **x**; and if the point **x** lies on a fixed line l, then the point $\mathbf{y} - \mathbf{z}$ lies on a fixed line perpendicular to l.

19

SOME ABSTRACT CONCEPTIONS

This chapter gives a brief introduction to one or two abstract conceptions which are important for more detailed study. They might, indeed, have appeared earlier, but there are some advantages in waiting until a body of experience has been accumulated first.

1. Reflexive relations. Let

$$S\{a, b, c, \ldots\}$$

be a given set of elements among which there is a relation denoted by R, in virtue of which the notation

$$a \mathrm{R} b$$

reads, 'a stands in the relation R to b'. For example, R might mean, 'is equal to', 'is the square of',....

DEFINITION. The relation R is said to be *reflexive* if, for all elements $a \in S$,

$$a \mathrm{R} a,$$

so that each element stands in the relation R to itself.

For example, if

$$\mathrm{R} \equiv \text{'is equal to'},$$

then

$$a \mathrm{R} a,$$

since, always,

$$a = a.$$

On the other hand, if

$$\mathrm{R} \equiv \text{'is less than'},$$

then

$$a \not\mathrm{R}\, a,$$

since it is never true that

$$a < a.$$

Examples

Determine which of the following relations are reflexive:

1. $S \equiv$ the set of triangles,
   $\mathsf{R} \equiv$ is similar to;
2. $S \equiv$ the set of integers (including zero),
   $\mathsf{R} \equiv$ differs by a multiple of 3 from;
3. $S \equiv$ the set of squares,
   $\mathsf{R} \equiv$ is congruent to;
4. $S \equiv$ the set of integers,
   $\mathsf{R} \equiv$ is the square of;
5. $S \equiv$ the set of rational numbers,
   $\mathsf{R} \equiv$ is equal to or less than;
6. $S \equiv$ the set of complex numbers,
   $\mathsf{R} \equiv$ has the same modulus as.

2. Symmetric relations.

DEFINITION. The relation R is said to be *symmetric* if, for all pairs of elements $a, b \in S$,

$$a \mathsf{R} b \Leftrightarrow b \mathsf{R} a.$$

For example, if

$$\mathsf{R} \equiv \text{'is equal to'},$$

then the relation

$$a = b$$

leads inevitably to

$$b = a$$

and conversely.

On the other hand, if

$$\mathsf{R} \equiv \text{'is less than'},$$

then

$$a < b$$
$$\not\Rightarrow b < a.$$

Examples

Determine which of the following relations are symmetric:

1. $S \equiv$ the set of triangles,
   $\mathsf{R} \equiv$ is congruent to;
2. $S \equiv$ the set of integers,
   $\mathsf{R} \equiv$ is prime to;
3. $S \equiv$ the set of integers,
   $\mathsf{R} \equiv$ is a factor of;

4. $S \equiv$ the set of circles in a plane,
R $\equiv$ lies wholly inside;
5. $S \equiv$ the set of ellipses in a plane,
R $\equiv$ lies wholly outside;
6. $S \equiv$ the set of complex numbers,
R $\equiv$ has the same real part as.

3. Transitive relations.

DEFINITION. The relation R is said to be *transitive* if, for $a, b, c \in S$,

$$a \mathsf{R} b \quad \textbf{and} \quad b \mathsf{R} c$$
$$\Rightarrow a \mathsf{R} c.$$

For example, if

$$\mathsf{R} \equiv \text{'is equal to'},$$

then the two relations

$$a = b, \quad b = c$$

lead inevitably to

$$a = c.$$

Again, if

$$\mathsf{R} \equiv \text{'is less than'},$$

then the two relations

$$a < b, \quad b < c$$

lead inevitably to

$$a < c.$$

On the other hand, if

$$\mathsf{R} \equiv \text{'is the square of'},$$

then the two relations

$$a = b^2, \quad b = c^2$$

do *not* lead to

$$a = c^2.$$

Examples

Determine which of the following relations are transitive:
1. $S \equiv$ the set of triangles,
R $\equiv$ is similar to;
2. $S \equiv$ the set of integers (with zero),
R $\equiv$ differs by a multiple of 3 from;
3. $S \equiv$ the set of integers,
R $\equiv$ is prime to;

4. $S \equiv$ the set of integers,
R $\equiv$ is a factor of;

5. $S \equiv$ the set of circles,
R $\equiv$ lies wholly outside;

6. $S \equiv$ the set of circles,
R $\equiv$ is concentric with.

4. Equivalence relations. A relation R which is reflexive, symmetrical and transitive is called an *equivalence relation.*

Immediate examples of equivalence relations are:

(i) $S \equiv$ the set of integers,
R $\equiv$ is equal to;

(ii) $S \equiv$ the set of triangles,
R $\equiv$ is similar to.

Examples

Determine which of the following are equivalence relations:

1. $S \equiv$ the set of integers,
R $\equiv$ is less than;

2. $S \equiv$ the set of integers (with zero),
R $\equiv$ differs by a multiple of 3 from;

3. $S \equiv$ the set of spheres,
R $\equiv$ is wholly outside;

4. $S \equiv$ the set of complex numbers,
R $\equiv$ has the same modulus as;

5. $S \equiv$ the set of square matrices,
R $\equiv$ has the same determinant as;

6. $S \equiv$ the set of integers,
R $\equiv$ is the reciprocal of;

7. $S \equiv$ the set of straight lines,
R $\equiv$ is perpendicular to;

8. $S \equiv$ the set of straight lines,
R $\equiv$ is parallel to or coincident with.

5. Equivalence classes. Consider two typical equivalence relations, say congruence among the set of triangles or equality of modulus among the complex numbers. Each of these relations divides the set into a number of *classes*: when one triangle is given, then there is automatically determined a whole class of triangles similar to it; when one complex number is given, then there is automatically determined a whole class of complex numbers having equal modulus. In so far as the relation R is concerned, one element suffices to fix the whole class.

This statement must now be made more precise:

THEOREM. *An equivalence relation* R *divides the elements of* S *into a number of distinct classes.*

Let a be any given element of S and p a 'variable' element. The relation

$$a \mathsf{R} p$$

is either true or false; if true, then p is said to *belong to a class* S_a whereas, if false, p is said *not to belong to* S_a. The immediate concern is to prove that *the totality of elements of* S_a *is determined whenever any one of its members* (not necessarily a) *is known.*

Suppose that the relation

$$a \mathsf{R} p$$

is true, so that $p \in S_a$. Take another element $q \in S_a$, for which also

$$a \mathsf{R} q.$$

Then $$a \mathsf{R} p \Rightarrow p \mathsf{R} a \quad \text{(symmetric)}$$

and $$p \mathsf{R} a \ \ \textbf{and} \ \ a \mathsf{R} q$$
$$\Rightarrow p \mathsf{R} q \quad \text{(transitive)}$$

so that the relation

$$p \mathsf{R} q$$

is also true. That is, q belongs to the class S_p determined by p. The result is that every element (such as q) belonging to S_a belongs equally to S_p, where p is any arbitrary member of S_a. Identical reasoning shows that every element of S_p belongs to S_a. Hence S_p is the same as S_a.

On the other hand, the various classes are distinct in the sense that

$$a \notin S_p$$
$$\Rightarrow \textit{no element of } S_a \in S_p.$$

Suppose that, on the contrary,

$$k \in S_a \quad \text{and} \quad k \in S_p$$

but $$a \notin S_p.$$

Then $$a \mathsf{R} k \quad \text{and} \quad p \mathsf{R} k,$$

so that $$p \mathsf{R} k \ \ \textbf{and} \ \ k \mathsf{R} a \quad (a \mathsf{R} k \text{ is symmetric})$$
$$\Rightarrow p \mathsf{R} a \quad \text{(transitivity)}$$
$$\Rightarrow a \in S_p,$$

in contradiction of the hypothesis. Hence no element of S_a can belong to S_p.

The equivalence classes are therefore distinct.

Illustration **1.** (i) Let S be the set of triangles and let the relation

$$a \mathrm{R} b$$

denote that a is similar to b. Then the equivalence classes consist of all triangles of given shape.

(ii) Let S be the set of integers and let the relation

$$a \mathrm{R} b$$

denote that the difference between a and b is a multiple of 3. Then the equivalence classes consist of the three subsets

$$\{\ldots, -4, -1, 2, 5, 8, \ldots\},$$
$$\{\ldots, -3, 0, 3, 6, 9, \ldots\},$$
$$\{\ldots, -2, 1, 4, 7, 10, \ldots\}.$$

(iii) Let S denote the set of straight lines in a plane and let the relation

$$a \mathrm{R} b$$

denote that a is parallel to, or coincident with, b. Then the equivalence classes consist of all the straight lines of given direction.

Examples

1. Let S denote the set of ordered pairs

$$a \equiv (a_1, a_2)$$

whose elements are positive integers, and let the relation

$$a \mathrm{R} b$$

mean that

$$a_1 b_2 = a_2 b_1.$$

Prove that R is an equivalence relation, and that the equivalence classes consist of pairs whose ratios have constant values. (Essentially, the equivalence classes represent the rational numbers in the sense that the equivalent pairs

$$(2, 3), (4, 6), (10, 15), (14, 21)$$

represent the rational number 2/3.)

2. Let O be a given point in a plane. Choose for S the set of points $\{A, B, C, \ldots\}$ in the plane, and let the relation

$$A \mathrm{R} B$$

denote that

$$AO = BO.$$

Prove that R is an equivalence relation and identify the equivalence classes.

3. Let P, Q be two given points in a plane. Choose for S the set of points $\{A, B, C, \ldots\}$ in the plane, and let the relation

$$A \mathsf{R} B$$

denote that

$$PA+AQ = PB+BQ.$$

Prove that R is an equivalence relation and identify the equivalence classes.

4. Let S denote the set of line segments of length l, and let the relation

$$a \mathsf{R} b$$

mean that the two segments are collinear. Prove that R is an equivalence relation and identify the equivalence classes.

5. Two matrices **a**, **b**, each of type $m \times n$, are said to be *equivalent* if non-singular square matrices **p**, **q** of orders m, n can be found such that

$$\mathbf{a} = \mathbf{pbq}.$$

Prove that this relation is an equivalence relation between $m \times n$ matrices.

Prove that the equivalence classes consist of matrices having the same rank.

6. Two square matrices **a**, **b** of order n are said to be *congruent* if a non-singular matrix **p** exists such that

$$\mathbf{a} = \mathbf{p}'\mathbf{bp}.$$

Prove that the relation is an equivalence relation between $n \times n$ matrices.

6. Homomorphism. Let S denote the set of ordered triplets

$$(x_1, x_2, x_3)$$

whose elements (named by *roman* letters) are real numbers; let Σ denote the set of ordered pairs

$$(\alpha_1, \alpha_2)$$

whose elements (named by *greek* letters) are also real numbers. The relation

$$x_1+x_2+x_3 = \alpha_1,$$
$$x_1+2x_2+3x_3 = \alpha_2$$

gives *a mapping of the elements of S onto the elements of* Σ, the word

onto

being reserved for a mapping in which each element of Σ appears at least once as (x_1, x_2, x_3) traverses S completely. In matrix notation, the mapping is

$$\mathbf{ax} = \boldsymbol{\alpha},$$

where

$$\mathbf{a} \equiv \begin{pmatrix} 1 & 1 & 1 \\ 1 & 2 & 3 \end{pmatrix}.$$

The important feature of the mapping, to which attention is now directed, is that, if $\mathbf{x}_1$, $\mathbf{x}_2$ are two ordered triplets (column matrices) of S giving rise to ordered pairs (column matrices) $\boldsymbol{\alpha}_1$, $\boldsymbol{\alpha}_2$ of Σ, so that

$$\mathbf{ax}_1 = \boldsymbol{\alpha}_1,$$
$$\mathbf{ax}_2 = \boldsymbol{\alpha}_2,$$

then any element in S of the form $k_1\mathbf{x}_1 + k_2\mathbf{x}_2$ is mapped onto the element $k_1\boldsymbol{\alpha}_1 + k_2\boldsymbol{\alpha}_2$ obtained, in analogous way, from the *same* scalars k_1, k_2 combining $\boldsymbol{\alpha}_1$, $\boldsymbol{\alpha}_2$. In other words, *the mapping*

$$\mathbf{ax} = \boldsymbol{\alpha}$$

preserves the structure of linear combinations in the sense that any linear combination

$$k_1\mathbf{x}_1 + k_2\mathbf{x}_2 + \ldots$$

is mapped upon the analogous linear combination

$$k_1\boldsymbol{\alpha}_1 + k_2\boldsymbol{\alpha}_2 + \ldots.$$

DEFINITION. *A mapping T of any set of elements*

$$S\{a, b, c, \ldots\}$$

onto a set

$$\Sigma\{\alpha, \beta, \gamma, \ldots\},$$

such that, say,

$$Ta = \alpha, \quad Tb = \beta$$

is called a homomorphism if a structure relation

$$c = a * b$$

in S is mapped into the corresponding structure relation

$$\gamma = \alpha \circ \beta$$

in Σ in such a way that

$$Tc = \gamma.$$

Thus $$T(a*b) = T(a) \circ T(b).$$

At the risk of multiplying technical terms, it may be remarked that a homomorphism between elements of the *same* set is called an *endomorphism.*

7. Isomorphism. A particularly important example of a homomorphism occurs when the elements of a set S are mapped onto the elements of a set Σ in such a way that each element of Σ arises from *one, and only one*, element of S. The homomorphism is then called an *isomorphism.*

For example, if the homomorphic mapping of §6 is 'truncated' to

$$x_1+x_2 = \alpha_1,$$
$$x_1+2x_2 = \alpha_2,$$

then α_1 and α_2 arise uniquely from the particular values

$$2\alpha_1-\alpha_2 = x_1,$$
$$-\alpha_1+\alpha_2 = x_2.$$

In the language of matrices, the mapping

$$\mathbf{bx} = \boldsymbol{\alpha}$$

of $\mathbf{x}$ onto $\boldsymbol{\alpha}$ is equally the mapping

$$\mathbf{b}^{-1}\boldsymbol{\alpha} = \mathbf{x}$$

of $\boldsymbol{\alpha}$ onto $\mathbf{x}$.

A mapping which is unique both ways is also said to be *one-one.*

Here, again, a special word, namely *automorphism* may be used when the isomorphism is between the elements of the same set.

***Illustration* 2.** Let S denote the set

$$S\{1, 2, 3, \ldots\}$$

of positive integers under the rule of elementary addition, and let S' denote the set

$$S'\{2, 2^2, 2^3, \ldots\}$$

under the rule of elementary multiplication.

The transformation

$$Ta = 2^a$$

gives a mapping between the two sets which is one-one. Further, it is an isomorphism, since

$$Ta = 2^a, \quad Tb = 2^b$$

and

$$T(a+b) = 2^{a+b}$$
$$= 2^a \times 2^b = (Ta) \times (Tb).$$

Thus when a, b are combined according to the rule (addition) in S, their maps 2^a, 2^b combine by the rule (multiplication) in S' which associates $a+b$ with $2^a \times 2^b$, and this is the basic property of an isomorphism.

Examples

1. Prove that, if the elements of a group are subjected to a homomorphic mapping, the image is also a group.

Prove also that the neutral element maps onto the neutral element and that inverses map onto inverses.

2. Prove that, if S is the set of all the integers, then the mapping T such that, for all $x \in S$,

$$Tx = 3x$$

is an endomorphism, the rule of combination in S being addition.

3. Prove that the following mappings of the set of points of a given plane into itself are automorphic:

(i) reflexions about a line,

(ii) reflexions about a point.

4. Suggest a definition for the *inverse* of an automorphism, and prove that it is also an automorphism.

5. Prove that the mapping upon itself of the positive rational numbers by means of the relation

$$y = 1/x$$

is *not* an automorphism.

6. Establish an isomorphism between the points of the x-axis in a plane π and the lines through the origin O' in a plane π'.

7. Establish an isomorphism between the circles of a coaxal system and the points of a straight line.

8. Three sets of elements S_1, S_2, S_3 are given. An isomorphic mapping P transforms the elements of S_1 into elements of S_2 and an isomorphic mapping Q transforms the elements of S_2 into the elements of S_3. Prove that the mapping of S_1 into S_3 denoted symbolically by QPS_1 is also an isomorphism.

9. Prove that an isomorphism can be set up between two groups which have the same structure table.

10. Establish an isomorphism between the positive integers and the negative integers.

11. Prove that an Abelian group cannot be isomorphic with a non-Abelian group.

12. Let R denote the set of all real numbers and S the set of ordered number-pairs of type (a_1, a_2) whose elements are real numbers. Prove that the mapping

$$(a_1, a_2) \to a_1$$

is a homomorphism but not an isomorphism.

Prove that, however, the mapping

$$(a_1, 0) \to a_1$$

is an isomorphism.

[The rule of combination for the number-pairs is to be

$$(a_1, a_2)+(b_1, b_2) = (a_1+b_1, a_2+b_2).]$$

13. Prove that, if $\{S\}$ is a group with a law of combination $*$, and if a is any fixed element of $\{S\}$, then the mapping A such that, for all $x \in \{S\}$,

$$Ax = a * x * a^{-1}$$

is an automorphism of $\{S\}$.

8. The isomorphism of two Euclidean spaces. Let R, R^* be two Euclidean spaces (p. 53) which, for convenience of statement, we take to be each of dimension 3. It is to be proved that *a one-one mapping T can be set up between these two spaces having the isomorphic properties that, if* $\mathbf{x}$, $\mathbf{y}$ *in R are mapped onto* $\mathbf{x}^*$, $\mathbf{y}^*$ *in* R^*, so that

$$T\mathbf{x} = \mathbf{x}^*,$$

$$T\mathbf{y} = \mathbf{y}^*,$$

then

(i) $T(\mathbf{x}+\mathbf{y}) = \mathbf{x}^*+\mathbf{y}^*$,

(ii) $T(k\mathbf{x}) = k\mathbf{x}^*$;

further, the mapping may be chosen so that

(iii) *the lengths* (*p.* 55)

$$x \equiv \sqrt{(x_1^2+x_2^2+x_3^2)}, \quad x^* \equiv \sqrt{(x_1^{*2}+x_2^{*2}+x_3^{*2})}$$

are equal.†

Suppose that the vectors of R are expressed in terms of the basis (compare p. 195) in R given by

$$\mathbf{e}_1 \equiv (1, 0, 0), \quad \mathbf{e}_2 \equiv (0, 1, 0), \quad \mathbf{e}_3 \equiv (0, 0, 1)$$

so that

$$\mathbf{x} \equiv x_1\mathbf{e}_1+x_2\mathbf{e}_2+x_3\mathbf{e}_3$$

and that the vectors of R^* are expressed in terms of the base in R^* given by

$$\mathbf{e}_1^* \equiv (1, 0, 0), \quad \mathbf{e}_2^* \equiv (0, 1, 0), \quad \mathbf{e}_3^* \equiv (0, 0, 1),$$

the stars implying selection within R^*; then

$$\mathbf{x}^* \equiv x_1^*\mathbf{e}_1^*+x_2^*\mathbf{e}_2^*+x_3^*\mathbf{e}_3^*.$$

Choose the particular one-one mapping

$$x_1^* = x_1, \quad x_2^* = x_2, \quad x_3^* = x_3,$$

or

$$\mathbf{x}^* = \mathbf{x}.$$

† Compare also Example 9 on p. 289

Then
$$T(\mathbf{x}+\mathbf{y}) = T\{x_1+y_1, x_2+y_2, x_3+y_3\}$$
$$= \{x_1^*+y_1^*, x_2^*+y_2^*, x_3^*+y_3^*\}$$
$$= \mathbf{x}^*+\mathbf{y}^*$$
and
$$T(k\mathbf{x}) = T\{kx_1, kx_2, kx_3\}$$
$$= \{kx_1^*, kx_2^*, kx_3^*\}$$
$$= k\mathbf{x}^*.$$
Also
$$x = \sqrt{(x_1^2+x_2^2+x_3^2)}$$
$$= \sqrt{(x_1^{*2}+x_2^{*2}+x_3^{*2})}$$
$$= x^*,$$
so that the lengths are equal. The conditions are thus all satisfied.

Example

1. The elements of a Euclidean space R are expressed in the form
$$x_1\mathbf{a}_1+x_2\mathbf{a}_2+x_3\mathbf{a}_3,$$
where $\mathbf{a}_1$, $\mathbf{a}_2$, $\mathbf{a}_3$ are the column vectors of an orthogonal matrix $\mathbf{a}$; the elements of a Euclidean space R^* are expressed in the form
$$x_1^*\mathbf{b}_1+x_2^*\mathbf{b}_2+x_3^*\mathbf{b}_3,$$
where $\mathbf{b}_1$, $\mathbf{b}_2$, $\mathbf{b}_3$ are the column vectors of an orthogonal matrix $\mathbf{b}$. Prove that the more general mapping
$$\mathbf{ax} = \mathbf{bx}^*,$$
where
$$\mathbf{x} \equiv \{x_1, x_2, x_3\}, \quad \mathbf{x}^* \equiv \{x_1^*, x_2^*, x_3^*\}$$
has the three isomorphic properties just established.

[Note, in particular, that
$$\mathbf{x}'\mathbf{x} \equiv (\mathbf{a}'\mathbf{bx}^*)'(\mathbf{a}'\mathbf{bx}^*)$$
$$\equiv \mathbf{x}^{*\prime}\mathbf{b}'\mathbf{a}\mathbf{a}'\mathbf{bx}^*$$
$$\equiv \mathbf{x}^{*\prime}\mathbf{x}^*.]$$

9. Isometries. In metrical geometry, the mapping of one space onto another will usually destroy equality of length between corresponding elements, though, in fact, many of the mappings which seem to occur most often in practice do succeed in conserving it. This idea must be made more precise.

DEFINITION. *A one-one mapping T between metrical spaces is called isometric if any two vectors* $\mathbf{x}$, $\mathbf{y}$ *of equal length map into two vectors* $T\mathbf{x}$, $T\mathbf{y}$ *also of equal length.*

Examples

1. Prove that the following mappings are isometries (pp. 129–131):
(i) rotation,
(ii) reflexion in a line,
(iii) reflexion in a point.

2. Prove that a shear (p. 132) is not an isometry.

3. Prove that an orthogonal mapping

$$\mathbf{x} = \mathbf{a}\mathbf{x}^*$$

is an isometry.

4. Prove that translation (a mapping $\mathbf{x} \to \mathbf{x} + \mathbf{a}$, where $\mathbf{a}$ is a constant vector) is an isometry.

5. Prove that the geometrical transformation known as *inversion* is not an isometry.

6. Prove that there is a unique isometry in a plane mapping a (non-isosceles) triangle ABC into a congruent triangle $A'B'C'$.

7. Prove that there are two isometries in a plane mapping a line segment PQ into an equal line segment UV.

8. Prove that if an isometry in a plane maps a point O into itself, then the isometry is either a rotation or a reflexion.

9. Prove that the isomorphism detailed in §8 is in fact an *isometry* between the two Euclidean spaces.

20

SOME MATRIX GROUPS

1. Additive group of matrices. Let S denote the set of $m \times n$ matrices

$$S \equiv \{\mathbf{a}\}$$

in which the elements a_{ij} are, say, complex numbers.

PROPERTY. *The set S is a group under the ordinary rule of matrix addition*:

The proof is immediate. The neutral element is the $m \times n$ zero $\mathbf{0}$ and the inverse of an element $\mathbf{a}$ is its negative $-\mathbf{a}$.

Example

1. Prove that, if the elements are real numbers, the set of matrices of the form

$$\begin{pmatrix} a & b \\ -b & a \end{pmatrix}$$

forms a group under addition as the law of combination, and that this group is isomorphic with the additive group of complex numbers.

2. Multiplicative group of matrices. Let S denote the set of matrices

$$S \equiv \{\mathbf{a}\}$$

whose elements a_{ij} are, say, complex numbers.

We examine whether these matrices can form a group under multiplication.

Note first that, if $\mathbf{a}$, $\mathbf{b}$ are any two matrices whatever of such a group, then *each* of the products

$$\mathbf{ab},\ \mathbf{ba}$$

must exist, and so *all the matrices of the group must be square.* Denote the order by n.

The treatment varies according to the existence or otherwise of singular matrices in the group.

1. Group of singular matrices. Suppose that the set S contains one singular matrix $\mathbf{a}$. Let $\mathbf{u}$ be any other element whatever. Then, for a group, a matrix $\mathbf{v} \in S$ can be found such that

$$\mathbf{u} = \mathbf{av},$$

and so

$$\begin{aligned}\det \mathbf{u} &= \det \mathbf{a} \det \mathbf{v} \\ &= 0.\end{aligned}$$

Hence *a multiplicative group of square matrices of order n which has one singular matrix has all its matrices singular.*

An elementary example is the set

$$\begin{pmatrix} 1 & 0 \\ 0 & 0 \end{pmatrix}, \begin{pmatrix} -1 & 0 \\ 0 & 0 \end{pmatrix}, \begin{pmatrix} i & 0 \\ 0 & 0 \end{pmatrix}, \begin{pmatrix} -i & 0 \\ 0 & 0 \end{pmatrix}$$

where the unit is the matrix

$$\begin{pmatrix} 1 & 0 \\ 0 & 0 \end{pmatrix}.$$

This emphasizes the fact that, *for a group of singular matrices, the neutral element cannot be the unit matrix.*

2. Group of non-singular matrices. Let S be a set of non-singular matrices $\{\mathbf{a}\}$ of order n, forming a group under multiplication.

Suppose that the neutral element of the group is $\mathbf{k}$. Then, by definition,

$$\mathbf{ak} = \mathbf{ka} = \mathbf{a}.$$

But $\mathbf{a}$ is non-singular, so that $\mathbf{a}^{-1}$ exists. Hence

$$\mathbf{a}^{-1}(\mathbf{ak}) = \mathbf{a}^{-1}\mathbf{a},$$

or

$$\mathbf{k} = \mathbf{I}.$$

Hence *in a group of non-singular matrices the neutral element is the unit matrix* $\mathbf{I}$.

It follows that *the group inverse of any element* $\mathbf{a} \in S$ *is the matrix inverse* $\mathbf{a}^{-1}$.

Examples

1. A set S consists of the matrices of the form

$$\begin{pmatrix} a & b \\ -b & a \end{pmatrix},$$

where a, b are real. Prove that, if a, b are not both zero, the set forms a group under multiplication as the law of combination, and that this group is isomorphic with the multiplicative group of non-zero complex numbers.

2. Prove that the set whose elements are the matrices

$$\begin{pmatrix} a & \bar{b} \\ -b & \bar{a} \end{pmatrix}$$

forms a group under multiplication, where a, b are complex numbers, not both zero, and $\bar{a}, \bar{b}$ are their conjugate complexes.

3. Prove that the set of matrices of rotation

$$\begin{pmatrix} \cos\theta & \sin\theta \\ -\sin\theta & \cos\theta \end{pmatrix}$$

forms a group.

4. Prove that the matrices

$$\begin{pmatrix} a & b & c & d \\ -b & a & -d & c \\ -c & d & a & -b \\ -d & -c & b & a \end{pmatrix},$$

where a, b, c, d are real numbers, not all zero, form a group.

5. Prove that the matrices of the set

$$\begin{pmatrix} a & b \\ 0 & c \end{pmatrix} \quad (ac \neq 0)$$

form a group under multiplication, and that this group is isomorphic with the set of elements $\{a\}$ in a mapping in which a matrix gives rise to the element in its top left-hand corner.

6. Show geometrically, or otherwise, that, if the matrices

$$\begin{pmatrix} 1 & 0 \\ 0 & 1 \end{pmatrix}, \begin{pmatrix} -1 & 0 \\ 0 & -1 \end{pmatrix}, \begin{pmatrix} 0 & 1 \\ -1 & 0 \end{pmatrix}, \begin{pmatrix} 0 & -1 \\ 1 & 0 \end{pmatrix}$$

are multiplied with one another and with

$$\begin{pmatrix} 1 & 0 \\ 0 & -1 \end{pmatrix},$$

they will generate a group of 8 elements, and write down the other three.

7. Show that the two matrices

$$\begin{pmatrix} \cos\theta & \sin\theta \\ \sin\theta & -\cos\theta \end{pmatrix}, \begin{pmatrix} -\cos\theta & -\sin\theta \\ -\sin\theta & \cos\theta \end{pmatrix}$$

are two of the elements of a group of order 4 in which the rule for combining elements is the ordinary rule for the multiplication of matrices. Give the other two elements.

Is this group isomorphic with the group of cyclic permutations the elements of which change the letters *abcd* into *bcda, cdab, dabc, abcd*?

8. Show that the set of matrices

$$\mathbf{a}(\phi) \equiv \begin{pmatrix} \cos\phi & \sin\phi \\ -\sin\phi & \cos\phi \end{pmatrix} \qquad (0 \leqslant \phi < 2\pi)$$

forms a group G, the combination of two elements being defined by the usual rule for the multiplication of matrices.

Give an example of a subgroup of G.

Explain what is meant by the statement that G is isomorphic to the group of complex numbers $e^{i\phi}$ for which the law of combination is multiplication.

9. Prove that the square matrices of order n form a ring when subjected to the usual rules of addition and multiplication.

Show that, in general, the conditions for a field are not satisfied. Do the non-singular 2×2 matrices form a field?

3. The orthogonality group. Let S be the set of orthogonal matrices $\{\mathbf{a}\}$ of order n having the property that

$$\mathbf{a}'\mathbf{a} = \mathbf{a}\mathbf{a}' = \mathbf{I}.$$

THEOREM. *These matrices form a multiplicative group.*

Let $\mathbf{a} \in S$, $\mathbf{b} \in S$ be any two orthogonal matrices. It is to be proved first that their products $\mathbf{ab}$, $\mathbf{ba}$ are in S—say $\mathbf{ab} \in S$.

Now
$$\begin{aligned}(\mathbf{ab})'\mathbf{ab} &= (\mathbf{b}'\mathbf{a}')(\mathbf{ab}) \\ &= \mathbf{b}'(\mathbf{a}'\mathbf{a})\mathbf{b} \\ &= \mathbf{b}'\mathbf{b} \qquad (\mathbf{a}'\mathbf{a} = \mathbf{I}) \\ &= \mathbf{I} \qquad (\mathbf{b}'\mathbf{b} = \mathbf{I}).\end{aligned}$$

Hence
$$\mathbf{ab} \in S.$$

Further the unit matrix is orthogonal, so that

$$\mathbf{I} \in S.$$

Finally, the inverse $\mathbf{a}^{-1}$ of any matrix $\mathbf{a} \in S$ is orthogonal since

$$\begin{aligned}(\mathbf{a}^{-1})'\mathbf{a}^{-1} &= (\mathbf{a}')^{-1}\mathbf{a}^{-1} = (\mathbf{a}^{-1})^{-1}\mathbf{a}^{-1} \\ &= \mathbf{a}\mathbf{a}^{-1} \\ &= \mathbf{I}.\end{aligned}$$

Hence
$$\mathbf{a} \in S$$
$$\Rightarrow \mathbf{a}^{-1} \in S.$$

The conditions for a group are thus satisfied.

4. A group of similar matrices. To show how this group arises, let **a** be a given non-singular square matrix of order n. A one-one correspondence is induced among the set of column vectors of order n by the relation

$$\mathbf{y} = \mathbf{ax}$$

with its inverse

$$\mathbf{x} = \mathbf{a}^{-1}\mathbf{y}.$$

Suppose now that the space of column vectors of order n is subjected to a mapping

$$\mathbf{x} = \mathbf{px}^*,$$

where **p** is a given non-singular $n \times n$ matrix. Under it, the mapping for **y** is

$$\mathbf{y} = \mathbf{py}^*,$$

and so the one-one correspondence (given in terms of **x**, **y** in the form $\mathbf{y} = \mathbf{ax}$) becomes, in terms of $\mathbf{x}^*$, $\mathbf{y}^*$,

$$\mathbf{py}^* = \mathbf{apx}^*,$$

or
$$\mathbf{y}^* = \mathbf{p}^{-1}\mathbf{apx}^*,$$

or
$$\mathbf{y}^* = \mathbf{a}^*\mathbf{x}^*,$$

where $\mathbf{a}^*$ is the matrix

$$\mathbf{a}^* = \mathbf{p}^{-1}\mathbf{ap}.$$

DEFINITION. If **a** is a given matrix and if **p** is any non-singular matrix, both of order n, then the matrix $\mathbf{p}^{-1}\mathbf{ap}$ is said to be *similar* to **a**.

The correspondences induced by **a** and $\mathbf{p}^{-1}\mathbf{ap}$ are essentially the same.

Let T be the set of non-singular matrices $\{\mathbf{a}\}$ defining a set of one-one correspondences, and let

$$S\{\mathbf{p}^{-1}\mathbf{ap}\}$$

be the set of matrices into which they are transformed under a *given* matrix **p**.

THEOREM. *The set of matrices*

$$\{\mathbf{p}^{-1}\mathbf{ap}\}$$

forms a multiplicative group.

Let $$\mathbf{p}^{-1}\mathbf{a}\mathbf{p} \in S, \quad \mathbf{p}^{-1}\mathbf{b}\mathbf{p} \in S$$

be two matrices of the system. Their product (say in that order) is

$$\begin{aligned}&(\mathbf{p}^{-1}\mathbf{a}\mathbf{p})(\mathbf{p}^{-1}\mathbf{b}\mathbf{p})\\ &= \mathbf{p}^{-1}\mathbf{a}(\mathbf{p}\mathbf{p}^{-1})\mathbf{b}\mathbf{p}\\ &= \mathbf{p}^{-1}(\mathbf{a}\mathbf{b})\mathbf{p}\\ &\in S,\end{aligned}$$

so that the products are in S.

Further the unit matrix is in S, since it can be expressed in the form

$$\mathbf{p}^{-1}\mathbf{I}\mathbf{p}.$$

Finally, the inverse of $\mathbf{p}^{-1}\mathbf{a}\mathbf{p}$ is also in S, for it is the matrix

$$\mathbf{p}^{-1}\mathbf{a}^{-1}\mathbf{p}$$

since
$$\begin{aligned}&(\mathbf{p}^{-1}\mathbf{a}^{-1}\mathbf{p})(\mathbf{p}^{-1}\mathbf{a}\mathbf{p})\\ &= \mathbf{p}^{-1}\mathbf{a}^{-1}(\mathbf{p}\mathbf{p}^{-1})\mathbf{a}\mathbf{p}\\ &= \mathbf{p}^{-1}(\mathbf{a}^{-1}\mathbf{a})\mathbf{p}\\ &= \mathbf{p}^{-1}\mathbf{p}\\ &= \mathbf{I}.\end{aligned}$$

The conditions for a group are thus satisfied.

Examples

1. Prove that the relation

a is similar to **b**

among $n \times n$ matrices is an equivalence relation.

2. Prove that, if

$$G\{\mathbf{a}\}$$

is a group of matrices of order n, with multiplication as the law of combination, then the set

$$\{\mathbf{p}^{-1}\mathbf{a}\mathbf{p}\},$$

for given non-singular $\mathbf{p}$, forms a group isomorphic with G.

3. Prove that similar matrices have

(i) equal determinants,

(ii) the same characteristic equation,

(iii) equal eigenvalues but not necessarily equal eigenvectors.

4. Prove that the set $S\{\mathbf{a}\}$ of all matrices commuting with a given matrix $\mathbf{k}$, so that

$$\mathbf{ak} = \mathbf{ka}$$

for all $\mathbf{a} \in S$, is a group under multiplication.

5. Prove that the set of matrices of unit determinant forms a group under multiplication.

Does the set of matrices of determinant -1 form a group under multiplication?

6. Show that the set of all real 2×2 matrices is a non-commutative ring M with unit element, and that those matrices in M which commute with a fixed matrix in M form a subring.

Prove that the only matrices in M which commute with *every* matrix in M are the matrices $\alpha\mathbf{I}$.

7. The matrices

$$R_n = \begin{pmatrix} \cos\frac{1}{6}n\pi & \sin\frac{1}{6}n\pi \\ -\sin\frac{1}{6}n\pi & \cos\frac{1}{6}n\pi \end{pmatrix},$$

for $n = 0, 1, \ldots, 11$, and the matrix

$$S_0 = \begin{pmatrix} 1 & 0 \\ 0 & -1 \end{pmatrix}$$

form part of a group of 24 elements, in which the rule of combination of elements is the usual rule for matrix multiplication. Find the remaining elements, and prove that it *is* a group.

Give at least one example of a subgroup.

SELECTED ANSWERS

Some of the examples are set merely to give practice in elementary manipulation and it is unlikely that the reader will be in doubt, in such cases, that his answers are correct. For this reason answers requiring much space for little apparent gain are not given. The sign — denotes such omissions. (The very first case, Chapter I, §1, indicates what is intended.)

CHAPTER 1

§1. Examples (pages 4, 5) — .

§3. Text examples, *second set* (page 9):

1. 9. **2.** 3, 7.

Revision examples (page 15):

1. (i) associative, commutative; (ii) all; (iii) associative, commutative; (iv) all; (v) all.

5. $M \cup N = M$; $M \cap N = N$.

6. P; P; $\{2,4,8,16\}$; $\{2,4,8,16\}$; $\{1,2,4,6,8,10,12,14,16\}$; $\{1,2,4,6,8,10,12,14,16\}$; M; P.

7. squares.

9. N, R, C; R, C; C. [$x = 1, x = 2$ are real *complex* numbers.]

11. 10.

CHAPTER 2

§5. Text Examples (page 24):

4. $\{\ldots, -2, -1, 0, 1, 2, \ldots\}$ under multiplication.

6. (i) (*a*) Neutral element, 0; inverses, $\{0,9,8,7,6,5,4,3,2,1\}$.
(*b*) Neutral element, 1; inverses, $\{-,1,-,7,-,-,-,3,-,9\}$.
(ii) Neutral element, $0.x^2+0.x+0$.
(iii) Neutral element, through $4k$ right angles.
(iv) Neutral element, **0**.

CHAPTER 3

§1. Text example (page 27):

1. Neutral element, 1; inverses of $1, i, -1, -i$ are $1, -i, -1, i$.

Revision examples (page 39):

3. (i) (*a*) yes, (*b*) no.
(ii) (*a*) no, (*b*) yes.
(iii) (*a*) yes, (*b*) no.
(iv) (*a*) no, (*b*) no.

17. $T^2 x = x \Rightarrow T$ is identity or $a = d$.
$T^3 x = x \Rightarrow T$ is identity or (various forms being possible)

$$\begin{cases}(bc-ad)(d-a) = 1,\\ bc = a^2-ad+d^2.\end{cases}$$

It is also possible for $b = c = 0$, $a = \omega$ or ω^2 with $d = -\omega$ or $-\omega^2$.

Subgroups (i) $(1,0,0,-1)$, $(5,7,3,4)$, $(4,7,3,5)$;
(ii) $(1,0,0,-1)$, $(2,-1,3,2)$;
(iii) $(1,0,0,-1)$, $(11,19,6,11)$;
(iv) $(1,0,0,-1)$ $(13,18,9,13)$.

CHAPTER 4

Revision examples (page 51):

1. 13, 13, $9\sqrt{2}$.

5. $\pi-\cos^{-1}(\frac{1}{3})$, $\cos^{-1}(5/\sqrt{57})$, $\cos^{-1}(7/\sqrt{57})$, where the inverse cosines are acute.

9. Solutions of

$$\frac{l+2m+2n}{\pm 3} = \frac{2l+3m+6n}{\pm 7} = \frac{3m+4n}{\pm 5}.$$

For example (positive signs), (1,3,4).

10. $p = \dfrac{\sqrt{3}-1}{2}$, $q = \dfrac{\sqrt{3}+1}{2}$; $u = \dfrac{\sqrt{3}-1}{2\sqrt{3}}$, $v = \dfrac{\sqrt{3}+1}{2\sqrt{3}}$, $w = \dfrac{2}{\sqrt{3}}$.

12. $3\mathbf{i}-8\mathbf{j}+4\mathbf{k}$; $(1105)^{-\frac{1}{2}}(30\mathbf{i}+3\mathbf{j}-14\mathbf{k})$.

13. $\mathbf{x} = \alpha^{-1}\mathbf{c}-(\mathbf{c}.\mathbf{b})\alpha^{-2}\mathbf{a}$; all vectors perpendicular to $\mathbf{b}$.

14. $k\{(\mathbf{a}.\mathbf{b})\mathbf{a}-(\mathbf{a}^2)\mathbf{b}\}$; $k'(1,-1,0)$; $k''(1,1,1)$.

15. $\sqrt{[\mathbf{c}^2-\{(\mathbf{c}.\mathbf{a})^2/\mathbf{a}^2\}]}$; $\{(\mathbf{c}.\mathbf{a})-(\mathbf{b}.\mathbf{a})\}/\sqrt{(\mathbf{a}^2)}$.

§9. Examples (page 57):

1. $2, 2, \sqrt{14}, \sqrt{14}; 0, \dfrac{3}{\sqrt{14}}, \dfrac{1}{\sqrt{14}}, 0, \dfrac{-2}{\sqrt{14}}, 0$.

2. $1/\sqrt{14}$, $1/\sqrt{8}$, $1/13$.

3. $-44/7$.

4. 0, 5, -8.

§10. Examples (page 60):

2. 11, 11, 11; $(5,-1,-7)$, $(-1,-7,5)$, $(-7,5,-1)$; -18.

3. (3,6,9).

6. $\{\mathbf{b}+(\mathbf{a}.\mathbf{b})\mathbf{a}+\mathbf{a}\wedge\mathbf{b}\}/\{1+\mathbf{a}^2\}$.

9. Taking $\mathbf{n}$ as a unit vector,

$$\cos\theta = \frac{\mathbf{p}.\mathbf{q}-(\mathbf{p}.\mathbf{n})(\mathbf{q}.\mathbf{n})}{\sqrt{[\{\mathbf{p}^2-(\mathbf{p}.\mathbf{n})^2\}\{\mathbf{q}^2-(\mathbf{q}.\mathbf{n})^2\}}}.$$

15. For general values of **a**, **b**, **c**, k, the restriction $\mathbf{b}\cdot\mathbf{c} = 0$ is essential. Then

$$\lambda = k/(\mathbf{abc}), \quad \mu = -\{(\mathbf{abc}) - k\mathbf{b}^2\}/(\mathbf{abc})(\mathbf{ab}), \nu = 0.$$

16. (i) $-\dfrac{\mathbf{b}.\mathbf{c}}{k(k+\mathbf{a}.\mathbf{b})}\mathbf{a} + \dfrac{1}{k}\mathbf{c}$; (ii) no solution unless also $\mathbf{b}.\mathbf{c} = 0$, and then $x = \lambda\mathbf{a} + (\mathbf{a}.\mathbf{b})^{-1}\mathbf{c}$, λ arbitrary.

17. $\mathbf{u} = \mathbf{b}\wedge\mathbf{c}/(\mathbf{abc})$, etc.

18. $\mathbf{a}+\mathbf{b} = 6\mathbf{i} - 4\mathbf{j} - 3\mathbf{k}$;

$$\mathbf{a}.\mathbf{b} = -23;$$
$$\mathbf{a}\wedge\mathbf{b} = -29\mathbf{i} - 24\mathbf{j} - 26\mathbf{k};$$
$$\mathbf{c} = \lambda\mathbf{a} + \mathbf{b} \text{ for arbitrary } \lambda;$$
$$\mathbf{x} = 2\mathbf{a}.$$

CHAPTER 5

Examples (page 70):

4. (i) $(7\lambda, -6\lambda, \lambda)$; (ii) $(-\lambda + 5\mu, \lambda, -2\mu, -\mu)$.

CHAPTER 6

§3. Revision examples (page 80):

4. Inverses of 1, 2, 3, 4, 5, 6 are 1, 4, 5, 2, 3, 6. (i) 4, (ii) 2.
6. $(2+\sqrt{2})(2-\sqrt{2}) = 2$.

CHAPTER 7

§2. Examples (page 84):

6. Subgroups (i) $(1, -1)$, (ii) $(1, \omega, \omega^2)$.

CHAPTER 8

§3. Examples (page 96): —.

§5. Examples (page 98):

1. —.
2. —. [For example, for mod 7 the answers, in addition to $x = 1$, are: $x^2 \equiv 1 \Rightarrow x = 6$; $x^3 \equiv 1 \Rightarrow x = 2$ or 4; $x^4 \equiv 1 \Rightarrow x = 6$.]

§8. Text examples (page 100): —.

§9 Examples, ***first set*** (page 104):

1. 2,222; 10,102.
2. 102,021; 2,012,021.
10. 12; 111.
11. 100,111; 10,222,011,112; 1,000.
12. 2,121; 210 + remainder 110.
20. 2,101,000.

Examples, ***second set*** (page 104):

1. 101,000.
10. 10,111.

11. 100,011; 110,001.
100. 100 + remainder 101.
101. 11,011; 100.
110. −100,000.

Examples, *fourth set* (page 106):

3. 1, 1.

CHAPTER 9

§6. Examples (page 112):

1. $42 = 14\times 1092-11\times 1386$.
2. $7 = 47\times 1001-60\times 784$.
3. $6 = 4\times 1230-21\times 234$.

Equations:

1. $x = -3+13\lambda,\quad y = 4-17\lambda$.
2. $x = -8+25\lambda,\quad y = 17-53\lambda$.
3. $x = -7+30\lambda, y = 11-47\lambda$.

Revision examples (page 115):

4. $(n+1)!+2, (n+1)!+3, (n+1)!+4, \ldots, (n+1)!+(n+1)$.
8. $210\lambda+206$.
11. (i) $x = 1+3\lambda,\quad y = 1-2\lambda$;
(ii) $x = 2+3\lambda,\quad y = -1-5\lambda$;
(iii) $x = -2+5\lambda,\quad y = 5-6\lambda$;
(iv) $x = 5+7\lambda,\quad y = -5-8\lambda$.
16. 245.

CHAPTER 10

§3. Examples (page 119):

1. $1-x+4x^2-4x^3$; $1-3x+2x^2-4x^3$; $x-x^2+x^3+x^4-4x^5$.
2. $1-16x^4$; $1-2x^2+x^4$; $1+4x+10x^2+20x^3+25x^4+24x^5+16x^6$.

§5. Examples (page 121):

1. x^2+x+1; $2x+4$.
2. $x-2$; $-x+1$.
3. x^2+3; $-2x^2+x-3$.
4. x^3-x^2; 0.
5. 1; $-2x^3+6x^2-7x-1$.

§6. Examples (page 123):

1. x^2+4x+3.
2. x^2+x+1.
3. x^2+5x+4.
4. 1.
5. No.

Revision examples (page 124):

1. (i) $1 \equiv U - xV$.
   (ii) $x^2-4x+3 \equiv \mathrm{V}$.
   (iii) $x^2-4x+4 \equiv \mathrm{V}$.
   (iv) H.C.F. is 1, where
   $5 = -(2x^2-2x+5)\,U+(2x^4-2x^3+7x^2-6x+5)\,V$.
2. $x^2-1 = (x^6+1)(x^6-1)-x^2(x^{10}-1)$.
4. $4A(x) = -(x^2-2)$, $4B(x) = x^4+2x^3+3x^2+4x+2$.
7. $x-3$; $0, 2, 3, 3, 4, 5, 6$.
9. x^2+x+1.
10. $P = \dfrac{1}{21}(4x+5)$, $Q = -\dfrac{1}{21}(4x-11)$.
11. $2x+5$.
13. $3x-2 = \frac{1}{312}\{(x-11)\,G-(x^2-7x-47)\,F\}$.
14. (i) $x+1 = (x^3+x)\,F-(x^5+x^3-1)\,G$,
    (ii) $(x-1)^2 = \{F-(nx-n-1)\,G\}/n^2$.

CHAPTER 11

CHAPTER 12

§1. Examples (page 141):

1. —
2. (i) $\begin{pmatrix} 1 & \frac{1}{2} & \frac{1}{3} & \frac{1}{4} \\ 2 & \frac{2}{2} & \frac{2}{3} & \frac{2}{4} \\ 3 & \frac{3}{2} & \frac{3}{3} & \frac{3}{4} \end{pmatrix}$,

   (ii) $\begin{pmatrix} 3 & 4 & 5 & 6 \\ 5 & 6 & 7 & 8 \\ 7 & 8 & 9 & 10 \end{pmatrix}$,

   (iii) $\begin{pmatrix} 1^2 & 2^2 & 3^2 & 4^2 \\ 1^2 & 2^2 & 3^2 & 4^2 \\ 1^2 & 2^2 & 3^2 & 4^2 \end{pmatrix}$.

§2. Examples (page 143):

1. $\begin{pmatrix} -1 & 0 & 0 & 0 \\ 0 & -1 & 0 & 0 \\ 0 & 0 & -1 & 0 \\ 0 & 0 & 0 & -1 \end{pmatrix}$.

2. $\begin{pmatrix} 0 & 0 & 0 & 0 \\ 1 & 0 & 0 & 0 \\ 0 & 1 & 0 & 0 \\ 0 & 0 & 1 & 0 \end{pmatrix}$.

3. $$\begin{pmatrix} -1 & 1 & -1 & -1 \\ 1 & -1 & -1 & -1 \\ -1 & -1 & -1 & -1 \\ -1 & -1 & -1 & -1 \end{pmatrix}.$$

4. $$\begin{pmatrix} 1 & -1 & 1 & -1 \\ -1 & 1 & -1 & 1 \\ 1 & -1 & 1 & -1 \\ -1 & 1 & -1 & 1 \end{pmatrix}.$$

5. $$\begin{pmatrix} 1 & 1 & 1 & 1 \\ 2 & 2^2 & 2^3 & 2^4 \\ 3 & 3^2 & 3^3 & 3^4 \\ 4 & 4^2 & 4^3 & 4^4 \end{pmatrix}.$$

6. $$\begin{pmatrix} 0 & -1 & -2 & -3 \\ 1 & 0 & -1 & -2 \\ 2 & 1 & 0 & -1 \\ 3 & 2 & -1 & 0 \end{pmatrix}.$$

§3. Examples (page 144):

1. $$\mathbf{a}+\mathbf{d} = \begin{pmatrix} 4 & 2 & 0 \\ 4 & 3 & 6 \end{pmatrix}, \qquad \mathbf{a}+\mathbf{e} = \begin{pmatrix} 1 & 2 & 4 \\ 6 & 4 & 9 \end{pmatrix},$$

$$\mathbf{d}+\mathbf{e} = \begin{pmatrix} 3 & 0 & -2 \\ 2 & -3 & 3 \end{pmatrix}, \quad \mathbf{b}+\mathbf{c} = \begin{pmatrix} 2 & 0 \\ -2 & 0 \end{pmatrix},$$

$$\mathbf{b}+\mathbf{f} = \begin{pmatrix} 4 & 2 \\ 1 & 1 \end{pmatrix}, \qquad \mathbf{c}+\mathbf{f} = \begin{pmatrix} 0 & -2 \\ -3 & 1 \end{pmatrix}.$$

§4. Examples (page 145):

2. $$\begin{pmatrix} 32 & 39 \\ 46 & 53 \end{pmatrix}.$$ 3. $$\begin{pmatrix} 1 & 0 \\ 0 & 1 \end{pmatrix}.$$

§5. Examples (page 146): —.

§7. Examples (page 150):

1. (i) $$\begin{pmatrix} p+2u & q+2v & r+2w \\ 3p+4u & 3q+4v & 3r+4w \end{pmatrix};$$

(ii) $$\begin{pmatrix} 2 & -2 & 3 & 3 \\ 28 & 12 & 10 & 6 \end{pmatrix};$$

(iii) $$\begin{pmatrix} -11 \\ 10 \end{pmatrix}.$$

(iv) $\begin{pmatrix} \alpha & \beta & \gamma & \delta \\ 2\alpha & 2\beta & 2\gamma & 2\delta \\ 3\alpha & 3\beta & 3\gamma & 3\delta \\ 4\alpha & 4\beta & 4\gamma & 4\delta \end{pmatrix}$;

(v) $(\alpha+2\beta+3\gamma+4\delta)$;

(vi) $\begin{pmatrix} 7 & 2 \\ 11 & 16 \\ 28 & 8 \\ 29 & 35 \end{pmatrix}$.

§10. Examples (page 153):

1. (i) $\mathbf{abc} = \begin{pmatrix} 297 & -156 \\ 725 & -382 \end{pmatrix}$, $\mathbf{bcd} = \begin{pmatrix} 323 \\ 206 \end{pmatrix}$

$\mathbf{abcd} = \begin{pmatrix} 735 \\ 1793 \end{pmatrix}$;

(ii) $\mathbf{abc} = \begin{pmatrix} 48 & 66 \\ -80 & -110 \end{pmatrix}$, $\mathbf{bcd} = (142, 48, 104)$,

$\mathbf{abcd} = \begin{pmatrix} 426 & 144 & 312 \\ -710 & -240 & -520 \end{pmatrix}$.

2. (i) $ax^2+2hxy+by^2$;
(ii) $\alpha(ax+hy+gz)+\beta(hx+by+fz)+\gamma(gx+fy+cz)$;
(iii) 0.

Revision examples (page 155):

14. $\begin{pmatrix} u^3 & 3u^2 & 3u & 1 \\ u^2v & 2uv+u^2 & v+2u & 1 \\ uv^2 & 2uv+v^2 & u+2v & 1 \\ v^3 & 3v^2 & 3v & 1 \end{pmatrix}$.

CHAPTER 13

§3. Examples (page 160):

2. $\mathbf{I}_3$.

3. (i) $\begin{pmatrix} 0 & 0 & -1 \\ 0 & -1 & 0 \\ -1 & 0 & 0 \end{pmatrix}$, (ii) $\begin{pmatrix} 0 & -1 & 0 \\ -1 & 0 & 0 \\ 0 & 0 & -1 \end{pmatrix}$,

(iii) $\begin{pmatrix} 0 & 0 & 0 \\ 0 & 0 & 0 \\ 0 & 0 & 0 \end{pmatrix}$, (iv) $\begin{pmatrix} -1 & 2 & -1 \\ 2 & -4 & 2 \\ -1 & 2 & -1 \end{pmatrix}$.

§4. **Examples** (page 161):

1. and **2.** The determinants are -20, 49, -1, and the adjoints are

$$\begin{pmatrix} 5 & -5 & -5 \\ -2 & -2 & 10 \\ -7 & 3 & -5 \end{pmatrix}, \quad \begin{pmatrix} 14 & 1 & 11 \\ 7 & 4 & -5 \\ 0 & 14 & 7 \end{pmatrix}, \quad \begin{pmatrix} 5 & 2 & -1 \\ 2 & 1 & -1 \\ 4 & 2 & -1 \end{pmatrix}.$$

3. $\mathbf{aa}' = \begin{pmatrix} 5 & 0 & 1 \\ 0 & 5 & -7 \\ 1 & -7 & 10 \end{pmatrix}, \quad \mathbf{a}'\mathbf{a} = \begin{pmatrix} 14 & -3 \\ -3 & 6 \end{pmatrix}$; 0, 75.

§5. **Examples** (page 162):

1. $\frac{1}{16}\begin{pmatrix} 3 & -1 & 5 \\ 5 & -7 & 3 \\ -1 & 11 & -7 \end{pmatrix}$.

2. $\frac{1}{14}\begin{pmatrix} 5 & -1 & 3 \\ 3 & 5 & -1 \\ -1 & 3 & 5 \end{pmatrix}$.

3. $\begin{pmatrix} 1 & 0 & 0 \\ 0 & \cos\theta & -\sin\theta \\ 0 & \sin\theta & \cos\theta \end{pmatrix}$.

4. None.

5. $\Delta^{-1}\begin{pmatrix} A & H & G \\ H & B & F \\ G & F & C \end{pmatrix}$,

where $\Delta = abc+2fgh-af^2-bg^2-ch^2$; $A = bc-f^2$; $F = gh-af$; etc.

6. None.

7. $\begin{pmatrix} \frac{1}{5} & 0 & 0 \\ 0 & \frac{1}{5} & 0 \\ 0 & 0 & \frac{1}{5} \end{pmatrix}$.

8. $\begin{pmatrix} \frac{1}{5} & 0 & 0 \\ 0 & -\frac{1}{4} & 0 \\ 0 & 0 & \frac{1}{3} \end{pmatrix}$.

9. (i) -6; (ii) 2, 3; (iii) 0, -1, 5.

§8. **Examples** (page 166):

1. $\mathbf{a}^{-1}\mathbf{b}$, $\mathbf{d}^{-1}\mathbf{a}^{-1}\mathbf{b}$.

2. Example $\begin{pmatrix} 1 & 0 \\ 0 & 0 \end{pmatrix}, \begin{pmatrix} 0 & 0 \\ 0 & 1 \end{pmatrix}$.

5. (i) $\frac{1}{2}\begin{pmatrix} -5 & 2 & 1 \\ 2 & 2 & -2 \\ 1 & -2 & 1 \end{pmatrix}$, (ii) $\frac{1}{39}\begin{pmatrix} 15 & -9 & -9 \\ 10 & 7 & -6 \\ 3 & 6 & 6 \end{pmatrix}$.

7. 0.

Revision Examples (page 170):

8. $\begin{pmatrix} 1 & -\lambda & \lambda^2 \\ 0 & 1 & -\lambda \\ 0 & 0 & 1 \end{pmatrix}, \begin{pmatrix} 1 & 0 & 0 \\ -\mu & 1 & 0 \\ \mu^2 & -\mu & 1 \end{pmatrix}$,

$$\begin{pmatrix} 1 & -\lambda & \lambda^2 \\ -\mu & 1+\lambda\mu & -\lambda-\lambda^2\mu \\ \mu^2 & -\mu-\lambda\mu^2 & 1+\lambda\mu+\lambda^2\mu^2 \end{pmatrix}.$$

9. (i) $\begin{pmatrix}1 & 0\\0 & 1\end{pmatrix}$, $\begin{pmatrix}4 & 0\\0 & 1\end{pmatrix}$, $\begin{pmatrix}1 & 0\\0 & 4\end{pmatrix}$, $\begin{pmatrix}4 & 0\\0 & 4\end{pmatrix}$;

(ii) Any p, q, r, s for which $p+s=5$ and $p^2-5p+4+qr=0$ [so that $s^2-5s+4+qr=0$ also].

15. $\begin{pmatrix}3\theta+\phi & 4\theta\\6\theta & -3\theta+\phi\end{pmatrix}$ for arbitrary θ, ϕ.

16. ± 1.

19. $\mathbf{x}=\begin{pmatrix}1 & 3\\2 & 7\end{pmatrix}$, $\mathbf{y}=\frac{1}{3}\begin{pmatrix}43 & -7\\118 & -19\end{pmatrix}$.

20. $\mathbf{a}^2=\begin{pmatrix}-1 & 0 & 0\\0 & -1 & 0\\0 & 0 & 4\end{pmatrix}$ $\mathbf{a}^{-1}=\begin{pmatrix}-1 & -1 & 0\\2 & 1 & 0\\0 & 0 & \frac{1}{2}\end{pmatrix}$.

§12. **Example,** *first set* (page 175):

1. (i) $\mathbf{a}=\begin{pmatrix}0 & 4 & -5\\-4 & 0 & 6\\5 & -6 & 0\end{pmatrix}$, $\mathbf{b}=\begin{pmatrix}0 & -3 & 2\\3 & 0 & -1\\-2 & 1 & 0\end{pmatrix}$;

(ii) $\mathbf{a}=\begin{pmatrix}0 & 0 & -1\\0 & 0 & -3\\1 & 3 & 0\end{pmatrix}$, $\mathbf{b}=\begin{pmatrix}0 & 1 & 2\\-1 & 0 & 0\\-2 & 0 & 0\end{pmatrix}$;

(iii) $\mathbf{a}=\begin{pmatrix}0 & 0 & 0\\0 & 0 & 1\\0 & -1 & 0\end{pmatrix}$, $\mathbf{b}=\begin{pmatrix}0 & -1 & 1\\1 & 0 & 0\\-1 & 0 & 0\end{pmatrix}$.

§12. **Examples,** *second set* (page 177):

1. $-\frac{25}{169}(3,4,-12)$, $\frac{2}{169}(3,4,-12)$, $\frac{55}{169}(3,4,-12)$, $-\frac{52}{169}(3,4,-12)$.

§12. **Examples,** *third set* (page 179):

1. $\begin{pmatrix}16m & 0 & 0\\0 & 16m & 0\\0 & 0 & 16m\end{pmatrix}$.

2. $\begin{pmatrix}4m & 0 & 0\\0 & 4m & 0\\0 & 0 & 8m\end{pmatrix}$.

3. $\begin{pmatrix} 20m & 0 & 2m \\ 0 & 20m & 4m \\ 2m & 4m & 20m \end{pmatrix}$.

4. (i) $(16m\omega/\sqrt{3})(1,1,1)$; $(4m\omega/\sqrt{3})(1,1,2)$; $(2m\omega/\sqrt{3})(11,12,13)$.
(ii) $8\text{m}\omega^2$; $\frac{8}{3}m\omega^2$; $12m\omega^2$.

5. $\mathbf{R} = \mathbf{I} - 2\mathbf{l}\mathbf{l}'$.

CHAPTER 14

§3. Examples (page 186):

1. $\begin{pmatrix} 10 \\ 28 \\ 46 \end{pmatrix}$.

2. $\begin{pmatrix} 1 & s \\ 0 & s \\ 0 & 0 \end{pmatrix}$.

3. $\begin{pmatrix} a^2+h^2+g^2 & ah+hb+gf, & ag+hf+gc \\ ha+bh+fg & h^2+b^2+f^2, & hg+bf+fc \\ ga+fh+cg & gh+fb+cf, & g^2+f^2+c^2 \end{pmatrix}$.

4. $\begin{pmatrix} ax+by \\ cx+dy \end{pmatrix}$.

5. $\begin{pmatrix} 1 & 2 & 3 & 4 \\ a & b & c & d \\ p & q & r & s \end{pmatrix}$.

6. $\begin{pmatrix} a+x & h+x & x \\ h+y & b+y & y \\ x & y & 0 \end{pmatrix}$.

7. $\begin{pmatrix} -h^2-g^2 & gf & hf \\ fg & -h^2-f^2 & hg \\ fh & gh & -g^2-f^2 \end{pmatrix}$.

8. $\left(\begin{array}{c|c} \mathbf{a} & \mathbf{b} \\ \hline \mathbf{c} & \mathbf{d} \end{array}\right)$, $\left(\begin{array}{c|c} \mathbf{I} & \mathbf{a}^{-1}\mathbf{b} \\ \hline \mathbf{0} & \mathbf{d}-\mathbf{c}\mathbf{a}^{-1}\mathbf{b} \end{array}\right)$,

$\left(\begin{array}{c|c} \mathbf{a} & \mathbf{0} \\ \hline \mathbf{c} & \mathbf{0} \end{array}\right)$, $\left(\begin{array}{c|c} \mathbf{a} & \mathbf{0} \\ \hline \mathbf{c} & \mathbf{d}-\mathbf{c}\mathbf{a}^{-1}\mathbf{b} \end{array}\right)$.

§4. Examples (page 189):

1. $(\mathbf{x}, \mathbf{y}, \mathbf{z})\begin{pmatrix} 3 & 2 & 1 \\ -1 & 1 & 0 \\ 1 & 4 & -1 \end{pmatrix}$.

Revision Examples (page 191):

3. (i) $\begin{pmatrix} 1 & 0 & 0 \\ 0 & 1 & 0 \\ \lambda & 0 & 1 \end{pmatrix}$, (ii) $\begin{pmatrix} 0 & 0 & 1 \\ 0 & 1 & 0 \\ 1 & 0 & 0 \end{pmatrix}$,

(iii) $\begin{pmatrix} 2 & 0 & 0 \\ 0 & -3 & 0 \\ 0 & 0 & 4 \end{pmatrix}$, (iv) $\begin{pmatrix} 1 & 0 & -7 \\ 0 & 1 & 0 \\ 0 & 0 & 1 \end{pmatrix}$.

CHAPTER 15

§2. Examples, ***second set*** (page 196):

1. (i) 3, (ii) 2, (iii) 3, (iv) 3, (v) 3, (vi) 2, (vii) 1, (viii) 2.
2. Yes.
3. {4, 4, 4}, {9, 1, 4}.

§4. Examples (page 200):

(Call the given vectors **x**, **y**, **z**, ... in order.)

1.
$$(\mathbf{x},\mathbf{y},\mathbf{z},\mathbf{u},\mathbf{v})\begin{pmatrix} 1 & 1 \\ 1 & 1 \\ 1 & 7 \\ -6 & 0 \\ 0 & -6 \end{pmatrix} = \mathbf{0}.$$

2.
$$(\mathbf{x},\mathbf{y},\mathbf{z},\mathbf{u},\mathbf{v})\begin{pmatrix} 0 & 0 \\ 2 & 2 \\ 1 & -2 \\ -1 & 0 \\ 0 & 1 \end{pmatrix} = \mathbf{0}.$$

[The vector **x** is independent of the others.]

3.
$$(\mathbf{x},\mathbf{y},\mathbf{z},\mathbf{u})\begin{pmatrix} 1 & 2 \\ -2 & -3 \\ 1 & 0 \\ 0 & 1 \end{pmatrix} = \mathbf{0}.$$

4.
$$(\mathbf{x},\mathbf{y},\mathbf{z},\mathbf{u},\mathbf{v})\begin{pmatrix} 1 \\ -3 \\ 1 \\ 1 \\ -1 \end{pmatrix} = \mathbf{0}.$$

§7. Examples, ***first set*** **(page 202):**

4. $(\lambda+2, \lambda+1, \lambda)$.

§7. Examples, ***second set*** **(page 204).**

Ranks 2, 2, 2, 2, 2, 2.
Nullities 1, 1, 2, 1, 1, 2.

CHAPTER 16

§4. Examples, ***first set*** **(page 208):**

1. 1, 2, 2, 3, 1, 2, 2, 3.

§4. Examples, ***second set*** **(page 211):**

1. $(-2\lambda+2\mu, 3\lambda-5\mu, \lambda, \mu)$.
2. $(3\lambda+3\mu-3\nu, -5\lambda-5\mu+5\nu, \lambda, \mu, \nu)$.
3. $(-5\lambda. 11\lambda, \lambda)$.
4. $(\lambda-2\mu, -2\lambda+\mu, \lambda, \mu)$.

Revision Examples (page 220):

2. (i) $(6-5\lambda, -2+3\lambda, \lambda)$; (ii) $k=26$ only; $(-2+\lambda, 8-2\lambda, \lambda)$; (iii) $(2-\lambda, 2-\mu, \lambda, \mu)$; (iv) $(2\lambda, 6-5\lambda, \lambda)$.

3. (i) Normal solutions unless $a+b+c=0$ or $a-2b+c=0$. In either of these cases, no solution unless $d=0$ also.

(ii) Normal solutions unless $a+b+c=0$. In this case, no solution unless $p+q+r=0$ also.

(iii) Normal solutions unless $a=-1$ or $\frac{5}{3}$.

When $a=-1$, no solution unless $b=1$ also, in which case solution $(\lambda+\mu-1, \lambda, 1, \mu)$; when $a=\frac{5}{3}$, the solution is given by

$$16x=12-15b+3\lambda,$$
$$16y=6+15b-21\lambda,$$
$$z=\lambda,\ 8t=5b+3.$$

(iv) Normal solutions unless $\lambda=-1$ or $\frac{1}{3}$.

When $\lambda=-1$, solution $(k-2, 1, k)$; when $\lambda=\frac{1}{3}$, no solution.

(v) For solution to exist, need $ab=-8$. Solution is then $(1+\frac{1}{2}b\lambda, \frac{1}{2}a-\frac{3}{2}\lambda, \lambda)$.

4. $k=2$; solution $\{\frac{1}{5}(8-2\lambda-\mu), \frac{1}{5}(-1+4\lambda-8\mu), \lambda, \mu\}$.

APPENDIX TO CHAPTER 16

Revision examples (page 226):

7. An example is
$$\begin{pmatrix} 1 & -2 & 1 & 0 \\ 2 & -3 & 0 & 1 \\ 3 & -5 & 1 & 1 \\ 4 & -7 & 2 & 1 \end{pmatrix}.$$

Only two rows are independent, so many forms are possible.

11. 2; $(16\lambda+5\mu, 5\lambda+9\mu, 17\lambda, 17\mu)$.
(i) $(1, -1, 2, -3)$; (ii) none.

12. (i)
$$\mathbf{p} \equiv \begin{pmatrix} 2 & -1 & 0 \\ -3 & 2 & 0 \\ -2 & -1 & 1 \end{pmatrix}, \quad \mathbf{q} \equiv \begin{pmatrix} 1 & 0 & -7 \\ 0 & 1 & 10 \\ 0 & 0 & 1 \end{pmatrix};$$

(ii)
$$\mathbf{p} \equiv \begin{pmatrix} 1 & -3 & 0 \\ -1 & 4 & 0 \\ -1 & -3 & 1 \end{pmatrix}, \quad \mathbf{q} \equiv \begin{pmatrix} 1 & 0 & 11 \\ 0 & 1 & -14 \\ 0 & 0 & 1 \end{pmatrix};$$

(iii)
$$\mathbf{p} \equiv \begin{pmatrix} -3 & 5 & 0 \\ 2 & -3 & 0 \\ 1 & -2 & 1 \end{pmatrix}, \quad \mathbf{q} \equiv \begin{pmatrix} 1 & 0 & -26 \\ 0 & 1 & 16 \\ 0 & 0 & 1 \end{pmatrix}.$$

CHAPTER 17

§1. Examples (page 234): —.

§2. Examples (page 236):

1. (i) $8y_1^2+36y_1y_2+45y_2^2$,
(ii) $45y_1^2+36y_1y_2+8y_2^2$,
(iii) $10y_1^2-50y_1y_2+65y_2^2$,
(iv) $25y_1^2+46y_1y_2+26y_2^2$,
(v) $73y_1^2+92y_1y_2+29y_2^2$,
(vi) $29y_1^2+8y_1y_2+4y_2^2$.

§3. Examples (page 238):

2. $\frac{1}{17}$.

3. $$\begin{pmatrix} 1/\sqrt{5} & 2/\sqrt{5} \\ -2/\sqrt{5} & 1/\sqrt{5} \end{pmatrix}.$$

§4. Examples (page 240):

1. The matrices of transformation are:

(i) $\begin{pmatrix} 1/\sqrt{2} & 1/\sqrt{2} \\ -1/\sqrt{2} & 1/\sqrt{2} \end{pmatrix}$, (ii) $\begin{pmatrix} 2/\sqrt{5} & -1/\sqrt{5} \\ 1/\sqrt{5} & 2/\sqrt{5} \end{pmatrix}$,

(iii) $\begin{pmatrix} 3/\sqrt{10} & -1/\sqrt{10} \\ 1/\sqrt{10} & 3/\sqrt{10} \end{pmatrix}$, (iv) $\begin{pmatrix} 2/\sqrt{5} & -1/\sqrt{5} \\ 1/\sqrt{5} & 2/\sqrt{5} \end{pmatrix}$,

(v) $\begin{pmatrix} 3/\sqrt{10} & 1/\sqrt{10} \\ -1/\sqrt{10} & 3/\sqrt{10} \end{pmatrix}$, (vi) $\begin{pmatrix} 2/\sqrt{13} & 3/\sqrt{13} \\ -3/\sqrt{13} & 2/\sqrt{13} \end{pmatrix}$,

(vii) $\begin{pmatrix} 5/\sqrt{29} & 2/\sqrt{29} \\ -2/\sqrt{29} & 5/\sqrt{29} \end{pmatrix}$.

§5. **Example** (page 244):

1. (i) $\begin{pmatrix} 1 & 1 & -4 \\ 1 & -1 & 0 \\ 0 & 0 & 2 \end{pmatrix}$, (ii) $\begin{pmatrix} 1 & 0 & 2 \\ 1 & 1 & -1 \\ 1 & -1 & -1 \end{pmatrix}$,

(iii) $\begin{pmatrix} 1 & 2 & 0 \\ -1 & 1 & 1 \\ -1 & 1 & -1 \end{pmatrix}$, (iv) $\begin{pmatrix} 0 & 1 & 1 \\ 0 & 1 & -1 \\ 1 & 0 & 0 \end{pmatrix}$.

(i) $4x_2x_3+2x_1x_2$ (note that it is singular), $2y_1^2-2y_2^2$;
(ii) $7x_1^2+5x_2^2+5x_3^2+2x_2x_3-2x_3x_1-2x_1x_2$, $15y_1^2+8y_2^2+48y_3^2$.
(iii) $x_1^2-x_2^2-x_3^2+2x_2x_3+2x_3x_1+2x_1x_2$, $-3y_1^2+12y_2^2-4y_3^2$.
(iv) $3x_1^2+3x_2^2+5x_3^2-2x_1x_2$, $5y_1^2+4y_2^2+8y_3^2$.

Note that *orthogonal* matrices satisfying the conditions are:

(i) $\begin{pmatrix} 1/\sqrt{2} & 1/\sqrt{2} & -4/3\sqrt{2} \\ 1/\sqrt{2} & -1/\sqrt{2} & 0 \\ 0 & 0 & 2/3\sqrt{2} \end{pmatrix}$, (ii) $\begin{pmatrix} 1/\sqrt{3} & 0 & 2/\sqrt{6} \\ 1/\sqrt{3} & 1/\sqrt{2} & -1/\sqrt{6} \\ 1/\sqrt{3} & -1/\sqrt{2} & -1/\sqrt{6} \end{pmatrix}$,

(iii) $\begin{pmatrix} 1/\sqrt{3} & 2/\sqrt{6} & 0 \\ -1/\sqrt{3} & 1/\sqrt{6} & 1/\sqrt{2} \\ -1/\sqrt{3} & 1/\sqrt{6} & -1/\sqrt{2} \end{pmatrix}$, (iv) $\begin{pmatrix} 0 & 1/\sqrt{2} & 1/\sqrt{2} \\ 0 & 1/\sqrt{2} & -1/\sqrt{2} \\ 1 & 0 & 0 \end{pmatrix}$.

and the quadratic forms then become
(i) $y_1^2-y_2^2$,
(ii) $5y_1^2+4y_2^2+8y_3^2$,
(iii) $-y_1^2+2y_2^2-2y_3^2$,
(iv) $5y_1^2+2y_2^2+4y_3^2$.

§6. **Revision Examples** (page 248):

3. $\begin{pmatrix} \frac{4}{5} & -\frac{3}{5} \\ \frac{3}{5} & \frac{4}{5} \end{pmatrix}$. Hyperbola $-x^2+4y^2=5$; real axis $\frac{1}{2}\sqrt{5}$.

4. Matrix $\begin{pmatrix} 2/\sqrt{5} & 1/\sqrt{5} \\ -1/\sqrt{5} & 2/\sqrt{5} \end{pmatrix}$; $\lambda=1$, $\mu=6$.

5. Matrix $\begin{pmatrix} 0 & -1/\sqrt{10} & 1/\sqrt{2} & 2/\sqrt{10} \\ 1/\sqrt{2} & 2/\sqrt{10} & 0 & 1/\sqrt{10} \\ -1/\sqrt{2} & 2/\sqrt{10} & 0 & 1/\sqrt{10} \\ 0 & 1/\sqrt{10} & 1/\sqrt{2} & -2/\sqrt{10} \end{pmatrix}$,

reducing to $x'^2+2y'^2+4z'^2+7t'^2$.

6. 3, 6; $\begin{pmatrix} 1 & 2 & 2 \\ 2 & 1 & -2 \\ 2 & -2 & 1 \end{pmatrix}$.

An *orthogonal* matrix is $\begin{pmatrix} \frac{1}{3} & \frac{2}{3} & \frac{2}{3} \\ \frac{2}{3} & \frac{1}{3} & -\frac{2}{3} \\ \frac{2}{3} & -\frac{2}{3} & \frac{1}{3} \end{pmatrix}$.

7. Matrix $\begin{pmatrix} 1/\sqrt{2} & 1/\sqrt{3} & 1/\sqrt{6} \\ 0 & 1/\sqrt{3} & -2/\sqrt{6} \\ -1/\sqrt{2} & 1/\sqrt{3} & 1/\sqrt{6} \end{pmatrix}$,

reducing to the form $2y_1^2+3y_2^2+6y_3^2$.

8. Orthogonal matrix $\mathbf{p} \equiv \begin{pmatrix} \frac{2}{\sqrt{5}} & \frac{1}{\sqrt{5}} \\ -\frac{1}{\sqrt{5}} & \frac{2}{\sqrt{5}} \end{pmatrix}$

For last part, see note; $\lambda = 2$, $\mu = 7$.

9. $\lambda = -1$, $\mu = 2$.

10. $\mathbf{x} = -\mathbf{I}, -3\mathbf{I}$,

or $\begin{pmatrix} -1 & 0 \\ \lambda & -3 \end{pmatrix}$, $\begin{pmatrix} -3 & \lambda \\ 0 & -1 \end{pmatrix}$,

or $\begin{pmatrix} \lambda-2 & (1+\lambda) \\ (1-\lambda)/\mu & -\lambda-2 \end{pmatrix}$, with λ, μ arbitrary.

11. $1, \omega, \omega^2$; $\{1, 1, 1\}$, $\{1, \omega, \omega^2\}$, $\{1, \omega^2, \omega\}$.

12. 2, 1, −1; $\begin{pmatrix} 0 & 0 & 1 \\ 1/\sqrt{2} & 1/\sqrt{2} & 0 \\ 1/\sqrt{2} & -1/\sqrt{2} & 0 \end{pmatrix}$.

Two quadratic forms; Examples, ***first set*** **(page 253):**

1. $-y_1^2+2y_2^2+3y_3^2$, $y_1^2+y_2^2+y_3^2$.
2. $y_1^2+y_2^2+y_3^2$, $y_2^2+2y_3^2$.
3. $y_1^2+y_2^2+y_3^2$, $-2y_1^2+y_2^2+\frac{1}{2}y_3^2$.
4. $y_1^2+y_2^2+y_3^2$, $y_1^2+2y_2^2$.
5. $y_1^2+y_2^2+y_3^2$, $y_1^2+2y_2^2-8y_3^2$.
6. $2y_1^2+y_2^2-y_3^2$, $y_1^2+y_2^2+y_3^2$.
7. $y_1^2+y_2^2+y_3^2$, $y_1^2-y_2^2$.
8. $y_1^2+y_2^2+y_3^2$, $-y_1^2+4y_2^2-3y_3^2$.

Two quadratic forms; Examples, *second set* (page 255), with x, y, z written for x_1, x_2, x_3:

$$\text{(i)}\ u = \left\{\frac{x+y+z}{\sqrt{3}}\right\}^2 + \left\{\frac{2x-y-z}{\sqrt{6}}\right\}^2 + \left\{\frac{y-z}{\sqrt{2}}\right\}^2,$$

$$v = 2\left\{\frac{x+y+z}{\sqrt{3}}\right\}^2 - \left\{\frac{2x-y-z}{\sqrt{6}}\right\}^2 - \left\{\frac{y-z}{\sqrt{2}}\right\}^2.$$

$$\text{(ii)}\ u = \left\{\frac{x+2y+z}{\sqrt{3}}\right\}^2 + \left\{\frac{2x+y-z}{\sqrt{6}}\right\}^2 + \left\{\frac{y-z}{\sqrt{2}}\right\}^2,$$

$$v = 2\left\{\frac{x+2y+z}{\sqrt{3}}\right\}^2 - \left\{\frac{2x+y-z}{\sqrt{6}}\right\}^2 - \left\{\frac{y-z}{\sqrt{2}}\right\}^2.$$

(iii) $u = (y+z)^2+(z+x)^2+(z+y)^2,$
$v = (y+z)^2-(z+x)^2+(x+y)^2.$

(iv) $u = (x+y+z)^2+(y+z)^2+z^2,$
$v = (x+y+z)^2+(y+z)^2+3z^2.$

CHAPTER 18

§2. Examples (page 257):

1. $4, \{1, 1, -1\}; 4\pm 2\sqrt{3}, \{\pm\sqrt{3}+1, -2, \pm\sqrt{3}-1\}.$
2. $0, \{2, 1, -2\}; 3, \{1, 0, 1\}; -7, \{1, 4, -1\}.$
3. $9, \{1, 2, 2\}; 3, \{2, -2, 1\}; -3, \{2, 1, -2\}.$
4. $9, \{1, 2, 2\}; 18, \{2\theta-2, -\theta, 1\}$ for arbitrary θ.
6. $0, \{1, -1, 1\}; \pm ai\sqrt{3}, \{1\pm i\sqrt{3}, -1\pm i\sqrt{3}, -2\}.$

§3. Examples (page 260):

1. $\lambda^3+(a^2+b^2+c^2)\lambda = 0.$
2. 1, 2, 4.

Revision examples (page 270):

9. hyperbola, line-pair, ellipse, rectangular hyperbola, circle; $\sqrt{2}$, 1.
10. $1, \{1/\sqrt{2}, 1/\sqrt{2}, 0\}; \tfrac{1}{2}(1\pm i)\sqrt{2}, \{\tfrac{1}{2}, -\tfrac{1}{2}, \pm\tfrac{1}{2}i\sqrt{2}\}$ but cannot normalize.

13. $$\begin{pmatrix} \dfrac{1-p^2}{1+p^2} & \dfrac{-2p}{1+p^2} \\ \dfrac{2p}{1+p^2} & \dfrac{1-p^2}{1+p^2} \end{pmatrix}.$$

15. $9, \{1, 2, 2\}; 27, \{2, -2, 1\}; -27, \{2, 1, -2\}.$

24. $$\begin{pmatrix} 0 & 1 & 0 \\ -1 & 0 & 1 \\ 0 & -1 & 0 \end{pmatrix}.$$

25. ± 1 and roots of $7\lambda^2+\lambda+7=0.$

27. $$\begin{pmatrix} 0 & c & s \\ -c & s^2 & -sc \\ -s & -sc & c^2 \end{pmatrix};$$

$1, \{0, s, -c\};\ \pm i, \{\pm i, c, s\};$

$$\begin{pmatrix} 0 & i & -i \\ s & c & c \\ -c & s & s \end{pmatrix}.$$

CHAPTER 19

§1. **Examples** (page 278):

1. Yes. **2.** Yes. **3.** Yes. **4.** No. **5.** Yes. **6.** Yes.

§2. **Examples** (page 278):

1. Yes. **2.** Yes. **3.** No. **4.** No. **5.** Yes. **6.** Yes.

§3. **Examples** (page 279):

1. Yes. **2.** Yes. **3.** No. **4.** Yes. **5.** No. **6.** Yes.

§4. **Examples** (page 280):

1. No. **2.** Yes. **3.** No. **4.** Yes. **5.** Yes. **6.** No. **7.** No. **8.** Yes.

CHAPTER 20

§2. **Examples** (page 291):

6. $$\begin{pmatrix} -1 & 0 \\ 0 & 1 \end{pmatrix}, \begin{pmatrix} 0 & 1 \\ 1 & 0 \end{pmatrix}, \begin{pmatrix} 0 & -1 \\ -1 & 0 \end{pmatrix}.$$

7. $$\begin{pmatrix} -1 & 0 \\ 0 & -1 \end{pmatrix}, \begin{pmatrix} 1 & 0 \\ 0 & 1 \end{pmatrix}.$$

Not isomorphic.

8. $$\begin{pmatrix} \cos\phi & \sin\phi \\ -\sin\phi & \cos\phi \end{pmatrix}, \begin{pmatrix} \cos\phi & -\sin\phi \\ \sin\phi & \cos\phi \end{pmatrix}, \begin{pmatrix} 1 & 0 \\ 0 & 1 \end{pmatrix}.$$

9. No. (The zero matrix is *singular*.)

§4. **Examples** (page 295):

5. No.

7. $$S_n \equiv \begin{pmatrix} \cos\frac{1}{6}n\pi & -\sin\frac{1}{6}n\pi \\ -\sin\frac{1}{6}n\pi & -\cos\frac{1}{6}n\pi \end{pmatrix}.$$

Subgroups: The matrices R_n; (R_0, S_0).

INDEX